Navier–Stokes–Fourier Equations

Radyadour Kh. Zeytounian

Navier–Stokes–Fourier Equations

A Rational Asymptotic Modelling Point of View

Springer

R. Kh. Zeytounian
Université des Sciences et
Technologies de Lille
LML – cité Scientifique
59655 Villeneuve d'Ascq Cédex
France
zeytounian@free.fr

ISBN 978-3-642-20745-7 e-ISBN 978-3-642-20746-4
DOI 10.1007/978-3-642-20746-4
Springer Heidelberg Dordrecht London New York

Library of Congress Control Number: 2011943087

Printed on acid-free paper

Springer is part of Springer Science+Business Media (www.springer.com)

Preface

Rationality and *Asymptotics* are the two *main concepts* associated with the *Modelling* in *Fluid Dynamics,* which have completely changed our *look* on the *Understanding* of *Navier–Stokes–Fourier* (*NS-F) equations*, governing the *viscous*, compressible and *heat conducting Newtonian baroclinic* and *non-adiabatic fluid flows*.[1]

This *Rational Asymptotic Modelling* (*RAM*) *Approach* have raised, on the one hand, further new interesting questions and potentialities for *Applied Mathematicians,* in their quest of rigorous existence and uniqueness results for the *Fluid Flow* problems.

On the other hand, this *RAM Approach* have opening up of new vistas for the derivation, by *Fluid Dynamicians,* various consistent simplified models related with *real stiff fluid flow* problems, as an *assistance* to *Numericians* embarked *on a computational simulations* of complex problems of engineering interest with the help of high speed computers.

In this book we touch (see, in particular, the *Chap. 6*) the "crucial" problem of a *practical* (rather than *formal, abstract*) "*Mathematics*" for a *consistent RAM Approach,* via a *"Postulate"* and, some "*key rules*" inspired from asymptotics.

This "*mathematics for the RAM*" is applied in a consistent way to modelling of various stiff problems of the: *aerodynamics (Chap. 7), Bénard thermal convection (Chap. 8)* and *atmospheric motions (Chap. 9).*

The main lignes of the aims of this book are set out in the "*Prologue*", and in the "*Overview*" a brief outline of the events related with my rather long "*RAM Adventure*", during thc years 1968–2009, is given.

The book is divided into *nine Chapters*, an *Epilogue*, a *list of References*, and a *Subject Index*.

In *Chap. 2*, the Newtonian (*Classic*) Fluid Dynamics is considered as a Mathematical-Physical Science and the reader can find in *four Sections* a *concise*

[1] These *NS-F equations* are, in fact, the equations usually named *"Navier–Stokes Compressible equations"* – and assumed often *barotropic equations*, by the Mathematicians !

material concerning the *main* theoretical *concepts* and *principles*, *equations* and associated *initial* and *boundary conditions*.

The *Chap. 3* is devoted to a *tentative description of a rational way* for the *obtention*, from *NS-F equations*, various *main model* equations and also to a *discussion* concerning their *nonuniform validity, near the initial time (*where the *initial data* are given) and *in the vicinity of a solid wall limiting the fluid flow* (where the *boundary conditions* for the *velocity vector* and *temperature* are given).

The *Chap. 4,* is entirely concerned with the application of *RAM Approach* for a *justification of Boussinesq model* equations, assuming that the *Mach number is a small parameter*.

The *Chap. 5* is an application of the *RAM Approach* to *large Reynolds numbers unsteady fluid flow*, which leads to a complicated *Five Regions Structure of unsteady NS-F full equations*.

The *Chap. 6*, is a central one and present a "*sketch of a Mathematical Theory for the RAM Approach*". As a basis for this "*practical*" Mathematics, in the realization of our *RAM Approach, the following "Postulate" is accepted as true, despite its simplicity:*

> If a leading order an approximate simplified model is derived from a NS-F fluid flow problem, then it is necessary that a RAM Approach be adopted to make sure that terms neglected in a such NS-F stiff problem really are much smaller than those retained in derived approximate simplified, no-stiff, leading-order consistent model problem.

The *Chap. 7*, is concerned with the two applications of the *RAM Approach* in "*Aerodynamics*". *First*, the derivation of a *through-flow model problem*, for a fluid flow in an *axial compressor*, when the *blades* in a row are *very closely* spaced. *Secondly*, the *low Mach number flow of a gas within a cavity* which is *changing* its *shape* and *volume with time*.

In *Chap. 8, The RAM Approach* concerns the famous *Bénard convection problem for a liquid layer heated from below*. In particular, the following *alternative* is demonstrated:

> Either the buoyancy is taken into account, and in this case the free-surface deformation effect is negligible and we rediscover the classical leading-order Rayleigh-Bénard shallow convection, unless viscous dissipation, rigid-free, problem, or the free-surface deformation effect is taken into account, and in this case at leading-order, for thin films, the buoyancy and viscous dissipation effects does not play a significant role in the so-called Bénard-Marangoni thermocapillary instability problem.
>
> But, if you have intend to take into account, in the case of a deep liquid layer, the viscous dissipation effect – according to Zeytounian – in equation for the temperature, then it is necessary to replace, the Rayleigh-Bénard shallow convection equations, by a new set of equations called deep convection equations with a "depth" parameter."

The last *Chap. 9* is devoted to atmospheric motions. *First*, we derive for 2D steady *lee-waves problem*, in a baroclinic, non-viscous and adiabatic atmosphere, from the Euler atmospheric equations, a *single, exact but rather, awkward equation*. This equation, *coupled* with an exact relation for the *density*, prove to be very convenient for a *RAM Approach* of lee-waves starting problem, when we consider the *low Mach* number case. *Secondly*, the *low-Kibel/Rossby number* asymptotic model is considered, and a *global quasi-geostrophic* (G*QG*) model is derived from *NS-F hydrostatic dissipative atmospheric equations*. Namely: the *QG single main equation model*, *initial condition* (*at time* $= 0$) via an *unsteady adjustment (Adj)* and *matching*, and *boundary condition (at the flat ground*) via the *Ackerblom's problem in a steady Ekman boundary layer (Ek) problem* and *matching*.

In *Epilogue* some concluding remarks are sketched briefly.

A postgraduate Course may involve most of the contents of this book, assuming perhaps a working knowledge of a classical university fluid dynamics Course.

Short Courses for training Applied Mathematicians and Numericiens and young Scientists in Industry and Research Laboratories can also be based on most of the contents of this book.

In fact, the material in this book, it seems me, is primarily suitable (maybe indispensable!) for use by the Scientists and Research Engineers working in the fields of Fluid Dynamics and having as a main motivation the numerical simulation of very stiff complex real fluid flows.

Finally, I thank Dr. Christoph Baumann, Engineering Editor, and the members of the Springer Engineering Editorial Department, where the camera-ready manuscript was produced in LaTeX and my English type-script was reread by a native English speaker.

Paris

Radyadour Khatchig
Zeytounian

Prologue

In the middle of fifties years of the twentieth century, with the works, at Caltech (California, *USA*), of Kaplun (1954 and 1957), Lagerstrom and Cole (1955), Kaplun and Lagerstrom (1957) and also Proudman and Pearson (1957) in *England*, asymptotics gave a new tremendous impetus on research in theoretical fluid dynamics.

Concerning, as to, in *France*, the asymptotics was introduced by *Paul Germain*, who was without (any) doubt the initiator of the application of asymptotics and modelling in France; Germain write, in "*Paul Germain's Anniversary Volume*" (2000):

> During the Istanbul 1952 International Congress, Paco Lagerstrom spoke to me about some questions which he thought to be of the utmost importance for the understanding of fluid mechanics and which might be ripe for solution at that time. One of them was the mathematical basis of the boundary layer concept, discovered by Prandtl, nearly 50 years before. Another one concerned the steady flow of an inviscid fluid as the limit of a class of corresponding flows of the same fluid, involving a vanishingly small viscosity, so that, in some sense, solutions of the Euler equations might be related to a class of solutions of Navier–Stokes equations, through some limiting process.... I began to foresee a new link between mathematics and fluid mechanics, provided by asymptotic techniques; as a matter of fact, no simply a link, but a way of thinking at an enormous variety of problems. Needless to say that all this was opening avenues without a clear vision of getting the right way in.

Indeed, Paul Germain has had an extremely fruitful scientific career in which he contributed, in particular, to modelling and asymptotics in fluid mechanics, by discovering *new problems*, *new ideas*, *new methods*, *new fields of applications*. But, as it is written (page 3, in 2000) by Germain:

> "My best contributions to fluids were due to discovering, at the end of 1955, Jean-Pierre Guiraud" – in fact, Guiraud first applied (in 1958) the method of matched asymptotic expansions to hypersonic flow past a blunt-nosed plate, with a matching in subtle fashion to an outer layer that is governed by hypersonic small –disturbance theory.

Ten years later, the Van Dyke's book (1964), at a time when *numerical fluid dynamics* was in its *infancy*, was at pains to demonstrate that perturbation, or asymptotic, techniques could be used to advantage to simplify problem to the extent that they were amenable to analytical treatment; the fluid dynamical literature abounds with testimony to the success of this approach (see, for instance, the Annotated Edition of Van Dyke's book (1975), by Van Dyke himself).

But, it is necessary to observe that, at the beginning of the sixties, in fact, idea of a *RAM Approach* was first envisaged in famous two works.

The first work being related with the problem of the kinetic heating during re-entry, which gave aeronautical engineers a strong impulse to improve the boundary layer theory. For this, the extension of higher (at least to second) order boundary layer theory to compressible and heat conducting flow has been carried out by Van Dyke (1962). This rational asymptotic (consistent) modelling Van Dyke approach, of the "*second order compressible and heat conducting boundary layer model*", derived from *NS-F* full problem (formulated below, in *Chap*. 2, of the present book) was an important scientific contribution in the NASA program for the re-entry, from a space station, the space shuttle, when it is necessary to take into account two complementary effects such as: *slipping of the flow* and a *temperature jump at the wall*.

The *second work*, is linked with the fact that, if the viscosity and heat conduction have to be taken into account outside the boundary layer, that means that the Rankine-Hugoniot equations which rule the jump across the bow-shock wave to be rewritten and one must take into account the *thickness of the shock*. Unfortunately, the proposed preceding evaluation was not completely satisfactory! Germain and Guiraud (1960), are convinced that the only way to get correct results was to apply matched asymptotic expansion (*MAE*) and for this the conservative form of the full *NS-F* equations has been taken into account. In such a case, viscous and heat conducting terms appear to have been taken care of – but one has to add a contribution from the inner expansion, the leading term of which gives the internal shock structure. We observe that to order Re^{-1}, the jump conditions are very easily written out when one knows the internal shock structure to leading order only. Germain and Guiraud gives (in a paper published only in 1966) explicit formulas in gas dynamics for the shock conditions up to order 1.

It is interesting to observe that, in relation to above work related with the shock waves in gas dynamics, Germain, in (page 11, in 2000), write:

> A solution of a mathematical particular schematisation S^0 of a physical situation is not acceptable if it cannot be obtained by the limit of a solution of a more refined schematisation S when S tends to S^0

This requirement being, in fact, a strong condition for a consistent application of the *RAM Approach*.

Already, from these above two early examples, it might have been clear that asymptotics were well suited to derive mathematical models amenable, via simulation, to numerical treatment rather than to obtain closed form of solutions – however, we observe again: *at the time, numerical fluid dynamics was almost nonexistent due to the lack of high speed computers*. But, in the course of time, advances in computer technology have led to the development of increasing accurate numerical solutions and have thereby diminished the interest in approximate analytical results.

It is now evident, and for me, as early as the 1975 year, when I work on my Survey Lecture for, *XIIth Symposium on Advanced Problems and Methods in Fluid Mechanics (*Bialowieza, 8–13 September 1975, Poland – published in 1976), that asymptotic techniques provide very powerful tools in the process of constructing mathematical consistent models for problems which are stiff, from the point of view of numerical analysis and simulation.

Whilst numerical, computational, fluid dynamics is now a mature discipline: *"For some time the growth in capabilities of numerical simulation will be dependent on, or related to, the development of RAM Approach."* The simple definition of this *RAM Approach* being:

> The art of modelling assisted, rationally, by the spirit of asymptotics.

Real flows (such as the *turbomachinery flows,* which are arguably among the most complicated known to man and are of great technological importance) are extremely complex and exhibit an enormous range of length and time scales whose resolution (a well known example is the problem related with the "*weather forecast*":

> of what the weather will be like tomorrow or for the next few days!

will probably remain well beyond the capabilities of any computer foreseeable future.

"Our RAM Approach *provide a rational and systematic method for obtaining the necessary simplified flow models, which, in most cases still have to be solved numerically.*" A such approach is an extremely worthwhile objective because most of the relevant engineering computations are based (*up to now*!) on relatively ad hoc models that are rife with internal inconsistencies – usually these ad hoc models turn out to be nonuniform validity- i.e., they break down in certain regions of the flow. The applications, in *Chaps. 7–9*, show that we are able to make significant progress, via the *RAM Approach*, toward developing a consistent basis for some of the more prominent engineering and geophysics/meteo models flows. It is necessary to observe that:

> At the twenty-first century massive computations are capable to bring so much even for understanding, but there seems to be no indication that they are in competition with asymptotics, both are useful and complementary.

Let me close this "*Prologue*" by a *remark*:

> Quite often the modelling of stiff fluid flow problems may be found by various empirical procedures or by an 'ad hoc' approach. But it seems me obvious that, the ultimate goal is to find the mathematical key which explains, not only the success of the modelling, but the validity and consistency of these procedures in practice during the numerical simulation – The RAM Approach being obviously a well adequate consistent method for a such realization!

Cited References

Kaplun, S.: ZAMP. **5**, 11–136 (1954)

Kaplun, S.: J. Math. Mech. **6**, 595–603 (1957)

Lagerstrom, P. A., Cole, J. D.: J. Rat. Mech and Anal. **4**, 817 (1955)

Kaplun, S., Lagerstrom, P. A.: J. Math. Mech. **6**, 585–593 (1957)

Proudman, I., Pearson, J.R.A.J.: J. Fluid Mech. **2**, 237–262 (1957)

Germain, P.: My discovery of mechanics. In: Maugin, G.A., Drouot, R., Sidoroff, F. (eds.) Continuum Thermomechanics (Paul Germain's Anniversary Volume), Kluwer Academic Publishers, pp. 1–27 (2000)

Guiraud, J.-P.: C. R. Acad. Sci., Paris, **246**, 2842 (1958)

Van Dyke, M.: Second-order compressible boundary-layer theory with application to blunt bodies in hypersonic flow. In: Riddell, F.R. (ed.) Hypersonic Flow Theory, vol. **7**, pp. 37–76. Academic Press, New York (1962)

Germain, P., Guiraud, J.-P.: Conditions de choc dans un fluide doué de coefficients de viscosité et de conductibilité thermique faibles mais non nuls. C. R. Acad. Sci., Paris, **250**, 1965–1967 (1960)

Germain, P., Guiraud, J.-P.: Conditions de choc et structures des ondes de choc dans un écoulement non stationnaire de fluide dissipatif. Journal de Maths. Pures et Appliquées. **45**, 313–358 (1966)

Zeytounian, R. Kh.:La Météorologie du point de vue du Mécanicien des Fluides. Fluid Dynamics Transactions. **8**, 289–352 (1976)

Contents

Chapter 1
NS–F Equations and Modelling: A French Touch

This Overview is a brief outline of the events related to my rather long "RAM Adventure" during the years 1968–2009. In 1968–1969 my discovery of asymptotics and rational modelling of fluid dynamics problems was, for me, a revelation, and the Rational Asymptotics Modelling (RAM) Approach to these problems, governed by the Navier–Stokes–Fourier (NS–F) equations,[1] has been my main scientific activity during the last 40 years – the systematic, logical and well argued consistent approach via asymptotics, in perfect harmony with my idea about mathematically applied, but not ad hoc, theoretical researches in fluid dynamics, without any modern abstract, sophisticated, functional analysis!

This Overview first presents a short account of my first contribution to RAM in fluid dynamics, related to a justification of Boussinesq equations used in Chap. 1 of the original, version of my doctoral thesis, written in Moscow during 1965–1966. I then relate various events concerning my collaboration with Jean-Pierre Guiraud, working on asymptotic modelling of fluid flows at the Aerodynamics Department of ONERA[2] during the 16 years up to 1986, which resulted in the publication of 26 joint papers in various scientific journals. Finally, a few remarks are presented concerning my preceding seven books (three in French and four in English), published during the years 1986–2009, on modelling in Newtonian fluid flows.

Below we use "Navier" equations in place of "Navier–Stokes incompressible" equations. In fact, as main fluid dynamics equations we have Euler, Navier, and NS–F equations. Concerning the so-called "Navier–Stokes (isentropic)" equations – often used by mathematicians in their rigorous investigations – in reality these NS equations are unable to describe any real fluid flows! Note also that in a RAM

[1] Concerning the term "Navier–Stokes–Fourier" equations used in this book – NS–F equations, governing classical, Newtonian, viscous, compressible and heat-conducting fluid flows – it seems to me that it is better adapted than the term commonly used (mainly by mathematicians), "Navier–Stokes compressible" equations.

[2] Office National d'Études et de Recherches Aérospatiales, Châtillon-92320 (France).

R. Kh. Zeytounian, *Navier–Stokes–Fourier Equations*,
DOI 10.1007/978-3-642-20746-4_1, © Springer-Verlag Berlin Heidelberg 2012

Approach, the Euler (vanishing viscosity case) and Navier (low compressibility case) equations are, in fact, derived consistently by limiting processes from NS–F full equations – but this is not the case for the NS (isentropic) equations!

Concerning my "Soviet Adventure" of 1947–1966... In 1954 I graduated from Yerevan State University with a Master of Sciences degree in pure mathematics (in the class of Sergey Mergelyan[3]); after which, during 1955–1956, I worked in the Institute of Water and Energy at the Armenian Academy of Sciences in Yerevan. I then had the opportunity for serious study in theoretical fluid dynamics, and in 1957 I chose dynamic meteorology as my main scientific research activity as a Ph.D. student in the Kibel Department of the Hydro-Meteorological Centre in Moscow.

Now, more than 50 years later, I am still proud to have been a student of Il'ya Afanas'evich Kibel[4] – an outstanding hydrodynamicist of the twentieth century who was active and creative throughout his entire career. Unfortunately, his life was too short. He died suddenly, at the age of 66, on 5 September 1970.

Mainly on the basis of my various publications in mesometeorology (linked with the lee waves downstream of a mountain in a baroclinic atmosphere and also with the free atmospheric local circulations above the various Earth sites) during the years 1957–1966, in the Kibel Department of the Hydro-Meteorological Centre in Moscow, in 1968 I had the opportunity to publish my first course in mesometeorology [1] for the engineering students at the École de la Météorologie in Paris.

In September 1966 I returned to Paris to write my thesis [2] on the basis of the results of research (1961–1965) into the lee waves 2D (non-linear) and 3D (linear) steady problems in non-viscous and adiabatic atmospheres, with the help of the Boussinesq approximation. In 1969 I was awarded the degree of Docteur d'État es Sciences Physiques by the University of Paris, which added to my Russian Ph.D. of 1960, from the University of Moscow and my SSSR Academy of Sciences Chief Scientific Research Worker degree in hydrodynamics and dynamic meteorology, obtained in 1964.

---

[3] Sergey Nikitovich Mergelyan (1928–2008) was an Armenian scientist – an outstanding mathematician, and the author of major contributions in Approximation Theory (including his well-known theorem in 1951). The modern Complex Approximation Theory was based mainly on his work (see, for instance, the book *Real and Complex Analysis* by W. Rudin; French edition, Masson, Paris, 1978). He graduated from Yerevan State University in 1947, and in 1956 played a leading role in establishing the Yerevan Scientific Research Institute of Mathematical Machines (YerSRIMM). He became the first Director of this Institute, which today many refer to as the "Mergelyan Institute".

[4] Il'ya Afanas'evich Kibel (1904–1970), Member of the SSSR Academy of Sciences, was one of the leading Soviet scientist in the field of theoretical hydromechanics. He is famous as the founder of the hydrodynamic method of weather forecasting, and for implementation of mathematical methods in meteorology. See his pioneer monograph, *An Introduction to the Hydrodynamical Methods of Short Period Weather Forecasting*, published in Russian in Moscow (1957), and translated into English in 1963 (Macmillan, London). Some of his well-known works on the meteo-fluid are published in *Selected Works of I. A. Kibel on Dynamic Meteorology* (in Russian, GydrometeoIzdat, Leningrad, 1984).

In the original version of my thesis (hand-written in Moscow during 1965–1966), in Chap. 1 the Boussinesq approximate equations were derived in an ad hoc manner (*à la* Landau – as in [3, §56]). However, Paul Germain (future juryman during my thesis defence in March 1969) was unfavourable towards this method of deriving Boussinesq approximate equations for convection in fluids, and I was obliged to completely rewrite that chapter! Germain considered that it is possible to derive these Boussinesq equations by an asymptotic rational/consistent process (but by what method?), and in a letter[5] written in Paris and dated 8 March 1968, he wrote that "... I should understand the justification of our starting equations?"

1.1 My First Contribution to the RAM Approach in Fluid Dynamics

This "justification problem" was for me a difficult challenge – 1 year before my 1969 thesis defence – and I was in an awkward situation! For some time I did not fully understand the question in Germain's letter! Finally, however, I chose a bastardized method via the so-called isochoric model equations, when the density ρ is a conservative unknown function along the fluid flow trajectories in time–space (t, **x**), such that

$$D\rho/Dt = 0, \text{ with } D/Dt = \partial/\partial t + \mathbf{u}\cdot\nabla, \tag{1.1}$$

where ∇ denotes the gradient vector and **u** the velocity vector – this constraint being often used in fluid dynamics when gravity plays an active role. This above conservative condition on ρ is, in fact, an incompressible condition. In particular, it is systematically considered in Yih's monograph [4]; and see also the book by Batchelor, [5], p. 75.

For **u**, ρ, temperature T and thermodynamic pressure $p = R\rho T$, where R is the thermally perfect gas constant, when we consider a non-viscous, compressible and adiabatic atmospheric motion, we have the following Euler non-dissipative system of three equations:

[5] Paul Germain wrote to me (in French!): "J'ai pu regarder les feuilles que vous m'avez adressées sur la mise en équation de votre problème. Je prends note du fait que vous ne passez plus par la forme intermédiaire des équations de la convection qui figurait dans les documents que vous m'aviez antérieurement donnés. Je ne suis néanmoins pas satisfait, car je ne vois toujours pas comment est justifiée la cohérence de vos approximations et pourquoi, alors que vous supposer les perturbations de vitesses petites, en particulier la quantité: $u^2 + w^2 - U_\infty{}^2$, afin d'obtenir des équations linéaires, vous ne linéarisez pas les conditions aux limites. Vous devez me trouver un peu 'tâtillon'. Mais si je dois faire partie du jury de votre thèse, c'est à titre de mécanicien des fluides et comme tel, je souhaiterais comprendre le bien fondé des équations de départ. Or depuis votre exposé au séminaire, j'éprouve toujours la même difficulté et les variantes que vous m'avez proposées ne m'éclairent pas."

$$D\rho/Dt + \rho\nabla.\mathbf{u} = 0, \tag{1.2a}$$

$$DS/Dt = 0, \tag{1.2b}$$

$$\rho\, D\mathbf{u}/Dt + \nabla p + \mathbf{g}\rho = 0. \tag{1.2c}$$

For the specific entropy we have the relation:

$$S = Cv \log[p/\rho^{\gamma}], \text{ with } \gamma = Cp/Cv, \tag{1.3}$$

the ratio of specific heats, and in (1.2c) the gravity force (with a measure g) is taken into account

It is well known that an incompressible fluid motion is obtained (from a compressible fluid motion) as a result of the following formal limiting "incompressible process":

$$\mathrm{Lim}_{is} = \gamma \to \infty \text{ with Cp fixed.} \tag{1.4}$$

With (1.4), in place of Eq. 1.2b, according to (1.3), we recover the above mentioned isochoricity condition ($D\rho/Dt = 0$) which leads, from the equation of continuity (1.2a), to the usual incompressibility constraint:

$$\rho = \text{constant} \Rightarrow \nabla.\mathbf{u} = 0.$$

Finally, in place of the Euler system of Eq. 1.2a–1.2c, with (1.3) – and as a consequence of (1.4) – we derive for the limit isochoric functions, $\mathbf{u}_{is}$, p_{is}, and ρ_{is}, the following simplified isochoric system of inviscid equations:

$$D\rho_{is}/Dt = 0, \quad \nabla.\mathbf{u}_{is} = 0, \tag{1.5a, b}$$

$$\rho_{is} D\mathbf{u}_{is}/Dt + \nabla p_{is} + \mathbf{g}\rho_{is} = \mathbf{0}. \tag{1.5c}$$

From (1.5a, b), $\nabla.\mathbf{u}_{is} = 0$, we have the possibility of introducing two stream functions, ψ and χ (as in [6]), such that in a 3D, steady case, $\partial\mathbf{u}_{is}/\partial t = 0$, $\partial S_{is}/\partial t = 0$, and $\partial\rho_{is}/\partial t = 0$, we obtain the following three relations:

$$\mathbf{u}_{is} = \nabla\psi \wedge \nabla\chi, \tag{1.6a}$$

$$\rho_{is} = \rho^*(\psi,\chi), \tag{1.6b}$$

$$(1/2)\mathbf{u}_{is}^2 + (p_{is}/\rho_{is}) + g\, z = I^*(\psi,\chi), \tag{1.6c}$$

where z ($\equiv$ **x. k** is directed above along the unit upward vector **k**) is the altitude, and the two functions, $\rho^*(\psi,\chi)$ and $I^*(\psi,\chi)$, are subject to a determination.

In particular, for the lee-waves problem, over and downstream of a mountain, this determination is performed via the boundary conditions at upstream infinity where, in a simple case, an uniform horizontal flow is assumed given.

From Eqs. 1.6a–1.6c we derive (again, according to [6]) two scalar equations for ψ and χ :

$$(\nabla \wedge \mathbf{u}_{is}).\nabla\psi = \partial I^*/\partial\chi + (p_{is}/\rho{*}^2)\partial\rho^*/\partial\chi; \quad (1.7a)$$

$$(\nabla \wedge \mathbf{u}_{is}).\nabla\chi = -\partial I^*/\partial\psi - (p_{is}/\rho{*}^2)\partial\rho^*/\partial\psi. \quad (1.7b)$$

If, now, $\mathbf{u}_{is}{}^{\infty} = U^{\infty}(z_{\infty})\mathbf{i}$ is the speed (along the axis of $x \equiv \mathbf{x}.\,\mathbf{i}$) far upstream of the mountain, which is simulated by the equation $z = \mu h(x)$, then at $x \to -\infty$, with $h(-\infty) \equiv 0$, the conditions are:

$$u_{is}{}^{\infty} = U^{\infty}(z_{\infty}),\ v_{is}{}^{\infty} = w_{is}{}^{\infty} = 0\ \rho_{is}{}^{\infty} = \rho^{\infty}(z_{\infty}); \quad (1.8a)$$

$$\psi = -\int_0^{z_\infty} U^{\infty}(z)dz = \psi_{\infty}(z_{\infty}), \quad (1.8b)$$

where $\mathbf{u}_{is}{}^{\infty} = (u_{is}{}^{\infty}, v_{is}{}^{\infty}, w_{is}{}^{\infty})$, and z_{∞} being, therefore, the altitude of a stream line in the basic non-disturbed two-dimensional far flow. In this particular, simple case, (1.8a,1.8b), the second stream function at infinity upstream is simply the plane (x, z), and $\chi_{\infty} \equiv y = \text{const}$.

We will suppose also, implicitly, that the solution of the considered lee-waves problem ought to be uniformly bounded at all points of the infinite plane (x, z). We assume also that $\psi = 0$ determines the wall of the mountain, and in a such case:

$$I^* = B(\psi) \text{ and } \rho^* = R(\psi) \quad (1.9a, b)$$

and in place of two Eqs. 1.7a, 1.7b, with the conditions (1.8a,1.8b), we derive a three-dimensional generalization of the 2D equation of Long, considered in his well-known paper [7]:

$$\nabla \wedge [\nabla\psi \wedge \nabla\chi].\nabla\psi = 0, \quad (1.10a)$$

$$\nabla \wedge [\nabla\psi \wedge \nabla\chi].\nabla\chi = -U^{\infty}(dU^{\infty}/d\psi) + (1/2)(d\log R/d\psi)(\nabla\psi \wedge \nabla\chi)^2 - (d\log R/d\psi)\{(1/2){U^{\infty}}^2 + g(z - z_{\infty})\}. \quad (1.10b)$$

In the case when (far upstream of the mountain):

$$U^{\infty} = (U^{\infty})^0 = \text{const} \Rightarrow \psi_{\infty}(z_{\infty}) = -(U^{\infty})^0 z_{\infty}, \quad (1.11a)$$

$$\rho^{\infty}(z_{\infty}) = \rho^{\infty}(0)\exp[-\beta z_{\infty}], \tag{1.11b}$$

and if we introduce (see (1.12)) the non-dimensional quantities (H is a characteristic meso-length-scale)

$$\xi = x/H, \zeta = z/H, \Psi = \psi/H(U^{\infty})^{0}, X = \chi/H, \tag{1.12}$$

we obtain, as in our thesis [2], in place of Eq. 1.10b, the following dimensionless equation:

$$\nabla \wedge [\nabla\Psi \wedge \nabla X].\nabla X + \mathcal{D}(\Psi + \zeta) = \lambda[(\nabla\Psi \wedge \nabla X)^2 - 1], \tag{1.13}$$

where

$$\mathcal{D} = \beta H^2[g/(U^{\infty})^{02}] \text{ and } \lambda = \beta(H/2), \tag{1.14a}$$

and we observe that the following relation

$$\lambda/\mathcal{D} = {Fr_H}^2 \tag{1.14b}$$

is true, where ${Fr_H}^2(= (U^{\infty})^{02}/[gH])$ is the square of a Froude number.
But, ${Fr_H}^2 \ll 1$ when $H \gg (U^{\infty})^{02}/g$, and this is indeed the case for the usual meteo data.

The relation (1.14b) shows that the term proportional to λ, in the main Eq. 1.13 must be small ($\lambda \ll 1$), because it is necessary (in (1.13)) that

$$\mathcal{D} = \lambda/{Fr_H}^2 = O(1), \tag{1.15}$$

as a ratio of two small parameters.

The parameter $\mathcal{D}$, being the main lee-waves parameter is the so-called Dorodnitsyn–Scorer parameter.

The relation (1.15) is, in fact, a similarity rule between two small parameters: λ and ${Fr_H}^2$ – the use of (1.15) being a key step in the derivation of our leading-order consistent Eq. 1.17 below.

Rigorously, the term proportional to λ can be neglected, in a first approximation, relative to the term with $\mathcal{D}$, which is assumed O(1), only when

$$\beta \ll 2/H, \tag{1.16}$$

and, in such a case, in leading-order approximate model Eq. 1.17, with subscript '$_B$':

$$\nabla \wedge [\nabla\Psi_B \wedge \nabla X_B].\nabla X_B + \mathcal{D}(\Psi_B + \zeta) = 0, \tag{1.17}$$

where for X_B we have as the equation (according to (1.10a)):

$$\nabla \wedge [\nabla \Psi_B \wedge \nabla X_B].\nabla \Psi_B = 0, \tag{1.18}$$

The effect of the compressibility is present only in the last term of (1.17) proportional to $\mathcal{D}$.

This above approximation is just the well-known Boussinesq approximation of 1903 [8]: "The derivatives of $\rho^{\infty}(z_{\infty})$ can be neglected except when they intervene in the calculation of the force of Archimedes."

In particular, if we assume (2D case) that:

$$X_B \equiv \eta(= y/H) \text{ and } \Psi_B \equiv \psi_p(\xi, \zeta), \tag{1.19}$$

we derive, from (1.17), a linear Helmholtz (*à la* Long [7]) equation:

$$\partial^2\psi_p/\partial\xi^2 + \partial^2\psi_p/\partial\zeta^2 + \mathcal{D}(\psi_p + \zeta) = 0. \tag{1.20}$$

But, if (1.20) is a linear equation (derived without any linearization!) the slip boundary condition, along the wall of our mountain, remains non-linear – the slip condition being down a curvilinear surface of the mountain,

$$\psi_p(\xi, \zeta = \kappa\, h^*(\xi)) = 0, \tag{1.21a}$$

with

$$\kappa = \mu/H \tag{1.21b}$$

and

$$h^*(\xi) \equiv h(H\xi). \tag{1.21c}$$

The above results are, in fact, the main part of my first theoretical contribution to the RAM Approach in fluid dynamics, obtained during the rewriting of my Doctoral thesis in Paris during 1968–1969.

I do not see, in reality, whether Paul Germain was completely satisfied with my new derivation. But however that may be, my efforts in writing a new Chap. 2 for my thesis were successful, and on 10 March 1969, after the defence of this thesis in the Faculty of Sciences of the University of Paris, I obtained the degree of Docteur d'État es Sciences Physiques – Paul Germain and Jean-Paul Guiraud being members of my thesis jury, with, as President of the Jury, Paul Queney,[6] Professor at the Sorbonne.

[6] The first theoretical investigations concerning 3D lee-waves problems in linear approximation was, in fact, carried out by Paul Queney. On the other hand, an excellent synthesis of theoretical developments on relief (lee) waves will be found in WMO Technical Note: "The Air flow over mountains", N° **34**, Geneva, 1960, by P. Queney et al.

In the last chapter of this book (in Sect. 9.1) the reader can find a more elaborate RAM Approach to fluid dynamics, for the 2D steady lee-waves problem, in the framework of low-Mach-number fluid flow (hyposonic) theory, which leads to a family of consistent, limiting leading-order model equations.

Concerning the full justification of the Boussinesq assertion and a satisfactory answer to Paul Germain's question – this justification of the Boussinesq approximate equations was for 5 years a major challenge for me, and I devoted considerable effort to the resolution of this problem.

Only in 1973, in the framework of low-Mach-number asymptotics, taking into account the existence of a hydrostatic reference state (function only of the altitude), did I well understand the way for a consistent non-contradictory RAM Approach.

In 1974 [9] these Boussinesq approximate equations for a viscous and non-adiabatic dissipative atmospheric motions were derived from the full unsteady NS–F dissipative equations.

To describe the atmospheric motions, which represent the departure from the hydrostatic reference state, I have considered the perturbations of pressure, density and temperature (these atmospheric perturbations being usually very small, relative to the hydrostatic reference state) and have rewritten (without any simplifications) the NS–F atmospheric equations relative to these thermodynamic perturbations and velocity vector.

This derived, very awkward, dimensionless system of equations is, in fact, a new (exact) form of the NS–F classical atmospheric equations well adapted for the application of our RAMA theory. In Chap. 4 we discuss a detailed RAMA of these Boussinesq approximate equations, inspired by my "Boussinesq's Centenary Anniversary paper" [10] of 2003, but for the sake of simplicity, only in the framework of a Euler non-viscous, compressible and adiabatic system of Eq. 1.2a–1.2c – this derivation being an instructive test problem for the formulation of our key steps in Chap. 6, devoted to the mathematics of the RAMA.

1.2 My Collaboration with Jean-Pierre Guiraud in the Aerodynamics Department of ONERA

In September 1967, thanks to the recommendation of Jean-Pierre Guiraud, I began a new career as a research engineer in the Aerodynamics Department of the Office National d'Études et de Recherches Aérospatiales, in Chatillon, near Paris. After working at ONERA for 5 years, in October 1972 I was – thanks to my Doctoral thesis (March 1969) – appointed Titular Professor of Fluid Mechanics at the University of Lille 1 – a position which I held until 1996.

I continued part of my theoretical researches in fluid-flow modelling as a 'Collaborateur Extérieur' at ONERA, and during 16 years there, from 1970 to 1986, for a full day once a week I worked with Jean-Pierre Guiraud in exchanging ideas and envisioning asymptotic modelling for various aerodynamics, stability/

turbulence and meteo problems. As a result of this collaboration, throughout this period we jointly published 26 papers in various scientific journals. (See, for instance, in the References, our 1986 paper [11], and references to other papers published either jointly or separately.) These works (published during 1971–1986) are devoted, with more or less success, to the application of the ideas that we discussed concerning various fields in fluid dynamics – all being motivated by the need for solving or understanding the basis of the solution of technological and geophysical problems involving fluid flows. These problems are related to:

Vortex flows in rotating machines (taking into account that the blades in a row are usually very close).

Rolled vortex sheets (in a region where the contiguous branches of the rolled sheet are so close to each other that they are very difficult to capture by numerical simulation).

Hydrodynamic stability (in a weakly non-linear domain, through perturbation techniques – the underlying mathematical theory being the so-called bifurcation theory).

Atmospheric flows (see Chap. 9 in this book, and our monograph, *Asymptotic Modeling of Atmospheric Flows* [12], published in 1990).

Flow at low Mach numbers (see our *Topics in Hyposonic Flow Theory*, Lecture Notes in Physics, vol. 672, 2006 [13]).

It was an extremely stimulating period of scientific research, for me. As far back as at the end of 1970 years it is evident that asymptotic techniques provide very powerful tools in the process of constructing working models for fluid dynamics problems, which are stiff, from the point of view of numerical analysis, coupled with a simulation via a powerful super-computer.

My approach differs from Van Dyke's exposition in [14], in the sense that: "Computational fluid dynamics is now a quite mature discipline, and for some time the growth in capabilities of numerical simulation will be dependent on, or related to, the development of rational asymptotic modelling approach – RAMA." If such is the case, then a simple definition of our RAMA is: "The art of a strongly argumentative, consistent, non-ad hoc and non-contradictory modelling assisted by the spirit of asymptotics."

It is my opinion that RAMA will remain for many years, or even decades – a quite powerful tool in deriving mathematically consistent models for numerical/ simulation fluid research. By "mathematical", I mean that the models derived by RAMA, the approximate consistent models, under consideration, should be formulated as reasonably well-posed initial and/or boundary value problems, in place of the starting full NS–F extremely complex and stiff problems (as, for example, turbomachinery flows, which are arguably among the most complicated known to man and are of great technological importance – see Sect. 7.1).

I again observe that our RAMA is an extremely worthwhile objective, because most of the relevant engineering computations are based on relatively ad hoc models that are rife with internal inconsistencies.

1.3 A Few Remarks Concerning My Preceding Books on Modelling in Newtonian Fluid Flows

Concerning the joint efforts of Guiraud and myself in our tentative writing of a book on the RAMA in fluid dynamics, I must say first that in 1977, after several attempts to persuade him, we both worked intensively, up to 1982, on a hand-written (in French) manuscript entitled "The laminar flows at high Reynolds numbers: an essay on the asymptotic modelling of Newtonian dynamics of fluids." The possibility of publication, after a rewrite in English, became a reality – at least for me!

Unfortunately, at that moment our opinions diverged concerning the opportunity of such publication of our finished manuscript in its 1982 form. Guiraud wished to pursue a deeper investigation of some delicate and difficult questions requiring time and additional research. Contrary to Guiraud, I was of the opinion that further investigations would be of no benefit and, in particular, would not provide anything else to support our initial objective: to show the effectiveness of our RAMA!

Finally, in 1986 and 1987 I published alone (but by common consent) a course in two volumes, in French, in the Springer series Lecture Notes in Physics (LNP): *Les Modèles Asymptotiques de la Mécanique des Fluides*, I [15] and II [16] – more or less inspired by the manuscript produced by myself and Guiraud in 1982.

As Titular Professor at the Université de Lille 1, beginning in 1972, I systematically used, throughout almost 10 years, various parts of our manuscript in my teaching of theoretical fluid mechanics as a first Course and second Course, respectively, for final-year (M.Sc.) undergraduate students and post-graduate research workers, and for students preparing a doctoral thesis.

In the beginning, in 1977, my goal was, in fact, a monograph devoted to RAMA for Newtonian fluid flows, and I had in mind the derivation of various models corresponding to parameters (not only to Reynolds number) characterizing various (high or low) physical effects – viscosity, compressibility, heat conduction, gravity, Coriolis force, unsteadiness, geometrical constraints, and so on.

The above-mentioned two-volume Course was my first experience in opening a new way into the difficult field of theoretical (analytical) fluid mechanics via the NS–F equations, offering fresh ideas together with a first systematic presentation of asymptotic approach in fluid dynamics for both students and young researchers. In a short critical review (*J. Fluid Mech.*, 1991, vol. 231, p. 691), the following opinion was expressed concerning this two-volume Course:

> The text is in French. Equations are hand-written but very clearly done. In many of the areas covered in these two volumes there is a conspicuous lack of suitable expository material available elsewhere in the literature, and Professer Zeytounian's notes are to be welcomed for filling these gaps until fuller and more specialized accounts appear in book form.

In addition, the following appeared in *Mathematical Reviews*, 1988:

> A reader having acquired a practical knowledge of the asymptotic methods which are presented and used here may certainly benefit by the advanced material about Navier–Stokes equations provided in the main body of these two volumes.

Later, in 1994, a third volume was published, also in French: *Modelisation asymptotique en mécanique des fluides newtoniens* [17] – and here it seems judicious to quote several sentences extracted from a review (in *Appl. Mech. Rev.*, vol. **49**(7), July 1996, p. 879) by J.-P. Guiraud:

> The purpose of this book is to present, through extensive use of dimensional analysis and asymptotic calculus, a unified view of a wide spectrum of mathematical models for fluid mechanics ... Usually, adequacy of a mathematical model is evaluated a priori through physical insight, experience, and inquiries about the topic. Here, the reader is proposed to become, on his own, an expert in adequacy, by systematic use of asymptotic approximation. Although a rather large spectrum of books on fluid mechanics and asymptotic methods may be found, it seems to this reviewer that the present one is rather exceptional by the extent and logical organization of the material ... A fascinating aspect is that the reader is led by the hand through a jungle of very different mathematical models, including Euler and Navier for incompressible fluid; Prandtl for boundary layer; Stokes, Oseen, and Rayleigh for various viscous effects; the usual regimes of aerodynamics, Boussinesq for the atmosphere and the ocean, primitive equations and quasi-geostrophic approximation for meteorology, gravity waves, amplitude equations of KdV or Schrodinger type, and low Mach number flows, including acoustics ... *Modelisation asymptotique en mécanique des fluides newtoniens* is a valuable book which is recommended both to individuals and libraries for the precise purpose indicated in the second sentence of this review. In principle, it is self-contained and might be a reference for students, engineers, and researchers who master computational aspects but want to be able to assess what kind of approximations are involved in the equations, as well as initial and boundary conditions with which they struggle.

Concerning, more precisely, the application of our RAMA to atmospheric motion, after my 1975 survey lecture,[7] published in 1976 [18], I decided to write a monograph devoted entirely to asymptotic modelling of atmospheric flows (see [12]), and in 1985 a manuscript (written in French) was ready. This manuscript was accepted by Prof. Dr. W. Beiglböck, of Springer-Verlag, Heidelberg, for publication in English. Unfortunately, the translation into English, by Lesly Bry, is infelicitous!

I think that my *Meteorological Fluid Dynamics* [19] is good preparation for the reading of [12], which was in fact published a little earlier than [19]. Here again, I quote part of a review of [12] (SIAM Review, vol. **33**(4), 1991, pp. 672–3) by Huijun Yang (University of Chicago):

> The present work is not exactly a 'course', but rather is presented as a monograph in which the author has set forth what are, for the most part, his own results; this is particularly true of Chaps. 7–13. (quoted from the Preface of this book):'In the book, the author viewed meteorology as a fluid mechanics discipline. Therefore, he used singular perturbation methods as his main tools in the entirety of the book ... The book consists of the author's more than 25 years work.' In the 32 references to his own work, fewer than one third were published in English, with the rest in Russian or French. Throughout the book, the reader can strongly feel the influence of Soviet works on the author. However, the author does

[7] Entitled *La Météorologie du point de vue du Mécanicien des Fluides*, written for the XIIth Symposium on Advanced Problems and Methods in Fluid Mechanics, Bialowieza (Poland), 8–13 September 1975.

> have his own character. The issues raised in the book, such as the initialization (initial layer) boundary layer treatment, and well-posedness and ill-posedness of the system, are very important problems facing researchers today in atmospheric sciences and other related sciences. The reader will find some valuable information on these issues ... The mathematically consistent treatment of the subject does give this book a unique place on shelves of libraries and offices of researchers ... This book is very different from recent books on the market [for example, Holton [20], Gill [21], Haltiner and Williams [22], Pedlosky [23], and Yang [24]]. I recommend that researchers in atmospheric dynamics and numerical weather prediction read this book to have an alternative view of deriving atmospheric flow models. Researchers in theoretical fluid mechanics might also be interested to see how singular perturbation methods can be used in atmospheric sciences. The book may be used as supplemental material for courses like numerical weather prediction or atmospheric dynamics. However, I do not think it is a suitable textbook for a regular class: as the author said in his Preface. 'I am well aware that this book is very personal – one might even say *impassioned*.'

This review seems rather favourable, but this does not seem to be the case with P. G. Drazin's review (*Journal of Fluid Mechanics*, 1992, 242):

> The author acknowledges that dynamic meteorology is too large a subject for him to attempt to cover completely. He 'has set forth what are, for the most part, his own results' in accord with his 'conception of meteorology as a fluid mechanics discipline which is in a privileged area for the application of singular perturbation techniques.' He applies the method of multiple scales or the method of matched asymptotic expansions to any equations he can find, systematically reviewing his own and his associates' extensive research. So it is a very personal view of meteorology, covering some areas of geophysical fluid dynamics with a formidable battery of notation. The nature of the subject demands a large and complex notation in any rational treatment. But the notation will allow only a few to benefit from this book. It is a barrier which I found hard to cross, not having the time and will to work through the book line by line, as a result with which I am familiar was difficult for me to follow. The approximations of dynamic meteorology are mostly singular. Nonetheless, their essentials are well understood by meteorologists now, albeit in a rough and ready way, and meteorologists are unlikely to be [influenced] by Zeytounian's approach. Yet the approximations of meteorology are subtle and deserve a more rational development than is commonly understood. This is the achievement of the book. The author derives many equations systematically, albeit not rigorously, from the primitive equations, rather than solving equations governing particular problems. For all these reasons, I feel that the book will be studied intensively by a few specialists but neglected by others.

Obviously, the last two sentences of this critical review are very controversial – especially the assertion of 'albeit not rigorously'. It seems clear that for Drazin, my "French touch" is not to everyone's taste! Indeed, the publication of this monograph in 1990 was possibly premature, despite the publication, in 1985, of my survey paper [25], *Recent Advances in Asymptotic Modelling of Tangent Atmospheric Motions*, devoted to an asymptotic rational theory of the modelling of atmospheric motions in a 'flat earth' closely related with the so-called β–plane approximation.

In [12], [19], and [25] my clear purpose was to initiate a process which does not seem to have sufficiently attracted the attention of scientists. This process involves the use of the RAMA for carrying out models; that is, for building approximate simplified (but consistent) well-posed (at least from a fluid dynamician's point of view) problems based on various physical situations and concerned with one or several high or low parameters.

I do not, of course, affirm that this is the only method, nor even the most efficient one, for deriving such problems in view of a numerical/computational simulation. I do, however, feel that when such a procedure is feasible it should be undertaken. As a matter of fact, the application of this approach makes it possible, in principle, to improve, at least, a second-order model problem obtained from the NS–F full problem, used by advancement in the associated asymptotic expansion.

After 1996 – having retired from the University of Lille 1, and with more time to write – I decided to return to my first (1977) idea concerning a monograph devoted to RAMA for Newtonian fluid flows and the derivation of models corresponding to various parameters (not only to Reynolds number) characterizing various (high or low) physical effects. I (partly) realized this objective in 2002 with my monograph [26] *Asymptotic Modelling of Fluid Flow Phenomena* – the first book in English devoted entirely to asymptotic modelling of fluid flow phenomena, dealing with the art of asymptotic modelling of Newtonian laminar fluid flows. In Chaps. 2–12 of this work I consider several important topics involved in the accomplishment of my objective in determining how simplified rational consistent asymptotic simplified models can be obtained for the most technologically important fluid mechanical problems.

According to Marvin E. Goldstein (of NASA Glenn Research Center) in his detailed, scrupulous, and rigorous review (SIAM Review, vol. **45**(1), pp. 142–146, 2003): "There are enough of the selected topics that accomplish the author's objective to make this book an important contribution to the literature." Goldstein also writes:

> Applied mathematicians have always found fluid mechanics to be a rich and interesting field, because the basic equations (i.e., the Navier–Stokes equations) have an almost unlimited capacity for producing complex solutions that exhibit unbelievably interesting properties, and because the dimensionless form of these equations contains a parameter (called the Reynolds number) which is usually quite large in technologically and geophysically interesting flows. This means that asymptotic methods can be used to obtain approximate solutions to some very interesting and important flow problems. These solutions usually turn out to be of non-uniform validity (i.e., they break down in certain regions of the flow), and matched asymptotic expansions have to be used to construct physically meaningful results. However, advances in computer technology have led to the development of increasingly accurate numerical solutions, and have thereby diminished the interest in approximate analytical results. But real flows (especially those that are of geophysical or engineering interest) are extremely complex and exhibit an enormous range of length and time scales whose resolution will probably remain well beyond the capabilities of any computer that is likely to become available in the foreseeable future. So simplification and modelling are still necessary, not only to meet the engineer's requirement for generating numbers but also for developing conceptual models that are simple enough to be analyzed and understood. The asymptotic scaling techniques and the reduced forms of the general equations that emerge from the matched asymptotic expansion process (as well as from other singular perturbation techniques) provide a rational and systematic method for obtaining the necessary simplified flows model, which, in most cases, still have to be solved numerically. The author states that the goal of this book is to promote the use of asymptotic methods for developing simplified but rational model for the Navier–Stokes equations which can then be solved numerically to obtain appropriate descriptions of the flow.

> This is an extremely worthwhile objective, because most of the relevant engineering computations are based on relatively ad hoc models that are rife with internal inconsistencies. To my knowledge, this is the first book devoted to accomplishing the author's stated objective, and it is, therefore, unfortunate that it is not as well executed as it could be.

To mitigate this last critical sentence, however, he continues:

> However, there are enough of the selected topics that accomplish this objective to make this book an important contribution to the literature. It contains an excellent chapter (Chap. 11) in which the classical model equations for large-scale atmospheric motion are derived in a fairly rational fashion. The author also devotes a full section (Sect. 6.6) to turbomachinery flows, which are arguably among the most complicated known to man and are of great technological importance. It is remarkable that he was able to make some progress toward developing an asymptotic basis for some of the more prominent engineering models of these flows.

Goldstein concludes: "It is this reviewer's hope that the deficiencies in this work will encourage others to write new and improved books with similar themes" – but unfortunately, it seems that for the present this is not the case! On the other hand, in 2006 and 2009 Springer published my two monographs, *Topics in Hyposonic Flow Theory* [13], devoted to hyposonic (low Mach numbers) flows, and *Convection in Fluids: A Rational Analysis and Asymptotic Modelling* [27], mainly related to the well-known Bénard convection problem in a layer of weakly expansible liquid heated from below.

Concerning the first of these, a considerable amount of stimulation and encouragement was derived from my collaboration with Jean-Pierre Guiraud, who, over a period of 20 years, has played an important role in asymptotic modelling of the various low-Mach-number flow problems presented in that book. The reader should take into account that is the first book devoted to hyposonic flow theory, and it is the author's hope that the various unavoidable 'deficiencies' (noted by Goldstein in his review of [13]) will persuade others to work on similar themes.

On the other hand, in [27] the main motivation was a rational analysis of various aspects (in particular, the influence of the viscous dissipation, free surface and surface tension) of the Bénard convection (heated from below) problem. It presents a careful investigation of three significant approximate models (see, for instance, Chap. 8 in the present book) related to the Bénard (1900) experiments, by which he discovered his well-known Bénard cells! It is evident that Professor Manuel G. Velarde was influential when I wrote my book on convection in fluids [27], as I benefited greatly from our collaboration and many discussions relating to Marangoni thermocapillary convection during my sojourn at the Instituto Pluridisciplinar UCM de Madrid in 2000–2004.

1.4 Conclusion

More than 50 years ago, with the works of Kaplun, Kaplun and Lagerstrom, and Proudman and Pearson, asymptotic techniques provided a new impetus for research in theoretical fluid dynamics. Twenty years later, a much more powerful revival was possible due to the dramatic influence of high-speed computers and the numerical analysis and simulation of fluid flow problems. The survey paper by Birkhoff [28] includes, through a series of case studies, a detailed assessment of the status, development, and future prospects of numerical/computational fluid dynamics.

During the early times, asymptotic techniques were mostly used in order to derive approximate solutions in closed form. Perhaps of more significance for the progress of understanding and also of research, however, was the use of asymptotic techniques in order to settle, on a rational basis, a number of approximate models which much earlier were often derived by ad hoc non-rational procedures. One of the most well-known examples is Kaplun's celebrated paper on boundary-layer theory, which provided a firm theoretical basis for some 50 years of boundary-layer research.

From this early example it is clear that my 1977 idea, relative to asymptotic techniques as a well-suited and invaluable tool for the derivation of mathematically consistent models (from full fluid dynamics equations – NS–F equations) which are amenable to numerical treatment rather than for obtaining closed form solutions, was a perspective of scientific activity in interaction with numerical simulation – even though, in the 1970s, numerical fluid dynamics was almost non-existent due to the lack of high-speed computers.

It is now evident that asymptotic techniques serve as very powerful tools in the process of constructing rational consistent mathematical simplified models for problems which are stiff, from the point of view of numerical analysis. Here, Chap. 6 is devoted to the mathematics of the RAM Approach, which seems a good basis for a practical use of this RAMA in simulation/computation via high-speed computers.

As a matter of fact J. P. Guinaud, who read a large part of the Chaps. 1 to 6 of the present book suggest me to quote what follows: "While having been absent from the Community of fluid mecanicians fifteen years from now, it was a pleasure for me to read the report by Zeytounian, of a long coworking with him. It is mere justice to mention that, during this active collaboration, a number of ideas were initiated by Zeytounian. My main contribution was the result of ten years of struggle with asymptotics before I had the good fortune to meet Zeytounian." A significant contradiction is obvious in the scientific activity of J. P. Guinaud. He never published any book whilst having written a number of documents, corresponding to many courses he teached manuscripts which were much more carefully written than simple notes to be distributed for the students. In particular, the Guirand Notes ("Topics in hypersonic flow theory", Department of Aeronautics and Astronantics, Stanford University; SUDAER n° **154**, may 1963, Stanford, California, USA) are Published in Russian by MiR, Moscow Editions, as a book in 1965.

Part I
Navier–Stokes–Fourier Equations: General Main Models

Chapter 2
Newtonian Fluid Dynamics as a Mathematical – Physical Science

There are two mathematical–physical descriptions of fluid dynamics.[1] The first of them is a microscopic description, from the Boltzmann equation for the (one-particle) distribution function $f(\mathrm{t}, \mathrm{x}; \xi)$:

$$\partial f/\partial \mathrm{t} + \xi.\nabla f = (1/\mathrm{Kn})Q(f;f),$$

where $f(\mathrm{t}, \mathrm{x}; \xi)$ is precisely the density probability of finding a molecule at the space-position $\mathbf{x}$ (at the time t), with the velocity ξ. The parameter Kn in the Boltzmann equation is the Knudsen number, which is the ratio of the mean free path (between collisions of molecules – a microscopic reference length, l°) and a typical (macroscopic) reference length, L°, of the classical continuum theory, which is the ratio of Mach (M) and Reynolds (Re) numbers:

$$\mathrm{Kn} = \mathrm{M}/\mathrm{Re},$$

where $\mathrm{M} = \mathrm{U}^{\circ}/\mathrm{c}^{\circ}$ is the constant Mach (dimensionless) number, based on the reference macroscopic velocity U° and the speed of sound c°, that characterizes the compressibility effect. The parameter $\mathrm{Re} = \mathrm{U}^{\circ}\mathrm{L}^{\circ}/\nu^{\circ}$ is the constant Reynolds (dimensionless) number that characterizes the viscosity (via the kinematic viscosity coefficient ν°) effect.

[1] In my 2001 review paper [29] – written by a fluid dynamicist for fluid dynamicists – the curious reader can find a contribution concerning the many theoretical mathematical investigations of Navier–Stokes–Fourier problems. My intent was to extract from the huge literature the basic results, ideas, and goals of this currently wide activity and to present the results to the readers of *Applied Mechanical Review*. I am sure that rigorous mathematicians will find in this paper many shortcomings, non-rigorous formulations, and so on. I think, however, that such a paper will stimulate further thinking by engineers and applied scientists, including some exchange of opinions, and so on, and that it is therefore needed. The distance between theoretical mathematicians and applied mathematicians and engineers has become too large! I hope that both old and new investigators interested in Newtonian fluid flow problems might learn much from it.

R. Kh. Zeytounian, *Navier–Stokes–Fourier Equations*,
DOI 10.1007/978-3-642-20746-4_2, © Springer-Verlag Berlin Heidelberg 2012

The right-hand side of the above Boltzmann equation, the collision operator Q, is typical of kinetic theory of gases in that it preserves mass, momentum, and, namely:

$$\int \psi Q d\xi, = 0,$$

where $\psi = (1, \xi_1, \xi_2, \xi_3, |\xi|^2)$ is a five-component (the so-called collisional invariants) vector. For a modern exposition of the kinetic theory of gases (dilute gas) see Cercignani, Illner, and Pulvirenty, 1994 [30].

The second continuum description is linked with the macroscopic length scale (which is the real scale for applications in fluid flows), and is governed by the three conservation equations of classical continuum mechanics: principle of conservation of mass (assuming that the fluid possesses a density function $\rho(\mathrm{t}, \mathbf{x})$), principle of conservation of linear momentum (adopting the stress principle of Cauchy; see Sect. 2.3.1), and the conservation of energy (we postulate that the total energy of a volume – the sum of its kinetic energy and its internal energy – is conserved; see Sect. 2.3.3).

The resulting three equations of continuum mechanics, which proceeds on the assumption that a fluid is practically continuous and homogeneous in structure (see, Serrin, 1959 [31]) are:

$$\mathrm{D}\rho/\mathrm{Dt} + \rho\nabla.\ \mathrm{u} = 0,$$

$$\rho\mathrm{D}\mathbf{u}/\mathrm{Dt} = \rho \boldsymbol{f} + \nabla.\boldsymbol{T},$$

$$\rho\mathrm{D}E/\mathrm{Dt} = \boldsymbol{T} : \boldsymbol{D} - \ \mathrm{div}\boldsymbol{q},$$

where $\mathbf{u}$ is the velocity vector, $\boldsymbol{f}$ is the extraneous force per unit mass (a known function of position $\mathbf{x}$ and time t), $\boldsymbol{T}$ is the Cauchy stress tensor, E is the specific internal energy, $\boldsymbol{q}$ is the heat flux vector, and the term $\boldsymbol{T}:\boldsymbol{D}$, in the energy equation for E, is a "dissipation" term involving the interaction of stress and deformation (second-order tensor $\boldsymbol{D}$). We observe that $\boldsymbol{T}:\boldsymbol{D}$ stands for the scalar product $T^{\mathrm{ij}}D_{\mathrm{ij}}$ of two second-order tensors (dyadics), and T^{ij} and D_{ij} are, respectively, the components of $\boldsymbol{T}$ and $\boldsymbol{D}$.

The problem of the derivation of the fluid dynamic equations (derived from above three continuum mechanics equations, with Cauchy stress tensor $\boldsymbol{T}$ and the heat flux vector $\boldsymbol{q}$, due to Navier–Stokes and Fourier constitutive equations – see Sects. 2.3.2 and 2.3.3) from the Boltzman equation for small Knudsen numbers (Kn $\downarrow$ 0) is shortly expounded in Sect. 5 of our review paper [29], where the reader can find various pertinent references concerning this fluid dynamics limit of kinetic equations, initiated by Hilbert in 1912.

2.1 From Newton to Euler

Sir Issac Newton, English mathematician and physicist, was the greatest single influence on theoretical physics until Einstein. In his major treatise, *Philosophia Naturalis Principia Mathematica* (1687) [32] he presented a mathematical description of the laws of mechanics and gravitation, and applied this theory to explain planetary and lunar motions. In the Second Law we read: "The body moves in such a way that at each moment the product of its acceleration vector by the density is equal to the sum of certain other vectors, called forces, which are determined by the motion taking place." That is:

$$\rho\ \mathbf{Du}/\mathrm{Dt}\ = \rho\mathbf{g} - \text{internal force per unit volume.} \tag{2.1}$$

A second part of Newton's *Principia* is related to the conservation of mass: "To each small solid body can be assigned a positive number *m*, invariant in time, called its mass." That is:

$$\mathrm{D/Dt}\left\{\int_V \rho dV\right\} = 0. \tag{2.2}$$

In (2.1), $\mathbf{u}$ is the velocity vector, $\mathbf{g}$ is the gravitational force per unit mass, and ρ is the density. The (Cartesian) components of the nabla, ∇, operator, in material (or substantial) derivative

$$\mathrm{D/Dt}\ = \partial/\partial\mathrm{t}\ + \mathbf{u}.\nabla,$$

are $\partial/\partial\mathrm{x_i}$, $\mathrm{i} = 1,2,3$, where the time is denoted by t, and $\mathbf{x} = (\mathrm{x}_1, \mathrm{x}_2, \mathrm{x}_3)$ is the position vector.

In (2.2), *dV* is a volume element in the neighbourhood of the point P, and to this volume element will be assigned a mass ρdV.

We observe that D/Dt is related to the Euler rule of differentiation, and t, $\mathbf{x}$ is the Euler time–space variable. To express (2.2) in the form of a differential equation, the differentiation indicated in this equation is carried out by transforming the integral suitably. In this case we derive the so-called equation of continuity (this derivation is, in fact, due to Euler in 1755 [33]):

$$\mathrm{D}\rho/\mathrm{Dt}\ + \rho\nabla.\mathbf{u} = 0. \tag{2.3}$$

This (compressible) equation of continuity (2.3) remains unaltered when viscosity is admitted.

2.1.1 Eulerian Elastic Fluid

In reality, fluid dynamics was first envisaged as a systematic mathematical–physical science in Johann Bernoulli's *Hydrodraulics* (1737) [34], in Daniel

Bernoulli's *Hydrodynamica* (1738) [35], and also in D'Alembert *Traité de l'Équilibre et du mouvement des fluides* (1744) [36]. However, the fundamental ideas expounded in these books were formulated mathematically as partial differential equations in an epochal paper by Euler (1755) [33] which firmly established him as the founder of rational fluid dynamics. Nevertheless, he considered only non-viscous (inviscid) fluid flows with the pressure a function only of the density (a so-called isentropic/barotropic fluid flow – the fluid being an elastic fluid). We observe that an inviscid fluid is one in which it is assumed that the internal force acting on any surface element *dS*, at which two elements of the fluid are in contact, acts in a direction normal to the surface element. At each point *P* (with coordinates x_i, $i = 1, 2, 3$) the stress, or internal force per unit area, is independent of the orientation (direction of the normal) of *dS*, and the value of this stress is called the pressure, p, at the point *P*. Therefore, the internal force per unit volume, appearing in Newton's equation (2.1), has x_i – component $(-\partial p/\partial x_i)$, $i = 1, 2, 3$. As a consequence, for an inviscid (non-viscous–Eulerian) fluid we determine, from (2.1), the classical Euler equation of motion (momentum equation):

$$\rho D\mathbf{u}/Dt = \rho\mathbf{g} - \nabla p. \tag{2.4}$$

Equations 2.3 and 2.4, which express Newton's principles for the motion of an inviscid fluid, are usually referred to as the Eulerian fluid flow (compressible) equations, and include one vector equation (2.4) and one scalar equation (2.3) for **u**, ρ and p (five unknowns).

It follows that one more equation is needed in order that a solution of the system of Euler equations be uniquely determined for given initial and boundary conditions. According to Euler, if we add to Eq. 2.3 and 2.4 the following specifying equation:

$$p = p(\rho) \tag{2.5}$$

which gives the relation between the pressure and the density, we shall have five equations (a closed system) which include all the theory of the motion of fluids.

By this formulation, Euler believed (255 years ago!) that he had reduced fluid dynamics, in principle, to a mathematical–physical science; but it is crucial to note that, in fact, Eq. 2.5 is not an equation of state, but specifies only the particular type of motion (so-called barotropic) under consideration, and in this case the fluid is just called an elastic fluid.

In my book (Zeytounian, 2002) [37], the reader can find a theory and applications of non-viscous fluid flows, and in the next chapter, devoted to a discussion of various general models derived from Navier–Stokes–Fourier equations, we obtain, for large Reynolds number $Re \gg 1$ – as a vanishing viscosity limit – the full unsteady Euler compressible non-viscous adiabatic and baroclinic equations for a thermally perfect gas (a trivariate fluid). (Concerning the NS–F equations see Sect. 2.3.)

These general models very often form the basis of various chapters in fluid dynamics treatises. It is obvious, therefore, that these treatises may be organized through some models which are best obtained by asymptotic modelling. As an example we mention the case of inviscid flows which are often considered as a model, used from the outset, and need to be embedded in the more general model of slightly (vanishing) viscous (laminar) or with slight friction (turbulent) flow, to which asymptotic modelling is applied. Incompressible flows are seldom considered as flow at small Mach numbers – which may lead to almost nonsensical conclusions, as when one deals with incompressible aerodynamics, because phenomena such as sound produced by quite low-speed flow cannot be understood other than by low-Mach-number (hyposonic) aerodynamics.

2.1.2 From Adiabaticity to Isochoricity

In many cases the specification of the type of flow is given in thermodynamic terms. The most common (and rather naive) assumption in the study of compressible fluids is that no heat output or input occurs for any particle. In this case, heat transfer by radiation, chemical processes, and heat conduction between neighbouring particles are excluded, and the fluid flow is called adiabatic.

In order to translate either assumption into a specifying equation, the First Law of Thermodynamics must be used, which gives the relation between heat input and the mechanical variables [J. R. von Mayer (1842)]. If the total heat input from all sources, per unit of time and mass, is zero, the First Law for an inviscid fluid can be written in the following form:

$$C_v DT/Dt + pD/Dt(1/\rho) = 0, \tag{2.6}$$

where C_v is the specific heat of the fluid at constant volume. The first term in (2.6) represents the part of the heat input expended for the increase in temperature T, and the second term corresponds to the work done by expansion. It is well known, also from thermodynamics, that for each type of matter a certain relation exists among the three (thermodynamic) variables, pressure p, density ρ, and temperature T:

$$f(p, \rho, T) = 0, \tag{2.7}$$

Thus the temperature can be computed when p and ρ are known. Naturally, the equation of state (2.7) is not a specifying equation, since it implies temperature as a new variable. Finally, Eqs. 2.3, 2.4, and 2.6, together with (2.7), form a closed system of six equations for the six unknowns: **u,** p, ρ, and T.

For a thermally perfect gas (naturally, a perfect gas is not necessarily inviscid), the equation of state (2.7) is explicit:

$$p = R\rho T, \tag{2.8}$$

where R is a constant depending upon the particular perfect gas. From (2.8) it follows that for a perfect gas the condition $p/\rho = \text{const}$ implies a fluid flow at constant temperature, or isothermal flow.

The specific entropy S of a perfect gas is then given by:

$$S = [R/(\gamma - 1)]\log(p/\rho^{\gamma}) + \text{const}, \tag{2.9}$$

where γ is a constant, having the value 1.40 for dry air. Thus the motion of a perfect gas with the condition $p/\rho^{\gamma} = \text{const}$, as a specifying equation, is isentropic (constant entropy motion or, since $\gamma > 1$, polytropic). The equation of state for a perfect gas in equilibrium, connected with the names of Boyle (see, Birch (1744) [38]), Mariotte, Gay Lussac, and Charles, has been widely known since 1800.

In precisely the modern form, it was used freely by Euler, but did not appear again in the hydrodynamical literature until used by Kirchhoff in his paper of 1868 [39].

In some presentations, no distinction is made between the term "perfect gas" and "ideal gas". Here the term "perfect gas" is defined precisely by the equation of state (2.8). The term "Eulerian fluid flow" is used for an inviscid (non-viscous) and non-heat-conducting flow, governed by the system of Eqs. 2.3, 2.4, and 2.6, with (2.7). According to (2.6) and (2.7) this Eulerian fluid flow is a baroclinic and adiabatic fluid flow. In Eq. 2.6 an expression for C_v in term of the variables T, p, and ρ is needed, but for a perfect gas, where the equation of state is (2.8), it is generally assumed that $C_v = R/(\gamma - 1)$ is a constant – R being the usual gas constant.

As a consequence, we derive, for such a perfect gas with constants C_v and C_p (= $\gamma\, C_v$), specific heats, the following conservation equation for specific entropy in the case of a thermally perfect gas:

$$D/Dt\ [\log(p/\rho^{\gamma}) = 0 \Rightarrow\ DS/Dt\ = 0. \tag{2.10}$$

Equation 2.10, however, holds only for an adiabatic flow of a perfect inviscid gas, when the entropy is constant for each particle but varies from particle to particle. Generally, a thermally perfect inviscid gas in adiabatic flow does not necessarily behave like an elastic fluid.

If we assume that in (2.10),

$$\gamma \text{ tends to infinity (incompressible limit case)}, \tag{2.11a}$$

such that $R = O(1)$, and that in such a case,

$$C_v\ \textit{tends to zero}\ \text{but}\ Cp \equiv R\ =\ O(1), \tag{2.11b}$$

then we derive, again, from (in place of) (2.10) the following evolution equation (conservation law) for density (isochoricity):

$$D\rho/Dt = 0 \Rightarrow \nabla \cdot \mathbf{u} = 0, \tag{2.11c}$$

As a consequence, for a Eulerian incompressible but non-homogeneous (isochoric) fluid flow we obtain the following system of three equations for the velocity $\mathbf{u}$, pressure p, and density ρ:

$$\rho\ D\mathbf{u}/Dt = \rho\mathbf{g} - \nabla p, \tag{2.12a}$$

$$D\rho/Dt = 0, \tag{2.12b}$$

$$\nabla \cdot \mathbf{u} = 0. \tag{2.12c}$$

This isochoric system of three Eqs. 2.12a–2.12c is very well investigated in Yih's 1980 book [4]. The above "incompressible limiting process" (2.11a,2.11b) presents the possibility of taking into account some compressible (second-order) effects of the order $O(1/\gamma)$ – which is the case, for instance, in the theory of lee waves downstream of a mountain!

2.2 Navier Viscous Incompressible, Constant Density Equations

The equation of motion of a viscous and incompressible homogeneous (with a constant density) fluid flow was first obtained by Navier in 1821 [40] and later by Poisson in 1831 [41]. The necessity of such a viscous equation (in place of the above Euler equation (2.4) was strongly linked with the d'Alembert theorem (paradox?): "An object moving with constant velocity $U\infty$ in a potential field (from Bernoulli equation and Lagrange theorem in an incompressible fluid the velocity-potential Φ must satisfy Laplace's equation) does not feel any force – neither drag nor lift."

Obviously, this result is in sharp contrast with experience! For instance, an aircraft could not fly. Suppose that, initially, the aircraft and the fluid (air) are both at rest, then the aircraft begins to move. Since vorticity cannot be produced (Lagrange – permanence of irrotational flow), the potential flow around the aircraft cannot produce any lift, so that flight is impossible. Such a paradox can be avoided if vorticity is present.

However, the problem remains of understanding how vorticity can be created in the system. The conservation of vorticity in an inviscid (incompressible) fluid, while reasonably far from the obstacle, is too drastic near the boundary of this obstacle!

A more accurate description of the interaction among the particles of the fluid and the obstacle leads us to introduce the Navier (viscous and incompressible) equation, which is a correction to the Euler (incompressible and non-viscous) equation of motion.

Such a new (Navier) equation can explain the effects, such as vorticity production, which are relevant near the boundary. This Navier equation has the following rather simple form:

$$\mathbf{Du_N}/\mathrm{Dt} + (1/\rho_\circ)\nabla p + \mathbf{g} = \nu_\circ \Delta \mathbf{u_N}, \quad (2.13a)$$

where $\nu_\circ$ is the constant kinematic viscosity, and $\Delta \equiv \nabla^2$ is the Laplace operator for the Navier velocity vector $\mathbf{u_N}$. The companion to (2.13a), equation of continuity is simply:

$$\nabla.\mathbf{u_N} = \mathrm{O}. \quad (2.13b)$$

Because ν° multiplies the derivative of highest order in Navier equation (2.13a), it cannot be inferred that the solutions of (2.13a), for very small values of ν°, reduces to Euler, below (2.14a), equation for an incompressible homogeneous and non-viscous fluid flow, with $\mathbf{u_E}$ as velocity vector:

$$\mathbf{Du_E}/\mathrm{Dt} + (1/\rho_\circ)\nabla p + \mathbf{g} = \mathrm{O}. \quad (2.14a)$$

with

$$\nabla.\mathbf{u_E} = \mathrm{O}. \quad (2.14b)$$

It is important (in particular, in the framework of our RAMA) to observe that the passage from compressible flow to incompressible flow, which filters the acoustic fast waves, is a strongly singular limit.

The reader can find in our 2006 monograph [13], devoted to hyposonic flow theory, various facets of the unsteady very slow flows at low Mach number, which are strongly related to a category of fluid flow problems, called "hyposonic", when $M << 1$.

In the above Navier incompressible viscous equation (2.13a) and also in the Euler incompressible non-viscous equation (2.14a), the term $(1/\rho_\circ)\nabla p$ is not an unknown quantity of the initial value problem. In fact, $\nabla(p/\rho_\circ)$ is the force term acting on the particles of fluid allowing them to move as freely as possible, but in a way compatible with the incompressibility constraint (2.13b) or (2.14b): $\nabla.\mathbf{u_N} = 0$.

Note particularly that for a Eulerian incompressible flow, $\mathbf{Du_E}/\mathrm{Dt} = 0$ admits solutions violating the condition: $\nabla.\, \mathbf{u_E} = 0$ at $t > 0$, even if the velocity divergence vanishes at $t = 0$. The pressure term in the above incompressible equations (2.13a) and (2.14a) is not an unknown quantity, because it can be determined when we have found the velocity field $\mathbf{u_N}$ or $\mathbf{u_E}$ – for instance, taking the divergence of the Euler equation (2.14a), we obtain a Poisson (elliptic) equation:

$$\Delta p = -\rho_\circ\{\nabla \cdot [(\mathbf{u_E} \cdot \nabla)\mathbf{u_E}] + \nabla.\mathbf{g}\},$$

and, knowing $\mathbf{u_E}$ and external force $\mathbf{g}$, we can find p by solving a Poisson equation with a Neumann boundary condition,

$$\partial p/\partial n = -\rho_\circ\{\nabla \cdot [(\mathbf{u_E} \cdot \nabla)\mathbf{u_E} + \mathbf{g}] \cdot \mathbf{n},$$

in a domain with a boundary (after that the Euler equation (2.14a) is projected on the outward unit normal **n**).

As a consequence of the above, it is sufficient to consider the Navier incompressible equation in terms of vorticity $\omega_N (= \nabla \wedge \mathbf{u}_N)$ assuming that **g** is conservative:

$$D\omega_N/Dt = (\omega_N \,.\, \nabla)\mathbf{u}_N + \nu_\circ \Delta \omega_N, \tag{2.15a}$$

with

$$\nabla.\,\mathbf{u}_N = 0. \tag{2.15b}$$

Obviously, when we assume that **g** is conservative, any potential flow,

$$\mathbf{u} = \nabla \Phi,$$

trivially satisfies the Navier equation (2.13a) in term of the vorticity ω_N!

However, to obtain a well-set boundary value problem, for a fixed $\nu_\circ > 0$, one must also (according to Stokes) replace the slip boundary condition on a (stationary) boundary of a fluid flow domain, for $\mathbf{u}_E$ in Euler non-viscous equation (2.14a):

$$\mathbf{u}_E.\mathbf{n} = 0, \tag{2.16a}$$

by the more stringent condition of no-slip boundary condition on a stationary boundary:

$$\mathbf{u}_N = 0, \tag{2.16b}$$

for $\mathbf{u}_N$ in Navier (2.13a).

Concerning this no-slip boundary condition (2.16b), it is interesting to note that in his 1904 lecture to the ICM, Prandtl stated:

"The physical processes in the boundary-layer (BL – Grenzschicht) between fluid and solid body can be calculated in a sufficiently satisfactory way if it is assumed that the fluid adheres to the walls, so that the total velocity there is zero – or equal to the velocity of the body. If the viscosity is very small and the path of the fluid along the wall not too long, the velocity will have again its usual value very near to the wall (outside the thin transition layer). In the transition layer (Übergangsschicht) the sharp changes of velocity, in spite of the small viscosity coefficient, produce noticeable effects."

Prandtl not only mentions the existence and nature of the thin boundary-layer and its connection with frictional drag, but derives heuristically the boundary-layer (so-called Prandtl) equations valid in a thin viscous layer close to the wall of the solid body. These BL Prandtl equations, however, are not valid near the time $t = 0$, where the initial data are given in the case of an initial-boundary value problem.

Prandtl – curiously – did not have any idea concerning this singular nature of his discovered BL equations in unsteady compressible case!

But it is also necessary to not overlook the important investigations of Lanchester (1907) in England, concerning the nature of the boundary-layer and explanation of separation (independently of Prandtl). (For a detailed discussion concerning the initial and boundary conditions, see Sect. 2.4.)

In Chap. 5, as a consequence of the singular nature of BL compressible equations, the unsteady full NS–F equations for large Reynolds number are analyzed in detail. For this, it is necessary to consider five regions and the related matching conditions.

2.3 Navier–Stokes–Fourier Equations for Viscous Compressible and Heat-Conducting Fluid Flow

According to Truesdell, in "*The Mechanical Foundations of Elasticity and Fluid Dynamics*" (1966) [42, p. 2]:

"Classical fluid dynamics describes the flow of media altogether without springiness of form, so that when released from all deforming forces except a hydrostatic pressure, they retain their present shapes; it is a partially linear theory, in which a uniformly doubled rate of deformation if dynamically possible would lead to doubled viscous forces."

2.3.1 The Cauchy Stress Principle

The derivation of the equation of motion for the velocity vector **u,** for real (viscous compressible and heat conducting) fluids is based on the following stress principle of Cauchy, 1828 [43]:

"Upon any imagined closed surface *S* (with outward normal ***n*** to *S*) there exists a distribution of stress vector $\Sigma(\mathbf{x},\ t;\mathbf{n}) = \mathbf{n}.T$, where ***T*** is the stress tensor, whose resultant and moment are equivalent to those of the actual forces of material continuity exerted by the material outside *S* upon that inside."

This statement of Cauchy's principle is due to Truesdell's paper of 1952 [44]; and as Truesdell remarks (1953) [45], the above well-known Cauchy principle

> . . . has the simplicity of genius. Its profound originality can be grasped only when ones realizes that a whole century of brilliant geometers had created very special elastic problems in very complicated and sometimes incorrect ways without ever hitting upon this basic idea, which immediately became the foundation of the mechanics of distributed matter.

As a consequence of his stress principle, Cauchy obtained a general equation of motion – the simple and elegant Eq. 2.17 below – which is valid for any fluid, and indeed for any continuous medium, regardless of the form which the stress tensor $\boldsymbol{T}$ may take.

$$\mathrm{D}\mathbf{u}/\mathrm{Dt} = \rho \boldsymbol{f} + \nabla.\boldsymbol{T}. \tag{2.17}$$

We observe that the above equation of motion, discovered by Cauchy in 1828, can be derived easily according to the principle of conservation of linear momentum: "The rate of change of linear momentum of a material volume V equals the resultant force on the volume."

The necessity for a clear-cut statement of the postulates on which continuum mechanics rests was pointed out by Felix Klein and David Hilbert, but the first axiomatic presentation is due to G. Hamel (1908) [46]. With $\Sigma(\mathbf{x}, \mathrm{t}; \mathbf{n}) = \mathbf{n}\,.\boldsymbol{T}$, for the stress vector Σ, the above principle is expressed by the statement:

$$\mathrm{D}/\mathrm{Dt}\left(\int_v (\rho \mathrm{D}\mathbf{u}/\mathrm{Dt})\right) dv = \int_v \rho \boldsymbol{f} dv + \int_v \mathrm{div}\boldsymbol{T} dv, \tag{2.18a}$$

applying the divergence theorem. Since v (a fixed volume) is arbitrary, we obtain the Eq. 2.17.

We observe also that the stress forces are in local equilibrium, and it is postulated that the stress tensor is symmetric:

$$T^{\mathrm{ij}} = T^{\mathrm{ji}}. \tag{2.18b}$$

2.3.2 Navier–Stokes Constitutive Equations: The Cauchy–Poisson Law

In our 2004 book *Theory and Applications of Viscous Fluid Flows* [47], in Sect. 1.4 of Chap. 2, the reader can find a detailed account of the constitutive equation of a viscous (*à la* Navier–Stokes) classical fluid, mainly inspired by Serrin (1959) [31].

Here we present only a short comment. A first important moment in the history of N–S constitutive equations is Stokes' idea (1845) [48] of "fluidity" which can be stated as four postulates:

1. $\boldsymbol{T} = F(\boldsymbol{D})$ and $\boldsymbol{D} = \boldsymbol{D}(\mathbf{u})$.
2. $\boldsymbol{T}$ does not depend explicitly on the position vector $\mathbf{x}$ (spatial homogeneity).
3. There is no preferred direction in space (isotropy).
4. When $\boldsymbol{D}(\mathbf{u}) = 0$, then $\boldsymbol{T} = -\,\mathrm{p}\boldsymbol{I}$ (Eulerian non-viscous fluid flow).

A medium whose constitutive equation (via stress tensor $\boldsymbol{T}$, which define or delimit the type of medium subject to study) satisfies these above four postulates is called a Stokesian fluid.

With the above four postulates, according to matrix algebra, and if we add the condition that the components of **T** be linear in the components of $\boldsymbol{D}(\mathbf{u})$, we deduce (Cauchy–Poisson law):[2]

$$\boldsymbol{T} = -(p + \lambda\,\mathrm{div}\mathbf{u})\boldsymbol{I} + 2\mu\boldsymbol{D}(\mathbf{u}). \tag{2.19}$$

The coefficients λ and μ (of viscosity) being of scalar functions of the thermodynamic state (considered in Sect. 2.3.3) and $\boldsymbol{I}$ is the unit tensor, with δ_{ij} (the so-called Kronecker symbol with $\delta_{kk} = 1$ and $\delta_{ij} = 0$ if $i \equiv j$) as components.

Indeed, the fully general expression is Poisson's (1831) relation [41] – but the name of Poisson is rarely quoted today. In Cauchy, (1828), [43], the term – $p\boldsymbol{I}$, in (2.19), is absent.

Dynamical (Navier) equation (2.13a) equivalent to those resulting from (2.19) when $\mathrm{div}\mathbf{u} = 0$ and, $\mu = \text{const}$, is due to Navier (1821) [40], and Saint–Venant (1843) [49], proposed (2.19) in the special case when $\mu = \mu_0 = \text{const}$ and $\mu_v \equiv \lambda + (2/3)\mu = 0$, which is the so-called (1845) "Stokes relation" [48].

This is the simple and elegant constitutive equation (2.19) for the viscous NS motion, discovered by Cauchy in 1828. It is valid for any fluid, and indeed for any continuous medium, regardless of the form which the stress tensor $\boldsymbol{T}$ may take.

The coefficients λ and μ (of viscosity) are scalar functions of the thermodynamic state (considered in the next Section).

Concerning the long controversy regarding the Stokes relation:

$$3\mu_v = 0, \tag{2.20a}$$

$3\mu_v$ being the bulk viscosity, in the classical theory of viscous fluids; see Truesdell (1966) [42].

The viscosities coefficients (shear/dynamic and bulk) and the thermal conductivity k (see Sect. 2.3.3, Fourier's law (2.30)) are known functions, subject to the thermodynamic restriction (Clausius–Duhem inequalities):

$$\mu \geq 0, \;\; \kappa \geq 0 \text{ and } \mu_v \geq 0. \tag{2.20b}$$

---

[2] For a perfect (absence of viscosity) fluid, the pressure has already appeared as a dynamical variable in Euler Eq. 2.4. Characteristic of the discipline of gas dynamics is the postulate that the thermodynamic pressure, introduced via functional relations among the state variables (see Sect. 2.3.3), is equal to this dynamical pressure. When the deformation $\boldsymbol{D}(\mathbf{u}) = 0$, for a perfect fluid, p is the thermodynamic pressure when the fluid is compressible, while p is simply an independent dynamical variable otherwise. For an incompressible perfect or viscous fluid (Navier, see Sect. 2.2) p is not an unknown quantity, because it can be determined when we have found the velocity field **u**. In some works (see [31]) a mean pressure, $p^* = -(1/3)\mathrm{Trace}\,\boldsymbol{T}$ is defined, and we have the following relation: $p - p^* = [\lambda + (2/3)\mu]\mathrm{div}\mathbf{u}$.

The dynamical equation (in components form, with indices, i = 1, 2 and 3) resulting from (2.17) with (2.19) is the Navier–Stokes (compressible) equation for the component u_i of the velocity **u**:

$$\rho\, Du_i/Dt + \partial p/\partial x_i = \partial/\partial x_j[\mu(\partial u_i/\partial x_j + \partial u_j/\partial x_i)] + \partial/\partial x_i[\lambda(\partial u_k/\partial x_k)], \tag{2.21}$$

since for the components of deformation tensor $\boldsymbol{D}(\mathbf{u})$ we have:

$$2D_{ij} = \partial u_i/\partial x_j + \partial u_j/\partial x_i. \tag{2.22}$$

A Stokesian fluid whose constitutive, NS compressible, equation is given by (2.19) is called a Newtonian fluid.

In Saint–Venant (1843) (and Stokes (1845)), the resulting dynamical equation, for u_i, in place of (2.21), is:

$$\rho Du_i/Dt + \partial p/\partial x_i = \mu_0\{\Delta u_i - (1/3)\partial/\partial x_i\{D\log\rho/Dt\}, \tag{2.23a}$$

if we take into account the equation of continuity:

$$\partial u_k/\partial x_k = -D\log\rho/Dt, \tag{2.23b}$$

and the Stokes relation (2.20a).

For an incompressible homogeneous fluid, again we derive the Navier dynamical equation (2.13a), with $\nu_0 \equiv \mu_0/\rho_0 = \text{const.}$

Equation 2.23a (where μ_0 is a constant), with the specifying equation $p = p(\rho)$ and continuity equation (2.23b), forms a closed system of equations – the so-called Navier–Stokes (compressible N–S equations) for u_i, p, and ρ, and governs a barotropic(?) viscous and compressible fluid flow, without an energy equation for the temperature T – which unfortunately do not have any physical (fluid dynamics) signification.

Slightly more complete general Navier–Stokes compressible N–S equations, for the unknowns u_i, p, and ρ, are obtained from (2.21), with again the specifying equation $p = p(\rho)$ and continuity equation (2.23b), if we assume that viscosities μ and λ do not depend on temperature T and are known functions of the density ρ only.

These above N–S equations do not emerge via a RAM Approach from the full unsteady NS–F equations and, in fact, do not have in reality any interest for fluid dynamicians!

We also obtain a simplified model of compressible viscous fluid flow if we assume in addition, instead of $p = p(\rho)$, that the pressure is identically constant in the fluid flow (isobaric fluid flow).

In this case, then, we arrive at the so-called Burger's model equations (as in Kazhikov (1994) [50]).

$$\mathrm{D}\log\rho/\mathrm{Dt} + \partial u_k/\partial x_k = 0, \tag{2.24a}$$

$$\begin{aligned}\rho\, \mathrm{D}u_i/\mathrm{Dt} &= \partial/\partial x_j[\mu(\rho)(\partial u_i/\partial x_j + \partial u_j/\partial x_i)] \\ &\quad + \partial/\partial x_i\,[\lambda(\rho)(\partial u_k/\partial x_k)],\end{aligned} \tag{2.24b}$$

which is a closed system of two equations for the velocity components u_i and the density ρ, when we assume that the viscosity coefficients, λ and μ, are a function only of ρ.

A final important remark concerning the above N–S compressible viscous barotropic system equations, (2.23a) and (2.23b) with $p = p(\rho)$, or Eqs. 2.21 and 2.23b with $p = p(\rho)$, for the unknowns u_i, p and ρ, which are mainly used by applied mathematicians in their rigorous mathematical analyses (see, for instance, P. L. Lions (1998) [51]) does not have any physical reality, mainly because just viscosity always generates entropy (baroclinity).

For this, in particular, the various rigorous mathematical results concerning the so called "incompressible limit", related to the limiting process $M \downarrow 0$, in the framework of above, compressible barotropic ($p = p(\rho)$) and viscous, two systems, seems (to me) very questionable.

In the short paper by Leray (1994) [52], this question is pertinently discussed.

2.3.3 *Thermodynamics and Energy Equation via Fourier Constitutive Equation*

But in general (in reality) the coefficients of viscosity are assigned or empirical functions of the positive variables ρ (density) and especially T (temperature), which both are present also in equation of state (2.7) or (2.8) for a trivariate realistic fluid.

Indeed, Euler and Lagrange not only failed to include viscosity effects in their equations of motion, forcing them adopt corresponding simplified (slip) boundary condition (2.16a), but also oversimplified their equation of state.

In real fluids, the pressure, p, is a function of two variables, ρ and T (for a trivariate fluid in a baroclinic motion, see the equation of state (2.7)). Obviously, again, it is necessary to associate with N–S compressible equations (2.21), (2.23b), and (2.7), for u_i, ρ, p and T, an energy equation, if we want to obtain a closed system of equations for our six unknown functions. For this, some thermodynamical assumptions are required.

For the real fluid flows – compressible, viscous, and heat-conducting – the mechanical energy is converted into heat by viscosity, and the heat of compression is diffused by heat conduction.

Here we consider only an homogeneous fluid when the local equation of state is according to the basic postulate of Gibbs (1875) [53]; see Truesdell (1952) [44], and Serrin (1959) [31]:

$$E = \mathrm{E}(\rho, S), \tag{2.25}$$

where E is the specific internal energy and S is the specific entropy.

In this case the temperature T and the thermodynamic pressure p are defined by the two relations:

$$\mathrm{T} \;=\partial E/\partial S \text{ and } \mathrm{p} \;= -\partial E/\partial(1/\rho). \tag{2.26}$$

Now, for any compressible fluid, by differentiating (2.25) along any curve on the energy surface (characterizing the fluid) we obtain:

$$\mathrm{D}E/\mathrm{Dt} \;=\; \mathrm{T}\ \mathrm{D}S/\mathrm{dt} - \mathrm{p}\ \mathrm{D}(1/\rho)\mathrm{Dt}. \tag{2.27}$$

But, for any homogeneous medium in motion, the conservation of energy is expressed by the equation of C. Neumann (1894):

$$\rho\,\mathrm{D}E/\mathrm{Dt} + \mathrm{p}\ (\partial \mathrm{u_k}/\partial \mathrm{x_k}) = -\partial \mathrm{q_i}/\partial \mathrm{x_i} + [2\mu\,\mathrm{D_{ij}} + \lambda\,\mathrm{D_{kk}}\ \delta_{ij}]\mathrm{D_{ij}}, \tag{2.28}$$

where the $\mathrm{q_i}$ are the Cartesian components of the heat flux vector **q**.

For the special case of a non-viscous incompressible fluid, the energy equation was given by Fourier (1833) [54], and for small motions of a viscous perfect gas by Kirchhoff (1868) [39], and in this case we also have $\mathrm{p} = \mathrm{R}\rho\mathrm{T}$, where R is the constitutive constant of the viscous thermally perfect gas.

We observe also, that for a medium suffering deformation the two Eqs. 2.27 and 2.28 express different and independent assumptions: the former, the existence of an energy surface, characterizing the fluid, and the latter, that mechanical and thermal energy are interconvertible. Indeed the First Law, Second Law, and so on, of thermodynamics is rather misleading terminology. (For a history of the origin of thermodynamics, see H. Poincaré (1892) [55], and Truesdell and Muncaster (1980) [56].)

Finally, from (2.27) it follows that in place of (2.28) we can write the following equation for the specific entropy:

$$\rho\mathrm{TDS/Dt} \;= -\partial \mathrm{q_i}/\partial \mathrm{x_i} + (2\mu\,\mathrm{D_{ij}} + \lambda \mathrm{D_{kk}}\,\delta_{ij})\mathrm{D_{ij}}. \tag{2.29a}$$

In particular, if the heat flux (via the vector **q**) rises solely from thermal conduction, then according to Fourier's law gives:

$$\mathrm{q_i} = -k\,(\partial \mathrm{T}/\partial \mathrm{x_i}), \tag{2.30}$$

where k is the thermal conductivity and $\mathrm{q_i}$ the components of **q**.

With (2.30) we obtain from (2.29a) the usual form of the energy equation:

$$\rho\mathrm{TDS/Dt} \;= -\partial(k(\partial \mathrm{T}/\partial \mathrm{x_k}))/\partial \mathrm{x_k} + (2\mu \mathrm{D_{ij}} + \lambda \mathrm{D_{kk}}\delta_{ij})\mathrm{D_{ij}}. \tag{2.29b}$$

For an adiabatic ($q_i = 0$) and non-viscous (inviscid) fluid with $\mu = 0$ and $\lambda = 0$:

$$DS/Dt = 0. \tag{2.31}$$

Therefore, if a non-viscous homogeneous fluid be in continuous motion devoid of heat flux, then the entropy of each particle remains constant. In particular, if the motion be steady, then the entropy is constant along each streamline.

But in general, the real motion is not isentropic – S is different of a constant in the each point of the flow and in time – but constant for each particle (along the trajectory) for a Eulerian fluid flow.

But even if the flow is isentropic, isentropicity in fluid remains valid only up to the first shock front encountered by the particles, after which it may well fail its isentropicity property.

2.3.4 Navier–Stokes–Fourier (NS–F) Equations

The three equations – continuity (2.23b), N–S for compressible motion (2.21), and energy (2.28) – with the two state relations

$$p = R\rho\,T \text{ and } E = C_v T, \tag{2.32a, b}$$

valid for a thermally perfect gas with constants specific heats, constitute the so-called Navier–Stokes–Fourier (NS–F) equations for a compressible, viscous and heat-conducting Newtonian fluid.

In this case it is assumed that the three constitutive (dissipative) coefficients are functions of ρ and T:

$$\lambda = \lambda(\rho,\ T), \tag{2.33a}$$

and

$$\mu = \mu(\rho,\ T) \tag{2.33b}$$

in (2.21) and (2.28), and

$$k = k(\rho,\ T) \tag{2.33c}$$

in (2.30).

The compact form of these NS–F equations is:

$$D\rho/Dt + \nabla\cdot\mathbf{u} = 0, \tag{2.34a}$$

$$\rho\,D\mathbf{u}/Dt + \nabla p + \rho g\mathbf{k} = \nabla.\Pi, \tag{2.34b}$$

$$\rho C_v DT/Dt + p \nabla \cdot \mathbf{u} = \nabla \cdot [k \nabla T] + \Phi, \tag{2.34c}$$

where Φ is the viscous dissipation function and gravity $\mathbf{g} = -g\mathbf{k}$ acts in the negative x_3 direction. In (2.34b, 2.34c):

$$\Pi = \lambda(\nabla \cdot \mathbf{u})I + 2\mu\mathbf{D}(\mathbf{u}) \tag{2.35a}$$

and

$$\Phi = 2\mu\text{Trace } [(\mathbf{D}(\mathbf{u})^2] + \lambda(\nabla \cdot \mathbf{u})^2, \tag{2.35b}$$

where:

$$\text{Trace } [(\mathbf{D}(\mathbf{u}))^2] = \mathbf{D}(\mathbf{u}) : \mathbf{D}(\mathbf{u}) = (1/4)[\partial u_i/\partial x_j + \partial u_j/\partial x_i]^2, \tag{2.36}$$

and $\mathbf{D}(\mathbf{u})$ is the rate-of-deformation tensor.

2.4 Initial and Boundary Conditions

Obviously, there has always been considerable interest in initial-boundary value problems for various systems of partial differential equations arising in Newtonian fluid dynamics.

This interest of fluid dynamicians stems primarily from efforts to create useful computational models of various processes for the purposes of simulation, prediction, and the detailed study of various fluid flow phenomena.

Naturally, the initial-boundary value problems for fluid dynamics equations should have been carefully investigate – but unfortunately, rigorous proof of the existence and uniqueness of solutions of these well-posed fluid dynamics problems requires very difficult mathematical investigation.

While initial-boundary value problems for these systems of equations are not easy to analyze, mathematical tools useful for such problems can be found in the works of Kreiss (1970, 1974) [57, 58], Belov and Yanenko (1971) [59], Oliger and Sundström (1978) [60], Majda (1984) [61], and Kreiss and Lorenz (1989) [62].

The solveability of these problems (a fundamental problem in rigorous mathematical theory of NS–F equations) is discussed in Chap. 8 of our *Theory and Applications of Viscous Fluid Flows* (2004) [47], and here we note only that the baroclinic and barotropic Eulerian equations are both symmetrical, hyperbolic systems, but isochoric and incompressible equations are not hyperbolic. This has a profound influence on the well-posedness of initial-boundary value problems for these systems of partial differential equations.

2.4.1 The Problem of Initial Conditions

The above NS–F unsteady equations (2.34a–2.34c) contain a total of five times derivatives for the components u_i of the velocity $\mathbf{u}$, density ρ, and temperature T. As a consequence, if we want to resolve a pure initial value, or Cauchy, problem (in the L^2 norm, for example), then it is necessary to have a complete set of initial conditions (data) for $\mathbf{u}$, ρ, and T:

$$t = 0 : \mathbf{u} = \mathbf{u}^{\circ}(\mathbf{x}), \rho = \rho^{\circ}(\mathbf{x}),\ T = T^{\circ}(\mathbf{x}), \tag{2.37}$$

where $\rho^{\circ}(\mathbf{x}) > 0$ and $T^{\circ}(\mathbf{x}) > 0$.

Moreover, when we consider a free-boundary problem or an unsteady flow in a bounded container, with a boundary depending on time, an initial condition for the (moving) boundary $\partial\Omega(t)$ has to be specified.

For an NS compressible but barotropic flow for the velocity $\mathbf{u}$ and density ρ, governed by Eqs. 2.21 and 2.23b with $p = p(\rho)$, as initial conditions we assume:

$$t= 0 : \mathbf{u} = \mathbf{u}_b{}^{\circ}(\mathbf{x}) \text{ and } \rho = \rho_b{}^{\circ}(\mathbf{x}). \tag{2.38a}$$

For the isochoric Euler equations (2.21) it is necessary to impose also

$$t = 0 : \mathbf{u} = \mathbf{u}_i{}^{\circ}(\mathbf{x}) \text{ and } \rho = \rho_i{}^{\circ}(\mathbf{x}). \tag{2.38b}$$

If the flow is continuous, when $\rho^{\circ}(\mathbf{x}) = \text{const}$, we have an incompressible flow (Eqs. 2.22 or 2.23b), and it is sufficient to assume only an initial condition for the velocity u:

$$t = 0 : \mathbf{u} = \mathbf{u}_i{}^{\circ}(\mathbf{x}). \tag{2.38c}$$

It is important to note that for both isochoric and incompressible divergence, free flows, it is necessary that

$$\text{the boundary integral} \int \mathbf{u} \cdot \mathbf{n}\, d\Omega \text{ vanish} \tag{2.39a}$$

and

$$\nabla \cdot \mathbf{u}_i{}^{\circ} = 0. \tag{2.39b}$$

Naturally, this last condition has no analogue for compressible (baroclinic or barotropic) flows because of the occurrence of the term $\partial\rho/\partial t$ in the continuity equation (2.23b).

Obviously, for the Laplace equation, which governs an incompressible, irrotational Eulerian unsteady flow (for example, waves on water), we do not have the possibility of imposing any initial conditions! But this Laplace equation is very appropriate for the investigation of waves on (incompressible) water, and in this case it is necessary to consider a free-boundary problem; that is, a problem for which the fluid (water) is not contained in a given domain but can move freely.

Usually, for this Laplace elliptic equation, one boundary condition is given (on the contour line containing the fluid) – but in the case when the boundary is known!

Two unsteady, dynamic and kinematic, conditions are needed (and also two initial conditions) at the free surface (interface) $x_3 = \eta(t, x_1, x_2)$, because the surface position $\eta(t, x_1, x_2)$ has to be determined as well as potential function ϕ (t, x_1, x_2, x_3).

For the free surface problem (for the function $\phi(t, x_1, x_2, x_3)$ and $\eta(t, x_1, x_2)$) governing the non-linear waves on water, we can consider two physical problems.

First is the so-called "signalling" (two-dimensional) problem, in which we have as initial conditions, when the water is initially at rest in a semi-infinite channel, the following conditions:

$$\phi(0, x_1, x_3) = 0 \text{ and } \eta(0, x_1) = 0, \text{ when } x_1 > 0, \tag{2.40a}$$

and at initial time t = 0 an "idealized wave-maker", at $x_1 = 0$, will generate a horizontal velocity disturbance, such that the initial condition is:

$$\partial\phi/\partial x_1 = W^\circ B(t/t^\circ), \text{ for } x_1 = 0 \text{ and } t > 0, \tag{2.40b}$$

where W° and t° are the characteristic velocity and time scales associated with the wave-maker idealized by the function $B(t/t^\circ)$.

A second category of the problem for water waves, in the infinite channel, is obtained by specifying an initial surface shape but zero velocity:

$$\text{for } t = 0 : \eta = a^\circ\zeta^\circ(x_1/l^\circ, x_2/m^\circ) \text{ and}$$

$$\phi(0, x_1, x_2, x_3) = 0, \tag{2.40c}$$

where l° and m° are the characteristic wavelengths (in the x_1 and x_2 directions) for the three-dimensional water wave motion.

In (2.40c) the scalar a° is a characteristic amplitude for the initial elevation of the free surface characterized by the function $\zeta^\circ(x_1/l^\circ, x_2/m^\circ)$. (Concerning the boundary conditions (kinematic and dynamic) for this free surface problem, see the next Sect. 2.4.2.)

For meteorological motions (considered in the Chap. 9), when we consider various approximate model equations – f°-plane equations, primitive equations, quasi-geostrophic equations, or Boussinesq equations – it is necessary, in fact (mainly because the filtering acoustic waves), to resolve associated unsteady adjustment problems for the formulation of consistent initial conditions for these simplified model equations.

2.4.2 Unsteady Adjustment Problems

For the study of a compressible fluid flow it is necessary to have for the determination of the solution of the corresponding evolution unsteady equations – Euler, Navier, and NS–F – a set of initial data for ρ, **u**, and T (the Cauchy problem). However, when we consider, for example, the incompressible equations ((2.13a) or (2.14a) with $\nabla.\mathbf{u} = 0$) for **u** and p, one is allowed to specify a set of initial conditions less in number than for the full compressible baroclinic equations. This is due to the fact that the "main" low-Mach-number limiting process (in fact, M tends to zero, with t and $\boldsymbol{x}$ both fixed), which leads to the approximate incompressible (model) equations, filters out some time derivatives – these corresponding to acoustic fast waves, because such waves are of no importance for low-speed aerodynamics and various atmospheric and oceanic motions – at least for steady flows.

When $\mathrm{Re}\uparrow\infty$ (large Reynolds number), from the Navier (incompressible and viscous) equations we derive the Prandtl boundary layer equations, and accordingly, for an unsteady flow the term $\partial u_3/\partial t$ disappears in the limiting momentum equation for the vertical (x_3-direction) component (u_3) of velocity! For low Reynolds number ($\mathrm{Re}\downarrow 0$) in the Stokes and Oseen limiting (steady) equations, the unsteady terms also disappear! Due to this, one encounters the problem of finding an answer to the following question: "What is the initial condition that is necessary to prescribe for **u** a solution of an incompressible equation, and in what way is this condition related to the starting initial conditions (with given data) associated with the exact, compressible equations?"

It is important to note that the exact initial conditions for the full compressible equations are not in general consistent with the estimates of basic orders of magnitude implied by the approximate model (without acoustic waves) equations. A physical process of time evolution is necessary to bring the initial set to a consistent level as far as the orders of magnitude are concerned.

Such a process is called "unsteady adjustment" of the initial data set to the approximate structure of incompressible equations under consideration. This process of adjustment, which occurs in many fields of fluid mechanics besides Boltzmann kinetic theory (first discussed by Hilbert in 1912), is short on the time-scale of approximate simplified equations, and ultimately, in an asymptotic sense, we obtain values for the consistent set of initial conditions suitable for the simplified equations.

When we consider the set of approximate simplified model equations, usually derived heuristically, with time–space fixed, then it is first necessary, for instance, to elucidate various adjustment problems – namely, concerning Prandtl boundary layer, Stokes and Oseen steady, Navier incompressible viscous, and Boussinesq equations.

A number of adjustment problems occur in meteorology for atmospheric motions (adjustment to hydrostatic balance) and to geostrophy (as in the case, for example, cited in Sect. 9.2), and the reader can find a detailed discussion of these adjustment problems in Chap. 5 of our *Meteorological Fluid Dynamics* (1991) [19].

However, it is important to note that, depending on the physical nature of the problems, we may have two kinds of behaviour when the rescaled (short) time goes to infinity. Either one may have a tendency towards a limiting steady state, or an undamped set of oscillations (as, for example, the inertial waves in the inviscid problem of spin-up for a rotating fluid; see Greenspan (1968; §2.4) [63]). The problem, considered in Sect. 7.2, is also very particular, and requires a special approach due to the persistence of acoustic oscillations. (For the terminology of the initial layer as adapted to this kind of singular perturbation problem, see Nayfeh (1973; p. 23) [64].) Finally, we note that usually, the process of the unsteady adjustment of the aerodynamical (or meteorological) fields is a result of the generation, dispersion and damping of the fast internal waves. According to method of matching asymptotic expansions (MMAE), the initial conditions for the limiting model equations are, in fact, matching conditions between the two asymptotic representations – the main one (with t fixed), and the local one (near $t = 0$), which is a necessary companion to main one!

In conclusion, we can say that the aim of the unsteady adjustment problem can be stated as follows: Clarify just how a set of initial data associated with an exact system of unsteady equations can be related to another set of initial data associated with a simpler, approximate model system of equations which is a significant degeneracy of the original system of exact equations considered at the start, but with the less time derivatives in this approximate model system.

In order to solve such a problem it is necessary to introduce an initial layer in the vicinity of $t = 0$, characterized by a short fixed time τ. Obviously such an unsteady adjustment problem is very important in meteorology for the formulation of a well-posed initial/Cauchy evolution in a time-prediction problem relative to, for example: "what the weather will be like tomorrow or for the next few days?"

Concerning the rigorous mathematical results of the singular limits in compressible fluid dynamics, see, for instance, the paper by Beiräro da Veiga (1994) [65], and also the various references in this paper. More recent papers have been published concerning the passage of compressible $\Rightarrow$ incompressible, by Desjardins, Lions, Grenier, Masmoudi, Hagstrom, Lorenz, and Iguchi; and for references see our *Topics in Hyposonic Flow Theory* (2006) [13]. Here we do not consider these contributions, but instead discuss some of these singular-limit problems (low-Mach asymptotics) which deserve a serious, consistent, fluid dynamics investigations via a RAM Approach (as in [13]).

Concerning the low-Mach asymptotic, we observe also that in the case of a flow affected by acoustic effects in a confined gas (internal flow within a bounded domain D(t)), over a long time when the wall ∂D(t) is started impulsively from rest, a multiple-time-scale technique is necessary, because acoustic oscillations remain undamped and the unsteady adjustment problem (with matching) does not work (see Sect. 7.2 on applications in aerodynamics).

2.4.3 Boundary Conditions for the Velocity Vector u and Temperature T

Several boundary conditions could be considered with respect to different physical situations.

If we consider, as a simple example, the motion of a fluid in a rigid container Ω (with a boundary $\partial\Omega$, independent of time), a bounded connected open subset of $\boldsymbol{R}^d$ (where $d > 1$ is the physical dimension), the different structure of the equations leads to the necessity of distinguishing between viscous (NS–F, NS, or Navier) and inviscid (Eulerian) fluids.

(a) For a viscous (NS–F or Navier) fluid: $\mu > 0$ and $\mu_v \equiv \lambda + (2/3)\mu > 0$
In this case, the physical effects due to the presence of the dynamic (shear) viscosity coefficient μ yield the validity of the steady no-slip condition:

$$\mathbf{u} = 0 \text{ on } \partial\Omega, \tag{2.41a}$$

(b) For a bulk-viscous fluids: $\mu = 0, \mu_v > 0$
Since only the bulk viscosity coefficient μ_v is different from zero, in this situation the slip boundary condition

$$\mathbf{u}.\mathbf{n} = 0 \text{ on } \partial\Omega \tag{2.41b}$$

where it is assumed here, and in what follows, that $\mathbf{n} = \mathbf{n}(\mathbf{x})$ denotes the unit outward normal vector to $\partial\Omega$.

(c) For an inviscid (Eulerian) fluid: $\mu = 0, \mu_v = 0$
Also in this case, the slip boundary condition (2.41b) is assumed.

As concerns the (absolute) temperature T, the boundary condition takes different forms in the two alternative cases, $k > 0$ and $k = 0$.

(d) Conductive fluids: $k > 0$
Several boundary conditions have physical meaning. Limiting ourselves to the most common cases, we can require :

$$\mathrm{T} = \mathrm{T_w} \text{ on } \partial\Omega(\text{Dirichlet}) \tag{2.42a}$$

$$k\partial\mathrm{T}/\partial\mathrm{n} = \Xi \text{ on } \partial\Omega(\text{Neumann}) \tag{2.42b}$$

$$k\partial\mathrm{T}/\partial\mathrm{n} + h(\mathrm{T} - \mathrm{T_0}) = \Xi \text{ on } \partial\Omega(\text{third type}), \tag{2.42c}$$

where $\mathrm{T_w} > 0$ and Ξ are known functions, and $h > 0$ is a given constant.

(e) Non-conductive (adiabatic case) fluids: $k = 0$

No boundary condition have to be imposed on temperature T if (2.41a) or (2.41b) are satisfied, since in these cases the temperature is not subjected to transport phenomena through the boundary.

According to Gresho (1992, pp. 47–52) [66], if $\mathbf{u}_N{}^0(\mathbf{x})$ is the initial (t = 0) velocity field for the Navier equation (2.13a), then in the domain Ω it is necessary to impose the above incompressibility constraint (2.13b):

$$\nabla \cdot \mathbf{u}_N{}^0(\mathbf{x}) = 0, \tag{2.43a}$$

and on the boundary $\partial\Omega = \Gamma(s)$:

$$\mathbf{n} \cdot \mathbf{u}_N{}^o(s) = \mathbf{n} \cdot \mathbf{w}(s, 0) = \mathbf{n} \cdot \mathbf{w}^0(s), \tag{2.43b}$$

where $\mathbf{w}(s, t)$ is the specified boundary condition for the Navier velocity vector which satisfies Eq. 2.13a.

2.4.4 Other Types of Boundary Conditions

In many situations (inflow–outflow problems) the velocity cannot be assumed to vanish on ∂Ω. This is the case, for instance, for the flow around an airfoil, where an inflow region is naturally present upstream (and an outflow region appears in the wake), or the flow near a rigid body, where the velocity can be assumed to vanish only on the boundary of the body. In these cases, several different boundary conditions may be prescribed.

Let us begin by considering the viscous case. Concerning the velocity-field, a (non-zero) Dirichlet boundary condition can be imposed everywhere, or, alternatively, only in the inflow region: that is, the subset of ∂Ω where $\mathbf{u} \cdot \mathbf{n} < 0$, whereas, on the remaining part of the boundary, the conditions

$$\mathbf{u} \cdot \mathbf{n} = U^+ > 0 \text{ and } (\mathbf{n} \cdot \mathbf{D}) \cdot \mathbf{t} = 0, \tag{2.44a}$$

have to be prescribed. (Here, $\mathbf{t}$ is a unit tangent vector on ∂Ω, and $\mathbf{D} = \mathbf{D}(\mathbf{u})$ is the rate of strain (deformation) tensor.)

Let us, moreover, remark that the condition

$$\mathbf{u} \cdot \mathbf{n} = 0 \text{ and } (\mathbf{n} \cdot \mathbf{D}) \cdot \mathbf{t} = 0, \tag{2.44b}$$

could also be considered, on the whole, ∂Ω. In this case, however, no inflow or outflow regions would be present.

More important is to analyse the boundary condition for the density ρ, since now it turns out that it is necessary to prescribe it on the inflow region. In fact, the first-order hyperbolic continuity equation (2.13a) can be solved by means of the theory of characteristics, and the boundary datum for ρ on the inflow region is indeed a (necessary) Cauchy datum for the density on a non-characteristic surface.

Let us note also, that if the heat conductivity coefficient k is vanishing and the fluid is inviscid, the same type of Dirichlet-inflow boundary condition has to be imposed on the temperature T, since in such a case Eq. 2.6 is also of the hyperbolic type for T. More complicated is the situation when the inviscid (Euler) case ($\mu = \lambda = k = 0$) is considered.

In fact, in this case the Eulerian system is a first-order hyperbolic one, and the number of boundary conditions, in the case of an "open boundary" (or a boundary located in the interior of a body or fluid), is different depending on whether the flow is

$$\text{subsonic} : |\mathbf{u}| < \mathrm{a},$$

or

$$\text{supersonic} : |\mathbf{u}| > \mathrm{a},$$

where

$$\mathrm{a} = [\gamma \ \mathrm{RT}\ ]^{1/2},$$

is the local sound speed for the perfect gas.

Take, for example, $d = 3$. An analysis of the sign of the eigenvalues of the associated characteristic matrix yields the following conclusion: The number of boundary conditions must be five or four on an inflow boundary, depending on whether the flow is supersonic or subsonic, and zero or one on an outflow boundary, again depending on whether the flow is supersonic or subsonic.

Obviously, in the case of an "open boundary" the normal velocity is non-zero on the boundary, except at certain points. In both cases, no obvious physical boundary conditions are known. We will not enter more deeply into this argument, and will only briefly discuss the inviscid case subjected to the slip boundary condition (2.41b), for which the boundary is a characteristic surface.

Further information on inflow–outflow boundary-value problems for compressible N–S and inviscid Euler equations can be found in two pertinent papers produced by Gustafsson and Sundström (1978)[67] and Oliger and Sundström (1978) [60]. Here, we note only that the solid-wall slip stationary boundary condition (2.41b) – the normal velocity

$$u_{\mathrm{n}} \equiv \mathbf{u} \cdot \mathbf{n} = 0,$$

should vanish at the boundary, is consistent with the number of inward characteristics (one).

The reader can find also in two papers by Viviand and Veuillot (1978)[68] and Viviand (1983)[69], a discussion of boundary conditions for steady Euler flow, considered as the limit (when time tends to infinity – the "pseudo-unsteady" method) of an unsteady flow (which does not have a precise physical meaning).

It is necessary to note that in numerical, computational, fluid dynamics, these problems of boundary conditions are very thoroughly considered by often taking into account the constraints related with the various particularity of the considered fluid flow problem and associated with numerical algorithms.

Another interesting set of boundary conditions appears when we consider the free-boundary problem; that is, a problem for which the fluid is not contained in a given domain but can move freely. In this case the vector $\mathbf{n} \cdot \boldsymbol{T}$ is prescribed on (interface) $\partial\Sigma$, where, moreover, $\mathbf{u} \cdot \mathbf{n}$ is required to be zero (stationary case) or equal to the normal velocity of the boundary itself (non-stationary case).

The value of $\mathbf{n} \cdot \boldsymbol{T}$ can be zero (free expansion of a fluid in the vacuum), or to (see, for example, the case of the well-known Bénard problem considered in Chap. 8):

$$-\,\mathrm{p}_e\mathbf{n} + 2\sigma K\mathbf{n} + \nabla_s\sigma, \text{ on interface,} \tag{2.45a}$$

where p_e is the external pressure, $\sigma = \sigma(\mathrm{T})$ is the surface tension (temperature-dependent, when the fluid is an expansible liquid), K is the mean interfacial curvature, and:

$$\nabla_s = \nabla - \mathbf{n}(\mathbf{n} \cdot \nabla), \tag{2.45b}$$

is the surface (projected) gradient at the interface, respectively.

But in this (viscous) case, it is necessary to also write a heat transfer condition across the interface, for an expansible and thermally conducting fluid (liquid),

$$k(T)\partial \mathrm{T}/\partial n + h_s\mathrm{T} \;=\; \text{prescribed function,} \tag{2.45c}$$

which is a Newton's cooling law, where the heat-transfer (constant) coefficient h_s is sometimes called the Biot number. We observe that rigorously, in a Bénard convection problem (with a temperature dependent tension in a free-surface) it is necessary to take into account two Biot numbers, respectively, for the conduction (motionless/no convective motion) state and convection state (see Chap. 8).

But the problem of "two Biot numbers asymptotics" is actually widely open (see, in Chap. 8, a discussion concerning this two Biot numbers problem in the framework of the Bénard problem for an expansible liquid layer on a solid horizontal flat surface and heated from below.) Since $\mathrm{d}\sigma/\mathrm{dT} \neq \mathrm{o}$, then, for the film problem it is necessary to take into account a Marangoni number proportional to the gradient:

$$(\mathrm{d}\sigma/\mathrm{dT})_{\mathrm{T}=\mathrm{T}^{\circ}},$$

where T° is a constant temperature.

In such a case we consider a thin film Bénard–Marangoni free-surface problem, which is fundamentally different from the classical Rayleigh–Bénard thermal instability problem. (See, for instance, in Velarde and Zeytounian (2002) [70], CISM Courses and Lecture (N0 428), pp. 123–90.)

Naturally, we are now imposing one more condition on the interface $\partial\Sigma$, since it is an unknown of the problem; in the non-stationary case an initial condition for the interface has to be added too (see (2.40a) or (2.40c)).

For an inviscid incompressible fluid (water) when we consider the wave on the water (in this case the problem for an irrotational flow is governed by the Laplace equation), an obvious physical simple condition is (if we assume that the surface tension is negligible), in place of (2.45a):

$$p = p_A, \textit{on interface between water and air above}, \tag{2.46}$$

where p_A denotes the air (constant, ambient) pressure on interface Σ and usually this ambient air (above the interface) is assumed passive (at temperature $T_A =$ const, pressure $p_A =$ const, with negligible viscosity and density).

In the case of a viscous liquid (thin film Marangoni problem, discussed in Chap. 8) the above condition (2.46) is replaced by a rather complicated explicit upper free surface condition (see Sect. 8.2.2).

Now, if the equation of the interface is $x_3 = \eta(t, x_1, x_2)$, in a Cartesian system of coordinates $(0, x_1, x_2, x_3)$, then from the Bernoulli incompressible integral we obtain the following dynamic condition on interface (for the wave on the water – the inviscid fluid problem) according to (2.46):

$$\partial\phi\partial t + (1/2)(\nabla\phi)^2 + g\,\eta = 0, \text{ on } x_3 = \eta(t, x_1, x_2), \tag{2.47}$$

and since the interface is a material wave surface we have also a kinematic condition:

$$\begin{aligned}\partial\phi/\partial x_3 = \partial\eta/\partial t + (\partial\phi/\partial x_1)\partial\eta/\partial x_1 \\ + (\partial\phi/\partial x_2)\partial\eta/\partial x_2, \text{on } x3 = \eta(t, x_1, x_2),\end{aligned} \tag{2.48}$$

Finally, if we assume that the water rests on a horizontal and impermeable bottom of infinite extend at $x_3 = -h_0$, where $h_0 =$ const is supposed finite, then we have the following simple (flat) bottom boundary condition for the Laplace equation:

$$\partial\phi/\partial x_3 = 0, \text{ on } x_3 = -h_0, \tag{2.49}$$

The Laplace equation for the potential ϕ, with (2.40a), (2.40b) or (2.40c) and (2.47)–(2.49), constitutes a well-posed problem for the investigation of the nonlinear unsteady waves on the water (see, for instance, Whitham (1974) [71], and the review paper by Zeytounian (1995) [72]).

It is important to note that each physical problem has specific boundary conditions related to the intrinsic nature of the problem. For example, in gas dynamics problems the boundary conditions are different if the fluid flow is subsonic ($M < 1$), supersonic ($M > 1$), transonic ($M \sim 1$) or hypersonic ($M >> 1$). If, for instance, the undisturbed basic flow is in the $x > 0$ direction and the body in question (in x, y plane) is located on the x-axis, with its leading edge at the origin and its trailing edge at $x = 1$ (with non-dimensional variables), then we can assume that the body shape is described by:

$$y = \delta\, h(x), \tag{2.50}$$

where the non-dimensional parameter δ is the maximum value of y of the body. The 2D steady velocity potential φ(x, y) is a solution of the steady 2D Steichen dimensionless equation:

$$\left[a^2 - M^2(\partial\varphi/\partial x)^2\right]\partial^2\varphi/\partial x^2 + \left[a^2 - M^2(\partial\varphi/\partial y)^2\right]\partial^2\varphi/\partial y^2$$
$$- 2M^2(\partial\varphi/\partial x)(\partial\varphi/\partial y)\partial^2\varphi/\partial x\partial y = 0, \tag{2.51}$$

with the following relation for the local sound speed:

$$a^2 = 1 + [(\gamma + 1)/2]M^2\{1 - [(\partial\varphi/\partial x)^2 + (\partial\varphi/\partial y)^2]\}. \tag{2.52}$$

In this case the slip condition is:

$$\partial\varphi/\partial y = \delta(dh(x)/dx)\partial\varphi/\partial x,$$
$$\text{when } x \in [0, 1], \text{ on } y = \delta\, h(x). \tag{2.53}$$

Far away, upstream, from the body the flow should be undisturbed, which requires:

$$\partial\varphi/\partial x \to 1 \text{ and } \partial\varphi/\partial y \to o \text{ as } x \to -\infty. \tag{2.54}$$

In most applications, the bodies of interest are thin and streamlined, so that generally δ is a small non-dimensional parameter ($\delta << 1$). We note here only that the classical linear, subsonic and supersonic theory is invalid when respectively:

$$[M^2 - 1]/\delta^{3/2} = O(1) - \text{transonic similarity}$$

$$\delta\, M = O(1) - \text{hypersonic similarity}$$

$$\eta\delta = O(1) - \text{far field similarity,}$$

where $\eta(= x + [M^2 - 1]^{1/2})\, y$ is a characteristic coordinate, such that:

$$\eta \sim \infty \text{ with M fixed.}$$

In the case of justification of the well-known Boussinesq (1903) [8] assertion (see, for instance, Chap. 4) concerning the convection in fluids [13]: "The derivatives of the density can be neglected except when they intervene in the calculation of the force of Archimedes."

It is also necessary to consider the hyposonic ($M << 1$) case, and in such a case, for the atmospheric motions it is necessary to take into account the following constraint:

$$\mathrm{M/Bo} = \mathrm{O}(1) - \text{hyposonic similarity,} \tag{2.55a}$$

where

$$\mathrm{Bo} = \mathrm{L}^\circ/(\mathrm{RT}^\circ/\mathrm{g}) \tag{2.55b}$$

a ratio of two lengths, is the so-called Boussinesq number (see our (1990) [12], p. 15).

The lee waves problem (related mainly with the dynamic influence of a mountain in a baroclinic, stratified, adiabatic atmosphere) is strongly influenced by the relief slip condition and also by the upstream flow conditions. In an unbounded atmosphere the radiation (in a simple Boussinesq model case) Sommerfeld condition for the Helmholtz equation at infinity (in altitude) plays an essential role. (See, in [13], various typical examples considered by Guiraud and Zeytounian.)

In the low Rossby model for atmospheric flow, the effect of the solid (earth) surface is taken into account (by matching) through the so-called viscous Ekman layer. Indeed, the viscous coefficients are so small that we should expect the boundary conditions to be close to those valid for the corresponding inviscid system. The viscous equations do, however, require additional boundary conditions, and as an effect, viscous boundary layers may occur at the boundaries.

Such boundary layers may sometimes be appropriate, as in the rigid wall situation (for example, the Ekman boundary layer). However, at open boundaries they are inappropriate.

In comparison to flows in interior or exterior domains, there are two new issues when the boundary extends to infinity. First, in addition to the usual initial and boundary conditions there needs to be some prescription of fluxes or pressure drops when the flow domain has several exits to infinity (as in (2.54)). Second, the solutions of interest often have infinite energy integrals, and recently a technique of integral estimates to deal with this problem has been developed. These estimates are called Saint Venant's type, because the method was first used in the study of Saint Venant's principle in elasticity.

Concerning, more precisely, the behaviour of an incompressible fluid velocity field at infinity, we note that in Dobrokhotov and Shafarevich (1996) [73], a simple method is given which makes it possible to determine an upper bound for the decay rate at infinity of an incompressible fluid velocity field of general form; that is, to determine a lower bound for the field itself.

This method is based on the use of simple integral identities which are valid for solutions of the Navier incompressible, viscous equations, in the external region which decrease quickly enough. For the equations in entire space, some of these identities were obtained by the two authors noted above.

The property of slow decay or spreading of localized fluid flow is a consequence of incompressibility, and is not associated with viscosity alone (in contrast to the case described by Serrin (1959) [31, 74]), so that it also holds good for an inviscid Eulerian fluid flow (in this case the reasons for spreading are related with the non-uniform external flow and non-linearity).

In fact, in order to compute in a bounded region a fluid flow modelled by a problem formulated on an infinite domain, one often introduces an artificial boundary Σ and tries to write on the domain $\Omega^* \subset \Omega$, bounded by Σ, a new problem whose solution is as close as possible to the original exact problem. When the solution of this new problem in Ω^* coincides with the restriction of the original problem, the boundary Σ is said to be transparent.

Here, we note also that the reader can find valuable information concerning this approach with applications to both inviscid and viscous fluid flows in various recently published papers in the leading journals devoted to numerical fluid dynamics (see, for instance, the recent issues of *Journal of Computational Physics*).

The general slip condition in an unsteady case:

$$\mathbf{n} \cdot (\mathbf{u} - \mathbf{u}_P) = 0, \tag{2.56}$$

is satisfied, in any case, for an impermeable solid wall, where $\mathbf{u}_P$ is the velocity of the moving wall. On the other hand, from the kinetic theory of gases, when the Knudsen number, *Kn* is small, we obtain

$$\mathbf{n} \wedge (\mathbf{u} - \mathbf{u}_P) = 0 \text{ on a moving wall.} \tag{2.57}$$

As a consequence of (2.56) and (2.57), we again deduce the no-slip condition (but for a moving wall):

$$\mathbf{u} = \mathbf{u}_P, \text{ on the moving wall.} \tag{2.58}$$

The above condition (2.57) is the so-called weak form of the no-slip condition on the moving wall.

Concerning the boundary condition for the temperature T on the wall, from the kinetic theory of gases, again when the Knudsen number Kn is small, we obtain:

$$\mathrm{T} = \mathrm{T}_P - \beta \mathbf{q} \cdot \mathbf{n}, \tag{2.59}$$

where β is a scalar function (related with the kinetic, Knudsen, sub-layer).

An interesting case of boundary condition is related to the so-called Prandtl–Batchelor condition (see, for instance, the papers by Batchelor (1956)[75] and Wood (1957)[76]).

For a 2D incompressible, steady Eulerian fluid flow, from Eq. 2.13a, when $v_\circ \equiv 0$, we derive the following equation for the 2D steady stream function $\psi(\mathrm{x}, \mathrm{y})$:

$$\nabla^2\psi = F(\psi), \tag{2.60}$$

where the function $F(\psi)$ is arbitrary! But, if the domain Ω where the flow is considered is a bounded connected open subset of $\mathbf{R}^2$, then we do not have the possibility of utilizing the behaviour condition at infinity for the determination of this function $F(\psi)$.

The key of this indeterminacy is strongly related with the vanishing viscosity problem. In fact, with the limiting process Re $\uparrow \infty$(or, $\nu_\circ \downarrow 0$) in the steady form of the Navier Eq. 2.13a. Namely, if we assume that the limit streamlines are closed in Ω, then according to Batchelor [75] we derive for the limit Euler stream line $\gamma_\circ$ (which is the one of the stream lines $\psi_\circ =$ constant) the following Prandtl–Batchelor condition:

$$dF(\psi_\circ)/d\psi_\circ = 0 \text{ and } F(\psi_\circ) = F_{\circ\circ} \equiv \text{const.} \tag{2.61}$$

As a consequence, the Eulerian vorticity, $\omega_\circ = -(1/2)F(\psi_\circ)$, for a steady incompressible 2D fluid flow, is constant in any region where the streamlines are closed.

From the matching performed by Wood [76] – with the corresponding Prandtl boundary-layer in the vicinity of $\partial\Omega$ – the value of this above constant, $F_{\circ\circ}$, is well determined.

On the other hand, in Guiraud and Zeytounian's short paper (1984) [77] a process for setting in motion a viscous incompressible liquid inside a 2D cavity is considered, and it is shown that the basic process occurs for a time of the order of $t = O(Re)$. Then a flow, *à la* Prandtl–Batchelor, with constant Euler vorticity is established after a time $t >> Re$.

In this same paper [77], a G–Z functional equation is derived which governs the distribution of the vorticity in the main stage of interest, and for the simple case of a cylindrical cavity it is shown that the vorticity tends towards its own steady-state value exponentially.

Finally, concerning the case of overspecified and underspecified boundary conditions, it is important to note that when for a given problem the number of boundary conditions is overspecified, the difference approximation (for a numerical calculation) may well be stable. However, the effective boundary conditions which influence the solution are, in general, difficult to determine, especially for problems in several space dimensions.

They may well be a complicated function of the conditions given and bearing little resemblance to them. An additional complication induced by over-specification is that the underlying solution being approximated is not generally continuous. In order to avoid the problems associated with the proper selection of boundary conditions, the order and type of the differential equations is often raised to obtain a problem that is easier to analyze and approximate.

For example, the Eulerian equations are usually modified by adding dissipative terms so that the number of boundary conditions is appropriate. Unfortunately, this idea seldom works. If a spurious boundary layer of appreciable size results, the effects are not unlike those for discontinuities (for a system of equations, the errors can propagate away from the discontinuity through other components of the solution), and unless the dissipative terms are very large, the error introduced at the boundary will again propagate into the interior.

Now, if the boundary conditions are underspecified there are no a priori estimates for the differential equations. In order for an approximation to be computable these must be a sufficient number of boundary conditions specified for the approximation. This cannot be fewer than the number required for the differential equation.

The well-posedness of the initial boundary value fluid flow problems follows, to some extent, from properly formulated initial and boundary conditions, and is strongly linked with the various facets (through the existence and uniqueness results) of the solvability of these fluid flow problems.

We recognize that in large part what might be called "mathematical topics in fluid dynamics" has remained closed to the mainstream of theoretical fluid dynamics and mathematical physicists, due in large part – as judiciously observed in the book by Doering and Gibbon (1995) [78] devoted to applied analysis of the Navier (Navier–Stokes incompressible) equations – to the technical nature of rigorous investigations, often phrased in the unfamiliar language of abstract (non-linear) functional analysis.[3]

The above summary of Chap. 2 presents the main theoretical concepts and principles, and also equations and associated initial and boundary conditions, of classical/Newtonian fluid dynamics. Various theoretical concepts can be found in our books devoted, respectively, to non-viscous (2002) [37] and viscous (2004) [47] fluid flows. In our survey paper on the well-posedness of problems in fluid dynamics (a fluid-dynamical point of view) (1999) [79] the problem is carefully considered, and an historical survey of some mathematical aspects of Newtonian fluid flows can be found in our (2001) [29] surveys.

[3] The curiour reader can find in "Handbook of Mathematical Fluid Dynamics, vol. 1 to 4", numerous papers related with rigorous mathematical results, of existence, unicity, regularity, well-posedness and limiting processes for solution of fluid flow problems, mainly by compactnesse-a very abstract functional approach!

Chapter 3
From NS–F Equations to General Main Model Equations

First of all, the main question is relative to the Rational Asymptotic Modelling Approach (RAMA) of Newtonian fluid flow problems. Our task is as follows.

Starting from a real technological or geophysical problem, mathematically formulated via the NS–F equations and associated initial and boundary conditions, we want to obtain a simplified and consistent model problem which is possibly resolved via a numerical/computational simulation with the help of a super-high-speed computer.

Our mathematical status (expounded, more precisely, in Chap. 6) is very naïve, and is based on a logical, non-contradictory, constructive process, according to a main heuristic postulate and some associated key steps for its realization.

The desirable, but difficult, requirement concerning the rigorous proof that the error admitted is really of the order suggested is simply abandoned, in consequence of two goals: first, to consider, as much as possible, real complicated problems; and second, to efficiently assist numericians in our numerical simulation.

In our survey paper with Guiraud (1986) [11], devoted to emphasizing the considerable support that mastering asymptotic tools may afford to researchers embarked on rational modelling of very difficult problems of fluid flows, the reader can find a guide for classification of various models in a RAMA.

Concerning the level[1] of "general models", our thesis below is that very often, various writings in fluid dynamics (especially in text-books for university undergraduates) are organised through several particular fluid flows.

---

[1] In addition to above-mentioned level of general models, it is necessary to consider also two particular levels: local models, in order to elucidate the behaviour in some localized region, and global specific models, which sometimes occur when the flow under consideration, as a whole, may be determined via asymptotic modelling.

R. Kh. Zeytounian, *Navier–Stokes–Fourier Equations*,
DOI 10.1007/978-3-642-20746-4_3, © Springer-Verlag Berlin Heidelberg 2012

3.1 Non-dimensional Form of the NS–F Typical Problem

In this book we work mainly with dimensionless time t and position vector **x**, and also with dimensionless velocity vector **u** and thermodynamic functions p, ρ, and T. Respectively, the physical time-space, velocity vector and thermodynamic functions (with dimensions) are reduced via the characteristic time t°, the characteristic length L°, the characteristic constant velocity U°, and characteristic constant thermodynamic values p°, ρ° and T°.

For simplicity I have used, for our above dimensionless quantities, **u**, p, ρ, and T, as function of dimensionless (t, **x**), the same notation as in Chap. 2.

3.1.1 Non-dimensional NS–F Equations and Reduced Parameters

First, in non-dimensional form, for a thermally perfect gas, we obtain as equation of state:

$$p = \rho\,T, \tag{3.1}$$

because p° = R ρ° T°, according to (2.8), written for the reference values.

In non-dimensional form we have for the time derivative, with respect to material motion (along trajectories):

$$S\,D/Dt = S\,\partial/\partial t + \mathbf{u}\cdot\nabla, \tag{3.2}$$

where

$$S = L^\circ/t^\circ U^\circ \text{ is the } \textit{Strouhal} \text{ number} \tag{3.3}$$

From (2.34a–c) with (2.35a, b) we derive the following three unsteady full NS–F non-dimensional equations:

$$S D\rho/Dt + \nabla\cdot\mathbf{u} = 0, \tag{3.4a}$$

$$\rho\,S\,D\mathbf{u}/Dt + \left(1/\gamma M^2\right)\nabla p + (Bo/\gamma M^2)\rho\mathbf{k} = (1/Re)\nabla\cdot\{\lambda(\nabla.\mathbf{u})I + 2\mu\mathbf{D}(\mathbf{u})\}, \tag{3.4b}$$

$$\begin{aligned}\rho\,S\,DT/Dt + (\gamma-1)\,p\,\nabla\cdot\mathbf{u} &= (\gamma/Pr\,Re)\{\nabla\cdot[k\nabla T]\}\\ &+ (\gamma-1)[\gamma M^2/Re]\{2\mu\,\mathrm{Trace}(\mathbf{D}(\mathbf{u}))^2 + \lambda(\nabla\cdot\mathbf{u})^2\}.\end{aligned} \tag{3.4c}$$

which form a closed set of three non-dimensional partial differential equations, for four dimensionless quantities, **u**, p, ρ and T, as functions of dimensionless time space variables (t, **x**), provided that one adds the non-dimensional equation of state (3.1) for an ideal (thermally perfect) gas.

The reduced parameters in the above non-dimensional NS–F equations (3.4a–c) are

$$\mathrm{Re} = \mathrm{L}^\circ \mathrm{U}^\circ / \nu^\circ \; - \; \text{the } \textit{Reynolds} \text{ number}, \tag{3.5a}$$

$$\mathrm{M} = \mathrm{U}^\circ / (\gamma \mathrm{R} \mathrm{T}^\circ)^{1/2} \; - \; \text{the } \textit{Mach} \text{ number}, \tag{3.5b}$$

$$\mathrm{Pr} = \mathrm{Cp}\, \mu^\circ / k^\circ \; - \; \text{the } \textit{Prandtl} \text{ number}, \tag{3.5c}$$

$$\mathrm{Bo} \; = \; \gamma(\mathrm{M}/\mathrm{Fr_L}^\circ)^2 - \text{the } \textit{Boussinesq} \text{ number}, \tag{3.5d}$$

where

$$(\mathrm{Fr_L}^\circ)^2 = \mathrm{U}^{\circ 2} / g\mathrm{L}^\circ, \tag{3.6}$$

is the square of the Froude number based on the characteristic length L°.

The Strouhal number takes into account the unsteady effects and the Reynolds number the viscous effects. In particular, in Chap. 5, the consideration of the limiting case, $\mathrm{Re} \uparrow \infty$, when $\mathrm{S} = \mathrm{O}(1)$, leads, in the framework of our RAMA, to a complicated asymptotic structure of the above NS–F system of equations (3.4a–c) which is very different from classical regular coupling:

$$\text{Euler (non} - \text{viscous fluid flow)} \Leftrightarrow \text{Prandtl (boundary layer fluid flow)}$$

This complicated, four regions, singular coupling of the full unsteady NS–F equations arise mainly because of the singular nature of the Prandtl boundary layer equations near $t = 0$, where the initial data are given.

The compressibility is strongly related with the Mach number, and the acoustics effects are, in fact, closely linked to the following similarity relation:

$$\mathrm{S\,M} \; = M^* \; = \; \mathrm{O}(1), \tag{3.7}$$

when

$$\mathrm{S} \gg 1$$

a strong unsteady-state effect, and

$$\mathrm{M} \ll 1$$

quasi-incompressibility.

In reality, only the low Mach number asymptotics presents the possibility of deriving in a consistent form, as a limiting simplified incompressible model of above NS–F system (3.4a–c), the associated Navier incompressible and viscous system of equations.

These Navier equations are again singular near t = 0, where the equations of acoustics are significant – but, unfortunately, this is really the case only if the fluid flow domain is not a bounded, time dependent, domain (a cavity, as is the case in an internal flow; see, for instance, Sect. 7.1).

In the Bénard thermal convection problem – a liquid layer heated from below, when the fluid is an expansible liquid such that the relation:

$$\rho = \rho(\mathrm{T}) \tag{3.8}$$

is often assumed, we can introduce the constant coefficient of cubical expansion:

$$\alpha^\circ = -(1/\rho^\circ)\left[\frac{d\rho}{dT}\right]_{T=T^\circ}. \tag{3.9}$$

Concerning the relation (3.8) it is necessary to observe that it is only consistent in the framework of a RAM Approach as a "leading-order" approximation of a trivariate liquid, $\rho = \rho\,(\mathrm{T},\ \mathrm{p})$, when the above cubical expansion (3.9) of the considered liquid is very small (more precisely, for $\alpha^\circ \Delta T^\circ \ll 1$) in the Grashof (Gr) number (3.10) below.

In this case, if $\Delta T^\circ$is the difference between the two temperatures in a layer of thickness d°, then the following similarity relation:

$$\alpha^\circ \Delta T^\circ/(\mathrm{Fr}_{d^\circ})^2 = \mathrm{Gr} = O(1), \tag{3.10}$$

plays a significant role for the derivation of classical Oberbeck–Boussinesq shallow thermal convection equations (see, for instance, Chap. 8). In (3.10) the so-called Grashof number is, in fact, a similarity parameter, and

$$\mathrm{Pr}\ \mathrm{Gr} = \mathrm{Ra} \tag{3.11}$$

is the Rayleigh number. The reduced dimensional parameter

$$\mathrm{Pr}\ \mathrm{Re} = \mathrm{Pe}' = \mathrm{L}^\circ \mathrm{U}^\circ/(k_0/\rho_0 \mathrm{Cp}) \tag{3.12}$$

is the Péclet number.

We observe that in the definition of Gr (3.10), in the numerator the parameter

$$\varepsilon = \alpha^\circ \Delta T^\circ \tag{3.13}$$

is the expansibility parameter – a small ($\approx 5 \times 10^{-3}$)parameter for many liquids, which plays the role as a main small parameter in our derivation (in Chap. 8) of an approximate limiting consistent model for shallow thermal convection (*à la* Rayleigh–Bénard). But for this it is necessary to assume also that

$$(\mathrm{Fr}_{d^\circ})^2 \ll 1 \text{ such that } \mathrm{Gr} = O(1)$$

In fact, in the framework of the full Bénard convection problem, three significant convection cases merit interest:

1. Shallow-thermal, Rayleigh-Bénard convection, without viscous dissipation and upper free surface deformation, when, $(\mathrm{Fr}_{d^\circ})^2 \ll 1$.
2. Deep-thermal, Zeytounian convection, with viscous dissipation, when, $1 \gg (\mathrm{Fr}_{d^\circ})^2 \approx \alpha^\circ g d^\circ/(\mathrm{Cv})^\circ$.
3. Thermocapillary, Marangoni convection, with a deformable upper free surface, when, $(\mathrm{Fr}_{d^\circ})^2 \approx 1$.

   A fourth case also deserves careful asymptotic investigation, linked with:
4. Ultra-thin film, when, $(\mathrm{Fr}_{d^\circ})^2 \gg 1$,

For this fourth case, many questions are still open and deserve serious discussion and investigations.

For $\mathrm{Pr} \ll 1$ with $\mathrm{Re} \gg 1$, but $\mathrm{Pé} = \mathrm{O}(1)$, the fluid motion is quasi-non-viscous but strongly thermally conducting, and the so-called high thermal conductivity model equations are valid, with a specified boundary condition for the temperature on the wall.

We observe that with $\mathrm{Pr} \ll 1$, and $\mathrm{Pé} = \mathrm{O}(1)$ we assume that

$$\mu \ll k_0/\mathrm{Cp} \approx \rho^\circ L^\circ U^\circ. \tag{3.14}$$

Strictly speaking, when $\mathrm{Re} \gg 1$ the viscosity effects remains important, mainly in a boundary layer near the wall which has thickness of the order of

$$\mathrm{H_{CL}} = L^\circ/(\mathrm{Re})^{1/2}, \tag{3.15a}$$

and in this case we have, again, a similarity relation, namely

$$\varepsilon^2\mathrm{Re} = \mathrm{Re}_\perp = \mathrm{O}(1), \text{ where } \mathrm{H_{CL}}/L^\circ = \varepsilon, \tag{3.15b}$$

when we assume that:

$$\varepsilon \downarrow 0 \text{ and } \mathrm{Re} \uparrow \infty. \tag{3.15c}$$

In (3.15b, c) the parameter ε is a long-wave parameter, and in the viscous fluid, when the limiting process (3.15b, c) is performed, the so-called long-wave approximation is very similar to classical (Prandtl, 1904) boundary-layer approximation.

For the atmospheric viscous and non-adiabatic motions (when, obviously, the Mach number is very low), we take into account (see Sect. 9.2) the Coriolis force, which is characterized by the Coriolis parameter,

$$f^\circ = 2\Omega^\circ \sin\phi^\circ, \tag{3.16a}$$

where $\Omega^\circ = |\Omega|$, and Ω is the angular velocity of the rotating earth's frame, and ϕ°a reference latitude (for $\phi^\circ \approx 45^\circ$, we have $a^\circ = 6{,}300$ km for the earth's radius).

In this case it is helpful to employ spherical coordinates λ, ϕ, and r, and for the gradient operator ∇ we can write:

$$\nabla = (1/\mathrm{r}\cos\phi)\partial/\partial\lambda\ \mathbf{i} + (1/\mathrm{r})\partial/\partial\phi\ \mathbf{j} + \partial/\partial \mathrm{r}\ \mathbf{k}. \tag{3.16b}$$

But it is more convenient to use a right-handed curvilinear coordinates system (x, y, z) lying on the earth's surface (for a flat ground we have $\mathrm{r} \approx \mathrm{a}^\circ$) at latitude ϕ° and longitude $\lambda = 0$; namely:

$$\mathrm{x} = \mathrm{a}^\circ\cos\phi^\circ,\ \mathrm{y} = \mathrm{a}^\circ(\phi - \phi^\circ),\ \mathrm{z} = \mathrm{r} - \mathrm{a}^\circ. \tag{3.16c}$$

Although x and y are, in principle, new longitude and latitude coordinates in terms of which the basic NS–F (written for the atmospheric motions with the Coriolis force term characterized by the Rossby number Ro) equations may be rewritten without approximation, they are obviously introduced with the expectation that for a small sphericity parameter,

$$\delta = \mathrm{L}^\circ/\mathrm{a}^\circ \tag{3.16d}$$

they will be the Cartesian coordinates of the f°-plane approximation.

For the atmospheric motions a fundamental parameter is the Kibel number, such that

$$\mathrm{Ki} = 1/\mathrm{f}^\circ\mathrm{t}^\circ \equiv \mathrm{Ro} = (1/\mathrm{f}^\circ)/(\mathrm{L}^\circ/\mathrm{U}^\circ), \tag{3.17}$$

when we assume that the Strouhal number $\mathrm{S} \equiv 1$ or $\mathrm{t}^\circ = \mathrm{L}^\circ/\mathrm{U}^\circ$.

On the other hand, for atmospheric motions in the whole troposphere, a very significant characteristic constant vertical altitude is

$$\mathrm{H_s} = \mathrm{RT}^\circ/\mathrm{g} \tag{3.18a}$$

and usually (for the weather prediction, in the hydrostatic approximation)

$$\varepsilon_\mathrm{s} = \mathrm{Hs}/\mathrm{L}^\circ \ll 1. \tag{3.18b}$$

Indeed, the viscous and non-adiabatic effect in the tropospheric motions are limited within a thin layer near the ground – the Ekman layer – which is characterized by the following Ekman number:

$$\mathrm{Ek_S} = \mathrm{Ki}/\mathrm{Re_s} = \nu^\circ/f^\circ\mathrm{Hs}^2, \tag{3.19a}$$

where, for $\mathrm{Re} \gg 1$,

$$\mathrm{Re_S} = \varepsilon_\mathrm{S}{}^2\mathrm{Re} = \mathrm{O}(1). \tag{3.19b}$$

The reader can find in my two books (1990) [12] and (1991) [19] various asymptotic models for the atmospheric motions. Section 9.2 presents an application of the RAM Approach to atmospheric motions when the Kibel number is a small parameter.

Finally, when we consider the oscillatory viscous flows (when the Strouhal number $S \gg 1$; see, for instance, Riley (1967) [80]), when the flow is induced by a solid body performing harmonic oscillations (assuming that the velocity of the body is $U^\circ \cos \omega$) in an unbounded viscous fluid which is otherwise at rest, with $t^\circ = 1/\omega$ being a typical time. In this case, in dimensionless Navier equations for an incompressible and viscous fluid we have two main parameters, α and β^2, such that:

$$1/S = \alpha = U^\circ \omega L^\circ \text{ and } Re/\alpha = \beta^2 \equiv \omega L^{\circ 2}/\nu^\circ. \tag{3.20}$$

In fact, Riley (1967) considers solely the situation corresponding to $\alpha \ll 1$, high Strouhal number, and for the case of $\beta^2 = O(1)$ it is assumed that the oscillatory fluid flow is strongly viscous: $Re \ll 1$.

3.1.2 Conditions for the Unsteady NS–F Equations

Roughly speaking, we can expect that the equations of motion for a viscous fluid are parabolic. However, a more detailed analysis of the structure of equations (3.4a) and (3.4b) for the velocity vector **u** and the density ρ shows that this seems correct concerning equation (3.4b) for the velocity vector **u**, but not quite correct for the system of Eqs. 3.4a, b, because the continuity equation with respect to density ρ is hyperbolic in the compressible case, even for non-trivial viscosity!

Thus, to be more exact, we can say that a system of two Navier–Stokes equations (3.4a, b), with a specifying relation $p = P(\rho)$ and viscosity coefficient

$$\mu = -(3/2)\,\lambda(\rho) \equiv \mu^*(\rho),$$

is a hyperbolic–parabolic or incompletely parabolic system according to the definition suggested by Belov and Yanenko (1971) [59] and Strikwerda (1977) [81], in the rigorous study of the mathematical properties of these equations.

The Fourier equation (3.4c), written in the dimensionless form, for a small characteristic Mach number (hyposonic flow, $M \ll 1$), with respect to temperature T, is parabolic.

Kreiss and Lorenz (1989) [62] presents pertinent information concerning the initial-boundary value problems for Navier–Stokes equations. On the other hand, Oliger and Sundström (1978) [60] discuss initial-boundary value problem for several systems of partial differential equations from fluid dynamics.

We observe also that both the viscosity and heat conduction terms (characterized mainly by the inverse of the Reynolds number, 1/Re), in dimensionless equations (3.4b) and (3.4c), are usually very small ($Re \gg 1$), but since these terms change the character of the NS–F partial differential equations (from parabolic to hyperbolic in the case of a Eulerian non-viscous adiabatic fluid flow), we are obliged to reinvestigate the boundary conditions.

Concerning the initial condition, at initial time $t = 0$ the situation is unchanged relative to hyperbolic Euler unsteady compressible non-viscous equations considered in Zeytounian (2002) [37], Chap. 8. In the NS–F equations (3.4a–c) we have three derivation in time for the velocity vector $\mathbf{u}$, density ρ, and temperature T.

As a consequence, if we want resolve a pure initial-value or Cauchy problem (in the L^2-norm, for example), it is necessary to have a complete set of initial conditions (data) for $\boldsymbol{u}$, ρ, and T (as was mentioned in Chap. 2).

We see (according to Sects. 2.4.3 and 2.4.4) that several boundary conditions could be considered with respect to different physical situations for the above NS–F system of Eqs. 3.4a–c.

If we consider, as a simple example, the viscous and heat-conducting fluid flow above a rigid, solid, steady wall, simulated by the equation

$$z = 0,$$

then, when $\mu > 0$ and $\mu_v \equiv \lambda + (2/3)\mu > 0$, it is necessary to assume the following no-slip condition:

$$\mathbf{u} = 0 \text{ on } z = 0. \tag{3.21}$$

for the velocity vector in non-dimensional NS–F equation (3.4b).

For the (absolute) temperature T, in the case of a heat conductive fluids, when $k > 0$, in non-dimensional NS–F equation (3.4c), just to limit ourselves to the most common cases, we can require, in dimensionless form, the following temperature condition:

$$T = 1 + \chi\Theta(t, P) \quad \text{on } z = 0. \tag{3.22}$$

where P is a position point on the wall $z = 0$. In (3.22), χ is a temperature reduced constant parameter − a rate for the wall temperature fluctuation, simulated by the function $\Theta(t, P)$ – a measure for the given known function $\Theta(t, P)$, which define the temperature field on the wall.

This parameter χ plays an important role during the RAM Approach via various similarity rule with the above reduced parameters (3.5a–d), and especially with the Mach number, M.

We observe that in the compressible case, from the continuity Eq. 3.4a, with (see (1.37)) initial condition $\rho = \rho^\circ(\mathbf{x}) > 0$ at $t = 0$, we can affirm (see, for example, Valli (1992) [82]) that one of the main mathematical problems (unlike the incompressible case) is to find a priori estimates assuring the non-degeneracy of the density ρ.

Today, 76 years later, despite tremendous progress in many aspects of the mathematically rigorous Navier (N–S incompressible) theory (see, for example, the book by Temam (2000) [83]), we have not yet answered the fundamental questions raised by Leray's pioneer papers (1933/1934) [84]. That is, we have not determined whether a solution that is initially smooth can develop a singularity at some later time, or whether singularities are an important feature of turbulence. This is also related to another question: "Are singularities really necessary to explain turbulence?". Lions book (1998) [51] is mainly devoted to new rigorous mathematical results for NS compressible (but isentropic) fluid flows.

3.2 General Main Model Equations

Here I reiterate that our main goal is modelling – a scientific activity which consists of deriving (according to the RAM Approach), for various technological and geophysical stiff fluid flow, consistent, approximate, simplified model problems, in a such way that they become amenable, on the one hand, to mathematical analysis, and on the other hand, to numerical simulation by a super-high-speed computers.

This is not an easy task – especially when numerical computation involves simultaneously, dominant and negligible effects in so-called "stiff problems". In such a complicated case, for the numerical simulation it is necessary to be able, via the RAM Approach, to derive model consistent problems where the "stiffness" is smoothed!

A such, the RAM Approach is remarkably illustrated (see Sect. 7.1) in the case of an internal flow in a turbomachinery row with a large number of blades – the space between blades being very small – when the stiffness is replaced by the smoothness via a source term in a simplified model, consistent, through-flow (without blades!), which occurs from redistribution of forces acted on the flow by the blades of the row.

Below I present a number of fluid flow problems which are linked with general main models, though we make no reference to the literature, as such reference is both unnecessary and rather arbitrary.

In many cases of practical interest, the parameters (3.5a–d) in NS–F non-dimensional equations make take on extreme values, either very large or very small – these extreme values of parameters being closely linked with various particular fluid flows, which are of great interest in understanding the profound nature of flow related with the full NS–F problem.

3.2.1 Some General Main Models

Inviscid Euler fluid flows, which are often considered as models, used from the outset, need to be embedded in the more general main model of slightly viscous (laminar – large Reynolds number – flow) or slightly frictional (turbulent?) flow, to which RAMA is applied when Re $\uparrow\propto$.

Creeping flows, with numerous applications in thin films (lubrication), micro-hydrodynamics, and so on, should be considered as flows at low Reynolds ($\mathrm{Re} \ll 1$) number. Entire books are devoted to creeping flows, but the role of the low Reynolds number, as the main small parameter, related with $\mathrm{Re} \downarrow 0$, is ignored except in the very few expository pages.

Incompressible flows are seldom considered as flows at small Mach number. This obviously can become almost nonsensical, as when one deals with incompressible aerodynamics and acoustics. Phenomena such as sound produced by quite low-speed flow cannot be understood other than as low-Mach-number (hyposonic, $\mathrm{M} \downarrow o$) aerodynamics.

Rotating flow, which dominates both industrial and geophysical nature applications, are indeed asymptotic models of flow at low Rossby number ($\mathrm{Ro} \downarrow 0$) or low Kibel number ($\mathrm{Ki} \downarrow 0$).

Large-scale models of flows, of current use in simulations of meteorological or oceanographic applications, are extracted from asymptotic modelling which explains the role of the hydrostatic balance.

A number of models for the flows in porous media, flow with suspensions, or turbulent flow, should be considered as models obtained through some kind of homogenization (see Sect. 6.4).

The Navier–Fourier model, which take into account the effects of a slight compressibility and the thermal effects, is derived from NS–F equations (3.4a–c), with the condition (3.22), when we assume that the temperature parameter χ is small, such that for $\mathrm{M} \ll 1$ we assume the following similarity rule:

$$\chi = \lambda^0 \mathrm{M}^2 \text{with } \lambda^0 = \mathrm{O}(1).$$

The Stokes and Ossen compressible model equations, when Re and M both tend to zero, such that

$$\mathrm{Kn} = \mathrm{M}/\mathrm{Re} \ll 1.$$

This above constraint is realized if we take into account the following similarity rule:

$$\mathrm{M} = \Lambda\, \mathrm{Re}^{1+a}, \text{ with } \Lambda = \mathrm{O}(1) \text{ and } a > 0, \text{ when } \mathrm{Re} \downarrow 0.$$

The model equations of non-linear acoustics when, in place of Re (given by (3.5a)), it is necessary to introduce an acoustic Reynolds number $\mathrm{Re}_{\mathrm{ac}}$, such that:

$$\mathrm{Re}_{\mathrm{ac}} = \mathrm{Re}/\mathrm{M}\ (\gg 1),$$

the characteristic length scale being

$$\mathrm{L}^\circ = 1/\mathrm{k}^\circ,$$

where

$$k^\circ = \omega/\sqrt{\gamma RT^\circ}$$

is the wave number, and ω is a reference frequency determined by the main frequency of the source signal.

The shallow-thermal, Rayleigh–Bénard, convection model is derived when simultaneously the Froude number (Fr) and the expansibility parameter (ε) tend both to zero such that the Grashof number $Gr = O(1)$.

Various meteo-fluid dynamics models linked with low Kibel number limiting process (as in Sect. 9.2).

Discussion of the RAM Approach applied to some of the above problems can be found in [12, 13, 17, 26], and [27]; and see also the above Overview, Sect. 1.3, concerning my earlier books on modelling in Newtonian fluid flows. In Chaps. 7–9, devoted to applications of the RAM Approach, we also consider some aerodynamics, thermal convection and meteo/atmospheric stiff fluid flows, and derive the associated consistent model problems.

3.2.2 A Sketch of the Various General Main Models

Below, the reader can find a sketch of the various consistent general main model equations issues, with the help of a RAM Approach, from the NS–F full unsteady (and assumed exact) system of equations.

A first observation concerning these above-mentioned general main models is relative to the fact that the later are derived from the RAMA when the space-time, $(\boldsymbol{x}, t)$, fluid domain is (implicitly assumed) fixed. As a consequence (often, unfortunately), these general main models equations turn out (in many cases) to be of non-uniform validity, as they break down in certain regions of the fluid flow. (Concerning this problem see Sect. 3.3.)

More precisely, these regions are usually strongly linked with the assumed initial and boundary conditions. These general main models break down near the time, $t = 0$, where the initial data are given, and in the vicinity of the wall, $z = 0$, where the conditions (3.21) and (3.22), for **u** and T, are assigned. Below is an overview relative to the above general main models arising from the full NS–F equations (Fig. 3.1).

The experienced reader will be aware of the fact that the partial differential equations of fluid dynamics (NS–F system of Eqs. 3.4a–c solely) are not sufficient – as is often the case, unfortunately, in mathematical fluid dynamics – for discussing fluid flow problems.

This seemingly anodyne remark not only has far-reaching consequences, but also presents the possibility of discovering various new intrinsic structures from NS–F equations.

A good illustration of such a possibility (it seems to me) is our recent RAM Approach, expounded in Chap. 5, applied to NS–F full unsteady equations in the framework of a large Reynolds number – the discovered five regions structure being a direct consequence of the singular nature of equations of the unsteady compressible boundary layer near the initial time (= 0) where the initial data for the unsteady full NS–F equations are given.

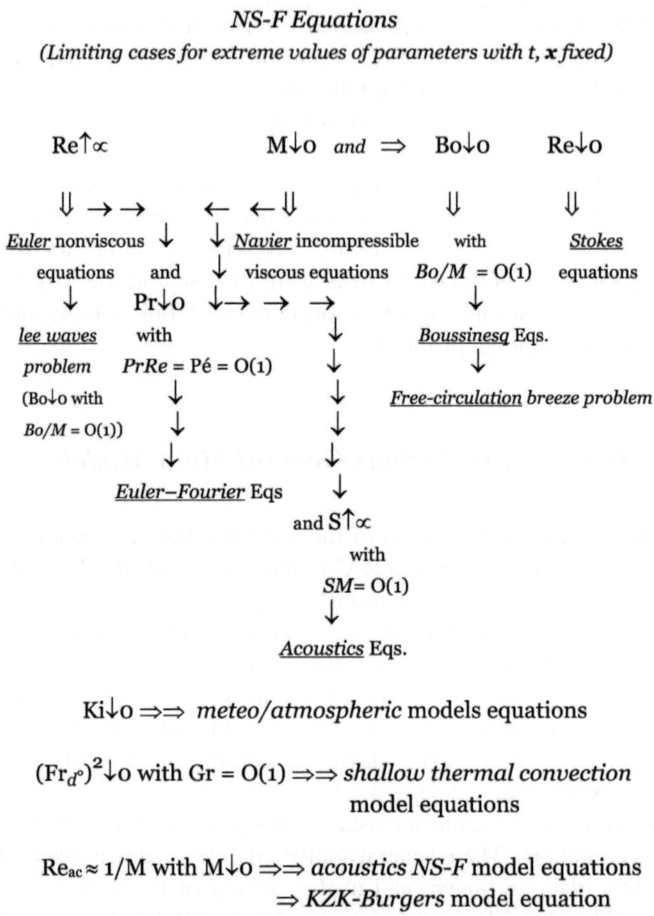

Fig. 3.1 From NS–F equations to some general main model equations

3.3 Non-uniform Validity of Main Model Equations and the Local Limiting Processes

The regions of non-uniform validity, near $t = 0$ and in the vicinity of $z = 0$ on the wall, of general main model equations, are mainly the results of the elimination of some terms with partial derivatives in t and z, in dimensionless NS–F full unsteady

Eqs. 3.4a–c during the main limiting processes, with t and $\boldsymbol{x}$ (= x, y, z) fixed, mentioned in Sect. 3.2.1.

3.3.1 Large Reynolds Case: A First Naive Approach

When the *Re* tends to infinity, with t and $\boldsymbol{x}$ fixed, we first consider the non-dissipative, Euler, main limiting process:

$$\mathrm{Lim}^{\mathrm{E}} \equiv \mathrm{Re} \uparrow \propto \text{ with t and } \mathbf{x} \text{ fixed} \tag{3.23}$$

in NS–F equations (3.4b) and (3.4c). The parameters Pr and M^2 being fixed, the dissipative terms in the right-hand side of both these equations disappears, and we recover the Euler compressible equations:

$$\mathrm{S}\ \mathrm{D}\rho_{\mathrm{E}}/\mathrm{Dt} + \nabla \cdot \mathbf{u}_{\mathrm{E}} = 0, \tag{3.24a}$$

$$\rho_{\mathrm{E}}\ \mathrm{S}\ \mathbf{D}\mathbf{u}_{\mathrm{E}}/\mathrm{Dt} + (1/\gamma \mathrm{M}^2)\nabla \mathrm{p}_{\mathrm{E}} + (\mathrm{Bo}/\gamma \mathrm{M}^2)\rho_{\mathrm{E}}\mathbf{k} = 0, \tag{3.24b}$$

$$\rho\ \mathrm{S}\ \mathrm{DT}_{\mathrm{E}}/\mathrm{Dt} + (\ \gamma - 1)\ \mathrm{p}\ \nabla \cdot \mathbf{u}_{\mathrm{E}} = 0, \tag{3.24c}$$

$$\mathrm{p}_{\mathrm{E}} = \rho_{\mathrm{E}}\mathrm{T}_{\mathrm{E}}, \tag{3.24d}$$

where

$$(\mathbf{u}_{\mathrm{E}}, \rho_{\mathrm{E}}, \mathrm{T}_{\mathrm{E}}) = \mathrm{Lim}^{\mathrm{E}}(\mathbf{u}, \rho, \mathrm{T}).$$

Obviously, for these non-viscous equations it is necessary to assume, in place of boundary wall condition (3.21), the slip condition:

$$\mathbf{u}_{\mathrm{E}}.\mathbf{k} \equiv \mathrm{w}_{\mathrm{E}} = 0 \text{ on } \mathrm{z} = 0, \tag{3.25}$$

This is the single boundary condition on the wall, for the above Euler unsteady compressible equations, because the term $\partial^2\mathbf{u}_{\mathrm{E}}/\partial \mathrm{z}^2$ is absent in the right-hand side of the limit Eq. 3.24b.

Concerning the initial conditions, it is again necessary to specify three initial conditions. However, the prescribed data in these three initial conditions, for Euler equations (3.24a–c), are all really the same than as for full unsteady NS–F equations (3.4a–c)? This is an interesting, but disquieting, question.

The singular nature of the Euler main limiting process in the vicinity of the wall $z = 0$ is directly related with the fact that condition (3.21) for an NS–F system of equations is replaced by the slip condition (3.25). In addition, the Euler equations (3.24a–c) take (unless dissipative terms) the form of a hyperbolic system.

This leads inevitably to the necessity of a RAM Approach for a slightly viscous and (when $\Pr = O(1)$) heat-conducting flow in a thin layer close to wall $z = 0$, and for this, according to a RAM Approach, it is necessary to consider a new limiting process: the Prandtl local limiting process in NS–F equations:

$$\lim{}^{\Pr} \equiv \ \mathrm{Re} \uparrow \propto \text{with t and x, y,} \quad \zeta = \ z/(1/\sqrt{\mathrm{Re}}) \ \text{fixed,} \tag{3.26a}$$

with

$$(\mathbf{v}_{\Pr}, \mathrm{w}_{\Pr}, \rho_{\Pr}, \mathrm{T}_{\Pr}) = \ \mathrm{Lim}^{\Pr}(\mathbf{v}, \ \mathrm{w}/(1/\sqrt{\mathrm{Re}}), \quad \rho\ , \mathrm{T}). \tag{3.26b}$$

From (3.26a, b), our small parameter, in a RAM Approach, is

$$\varepsilon = 1/\sqrt{\mathrm{Re}} \ \ll 1 \tag{3.27}$$

since $\mathrm{Re} \gg 1$.

In such a case in the leading-order we derive the well-known Prandtl boundary-layer equations. A detailed derivation of these Prandtl boundary-layer equations is presented in [26], Chap. 7. Here, in Chap. 5, we return to the case of a large Reynolds number fluid flow in the framework of a four-regions structure of NS–F full unsteady equations at high Reynolds number.

Here, if we simply assume that the dissipative coefficients λ, μ, and k are constant in dimensionless equations (3.4b, c) for $\mathbf{u}$ and T (in fact, equal to one), then we obtain, with (3.26a, b), the following Prandtl BL unsteady equations (with the horizontal gradient operator $\mathbf{D} = (\partial/\partial \mathrm{x}, \partial/\partial \mathrm{y})$):

$$\mathrm{S}\ \partial\rho_{\Pr}/\partial \mathrm{t} \ + \mathbf{D}.(\rho_{\Pr}\mathbf{v}_{\Pr}) + \partial(\rho_{\Pr}\mathrm{w}_{\Pr})/\partial\zeta = 0, \tag{3.28a}$$

$$\begin{aligned} &\mathrm{S}\ \rho_{\Pr}[\partial\mathbf{v}_{\Pr}/\partial \mathrm{t} + (\mathbf{v}_{\Pr}.\mathbf{D})\mathbf{v}_{\Pr} + \mathrm{w}_{\Pr}\partial\mathrm{w}_{\Pr}/\partial\zeta] \\ &+ (1/\ \gamma \mathrm{M}^2)\mathbf{D}\mathrm{p}_{\Pr} = \partial^2\mathbf{v}_{\Pr})/\partial\ \zeta^2, \end{aligned} \tag{3.28b}$$

$$\partial \mathrm{p}_{\Pr}/\partial\zeta = 0, \tag{3.28c}$$

$$\begin{aligned} \mathrm{S}\ \rho_{\Pr}[&\partial\mathrm{T}_{\Pr}/\partial \mathrm{t} + (\mathbf{v}_{\Pr}.\mathbf{D})\mathrm{T}_{\Pr} \\ &+ \mathrm{w}_{\Pr}\partial\mathrm{T}_{\Pr}/\partial\zeta] + (\ \gamma - 1)\mathrm{p}_{\Pr}[\mathbf{D}.\mathbf{v}_{\Pr} + \partial\mathrm{w}_{\Pr}/\partial\zeta] \\ &= (\ \gamma/\Pr)\partial^2\mathrm{T}_{\Pr})/\partial\zeta^2 + \ \gamma(\ \gamma - 1)\mathrm{M}^2|\partial\mathbf{v}_{\Pr}/\partial\zeta|^{2,} \end{aligned} \tag{3.28d}$$

$$\mathrm{p}_{\Pr} = \rho_{\Pr}\mathrm{T}_{\Pr}, \tag{3.28e}$$

where in (3.28b) and (3.28d) some dissipative terms are again present.

It is well known that the above two limiting systems (Euler and Prandtl) of Eqs. 3.24a–d and 3.28a–e are related to the following matching condition (the concept of matching will be discussed in more detail in Sect. 6.4.2):

$$\lim_{\zeta\uparrow\infty}[\mathrm{Lim}^{\mathrm{Pr}}] = \lim_{z\downarrow o}[\mathrm{Lim}^{\mathrm{E}}], \tag{3.29}$$

As a first consequence of (3.29) we recover the slip condition (3.25), which is, in our RAM Approach, a mathematically logical consequence of the passage from an incompletely parabolic (hyperbolic–parabolic) NS–F system to a hyperbolic Euler system.

In Chap. 5, devoted to a "deconstruction"[2] of the full NS–F unsteady equations, when Re $\uparrow\infty$, the consequence of the strong (BL–Prandtl) degeneracy (3.28c) is analyzed in the relation of initial conditions for a well-posed initial-boundary unsteady problem for the Prandtl BL equations.

3.3.2 Low Mach Number Case: The Navier System of Equations and Companion Acoustics Equations

The Navier equations govern incompressible viscous fluid flows, and are formally derived from the full NS–F equations (3.4a–c), when M $\downarrow$ 0.

Here, for aerodynamics, we assume that Bo $= 0$. The case when Bo is not zero (the gravity force being active) is considered in Chap. 4, in the framework of a rational justification of well-known Boussinesq approximation.

More precisely, we consider the following Navier main limiting process:

$$\begin{aligned}&\mathrm{Lim}^{\mathrm{M}} \equiv \mathrm{M}\downarrow 0\\&\text{with t and } \mathbf{x} \text{ fixed,}\end{aligned} \tag{3.30}$$

and we assume, again, that the dissipative coefficients λ, μ, and k are equal to one in the non-dimensional equations (3.4b, c). During the above $\mathrm{Lim}^{\mathrm{M}}$ (3.30) we assume that S, γ, Re, and Pr are all O(1).

First, according to (3.4b), in a very naive way we assume the following asymptotic expansion for the pressure p:

[2] From the French "déconstruction" – a word invented by Jacques Derrida (see the book by Charles Ramond, *Le vocabulaire de Derrida*, Ellipses, Paris, 2001), which presents us with the possibility of understanding the intrinsic structure of the NS–F equations, and reveals the presence of a profound unity in the puzzle of the partial models of fluid dynamics problems. To "deconstruct" the NS–F system of equations we have the possibility of unifying these partial models, used in Newtonian fluid dynamics, according to our RAM Approch, using the various limiting processes linked with the reference dimensionless parameters in equations, conditions, and geometry of considered non-dimensional fluid flow problems. From a such process we re-establish a well-ordered and unified family of partial fluid flows.

$$p = p_0(t) + M^2 p_N(t, \mathbf{x}) + ..., \tag{3.31}$$

and we enquire how one can obtain information concerning the function $p_0(t)$. Scrutinization of the starting NS–F equations, and trying more sophisticated expansion processes (in place of (3.31)) proves to be of no help. However, a way out appears if (3.31) holds in the whole of a domain of space where p is known to leading order, with respect to the smallness of M, as far as its dependence on x is concerned. This, obviously, occurs when (3.31) holds in a neighbourhood of infinity where the pressure is constant (especially in the case of a steady fluid flow past a finite body – external aerodynamic problem), and this leads, then, to $p_0(t) = 1$.

On the other hand, let us assume that the gas is contained in a container Ω bounded by an impermeable but eventually deformable (with time t) wall, so that the volume occupied by the gas is a given function of time: namely, $V_0(t)$. An obvious way to proceed is to assume that the density and temperature go to definite limits: $\rho_0(t, \mathbf{x})$ and $T_0(t, \mathbf{x})$, according to (3.30), Lim^N. It is a very easy matter to derive, from Eq. 3.4c, an equation satisfied by $T_0(t, \mathbf{x})$. Such an equation involves the unknown function $p_0(t)$, and it has an obvious solution $T_0(t, \mathbf{x}) \equiv T_0(t)$, which holds provided that the two unknown functions, $\rho_0(t, \mathbf{x})$ and $T_0(t, \mathbf{x})$, meet the requirement that

$$T_0/(\rho_0)^{\gamma-1} \text{ be independent of time} = 1. \tag{3.32a}$$

It is then easy to reach the conclusion that

$$\rho_0(t) = p_0(t)^{1/\gamma}, \tag{3.32b}$$

and, conservation of the mass for the whole of the gas contained within the container gives

$$p_0(t) = [V_0(t)]^{-\gamma}, \tag{3.32c}$$

so that we have found our way out of the indeterminacy concerning the leading term in (3.31). Of course, our argument relies on $T_0(t)$ being independent of space, and we have to discuss the adequacy of that. It is obviously a matter of conduction of heat within the gas. Such a phenomenon might have two origins – one of which is dissipation of energy within the gas; but consideration of Eq. 3.4c tell us that this enters into account at a rate of $O(M^2)$ and is negligible as far as $T_0(t)$ is concerned. A second origin for the conduction of heat is through the variations of temperature on the wall or from heat transfer through it.

Let us return to Eqs. 3.4a, b and assume that $\mathbf{u}$ goes to $\mathbf{u}_N(t, \mathbf{x})$ according to (3.30). We put:

$$p_N/\gamma\rho_0(t) = \text{Lim}^N\{[p - p_0(t)]/\gamma\rho_0(t)M^2\} = \pi, \tag{3.33}$$

This π is a fictitious pressure perturbation, and we derive the following quasi-incompressible system of equations for Navier velocity vector $\mathbf{u}_N$ and pressure perturbation π:

$$\nabla.\mathbf{u}_N = -(S/\rho_0(t))\ d\rho_0(t)/dt, \tag{3.34a}$$

$$[S\ \partial/\partial t\ + \mathbf{u}_N.\nabla]\mathbf{u}_N + \nabla\ \pi = [\ \mu(T_0)/\rho_0(t)Re]\nabla^2\mathbf{u}_N, \tag{3.34b}$$

$$\rho_0(t) = 1/V(t), p_0(t) = [V(t)]^{-\gamma}, T_0(t) = [V(t)]^{1-\gamma}. \tag{3.34c}$$

The above system of Eqs. 3.34a, b with 3.34c is a rather slight variant of the Navier equations (see also Sect. 7.2) which consists of a fluid flow with time-dependent viscosity and no-divergenceless velocity vector, rather than a divergenceless motion with constant in space.

The usual set of Navier equations is obviously obtained for a constant-volume container, and in this case ($p_0(t)$ and $\rho_0(t)\ \equiv\ 1$, in (3.33)) we can write, in place of (3.34a, b), the following classical Navier (NS incompressible) equations for a divergenceless velocity vector $\mathbf{u}_N$ and fictitious pressure perturbation π:

$$\nabla.\mathbf{u}_N = 0, \tag{3.35a}$$

$$[S\ \partial/\partial t\ + \mathbf{u}_N.\nabla]\ \mathbf{u}_N + \nabla\pi = (1/Re)\nabla^2\mathbf{u}_N. \tag{3.35b}$$

On the one hand, for the full unsteady NS–F compressible and heat-conducting equations (3.4a–c), with Bo $= 0$, aerodynamics case, if we want to resolve a Cauchy problem it is necessary to impose three initial conditions for ρ, $\mathbf{u}$, and T.

On the other hand, when considering the above Navier general main model equations (3.35a, b), we must give only the initial value for $\mathbf{u}_N$, and this Navier initial condition $(\mathbf{u}_N)^\circ$ is such that:

$$\nabla.(\mathbf{u}_N)^\circ = 0. \tag{3.36}$$

This shows that the Navier incompressible system (3.35a, b) – where acoustic waves are filtered – is certainly not valid (singular) near $t = 0$.

As a consequence, it is necessary to derive a new model of limiting equations consistent in an initial (local) thin region with a "short time". We therefore introduce

$$\tau =\ t/M\ =\ O(1),\ \text{when } M \downarrow 0, \tag{3.37a}$$

and consider the following local acoustic limiting process:

$$\lim{}^{Ac} \equiv M \downarrow 0 \text{ with } \tau \text{ and } \mathbf{x} \text{ fixed}. \tag{3.37b}$$

This short time τ is well suited for studying the transient behaviour – an unsteady adjustment problem related to a process of matching between main (for instance, (3.31)) and local,

$$p = 1 + M^a p_{ac} + \ldots, \tag{3.38a}$$

expansions.

In fact, for the case of an external aerodynamics – when the matching is possible – we must deduce from NS–F equations (3.4a–c) a local consistent system of acoustics equations which realize the above-mentioned matching.

For this, the following asymptotic expansions are considered, with (3.38a) for p, **u**, ρ, and T, in the framework of unsteady NS–F dimensionless equations (3.4a–c):

$$\mathbf{u} = \mathbf{u}_{ac} + \ldots, \tag{3.38b}$$

$$\rho = 1 + M^b \rho_{ac} + \ldots, \tag{3.38c}$$

$$T = 1 + M^c T_{ac} + \ldots, \tag{3.38d}$$

with a, b, and c being three scalars to be determined.

We now consider the above (3.37a, b) local acoustic limiting process, when again, S, γ, Re, and Pr are all O(1).

Various detailed investigations by Zeytounian and Guiraud (see, for instance, [85] and [86]) show that the above asymptotic local expansions (3.38a–d) are consistent only with the situation corresponding to an unbounded fluid flow, outside a solid bounded body **Ω**, starting in motion impulsively – mimicking a catapulting process.

A poor justification for working this way is that this is a classical problem in inviscid incompressible fluid dynamics (in fact, this transient behaviour is essentially characterized by the weak compressibility) in the NS–F equations, when we consider the local (in time) limiting process (3.37a, b) with (3.38a–d).

First, from (3.4a), at leading order we derive a least-degenerated (concerning the least/significant degeneracy concept; see Sect. 6.4.3) continuity equation when a = 1 and b = 1:

$$S\, \partial\rho_{ac}/\partial\, \tau + \nabla.\mathbf{u}_{ac} = 0. \tag{3.39a}$$

It is then easy to show that the more consistent limiting system of equations, which is derived from the full dimensionless NS–F equations (3.4a–c) with (3.37a, b) and (3.38a–d), is the linear acoustics system of equations, if we assume that a = c = 1. In such a case, with (3.39a) we obtain the following three equations:

$$S\, \partial\mathbf{u}_{ac}/\partial\tau + (1/\,\gamma)\nabla p_{ac} = 0, \tag{3.39b}$$

$$S\ \partial T_{ac}/\partial\tau + (1 - \gamma)\nabla.\mathbf{u}_{ac} = 0, \tag{3.39c}$$

$$p_{ac} = \rho_{ac} + T_{ac}. \tag{3.39d}$$

If we consider the transient behaviour of an NS–F fluid flow which is set into motion from rest by the displacement of a solid body, an unbounded medium, as an initial condition for above acoustics equations we write:

$$t = 0 : \mathbf{u}_{ac} = 0, T_{ac} = 0,\ \rho_{ac} = 0. \tag{3.40}$$

For the determination of the unknown Navier initial condition $(\mathbf{u}_N)^\circ$ it is necessary to consider an unsteady adjustment problem (as described in Sect. 2.4.2) with a matching process, such that:

$$\lim_{\tau\downarrow\infty}\mathbf{u}_{ac} = \lim_{t=0}\mathbf{u}_N \equiv\ (\mathbf{u}_N)^\circ. \tag{3.41}$$

Here I note that in the book by Wilcox (1975) [87] the reader can find a scattering theory which makes possible the analysis of the behaviour of the above equations of acoustics (3.39a–d), with (3.40), when $\tau \downarrow \infty$, and in Zeytounian (2000) [88] this matching has been realized.

Unfortunately, this is not the case when the gas is contained in a bounded container Ω, with an impermeable but eventually deformable (with time t) wall, so that the volume occupied by the gas is a given function of time, $V_0(t)$ – a problem first considered by Guiraud and Zeytounian in (1980) [88].

For instance, in a bounded fluid flow, inside a solid bounded body Ω, with a wall $\Sigma(t)$ deformable with time t, the above matching relation (3.41) often does not work (see Sect. 6.4.2, concerning "matching"). In Sect. 7.2, devoted to applied aerodynamics problems, we return to the consideration of a such case. In particular, this case may be related to compressible gas flow in the compression phase of an internal combustion engine.

3.3.3 *Oberbeck–Boussinesq Model Equations for the Rayleigh–Bénard Shallow Thermal Convection*

When we consider a weakly expansible (with a low, ε, expansibility parameter (3.13)) liquid, such that the following simplified equation of state is valid (at least, in leading-order, when the expansibility parameter, ε, defined by (3.13), tends to zero):

$$\rho = \rho\ (T), \tag{3.42a}$$

then, in place of the NS–F system of equations (2.34a–c) with (2.35a, b), we have in equation (2.34c) as C_v the coefficient (as a consequence of (3.42a)):

$$C(T) = DE_l(T)/Dt. \tag{3.42b}$$

because the specific internal energy $E = E_l(T)$ is a function of temperature T only:

$$DE_l/Dt = C(T)\ DT/Dt. \tag{3.42c}$$

For the derivation of a leading-order (relative to $\varepsilon \ll 1$) Oberbeck–Boussinesq model – as a simplified consistent Rayleigh–Bénard (RB) shallow thermal convection model problem – we can assume (due to a low squared Froude number (see (3.10)) hypothesis) that

$$\rho(T) \approx \rho_{d^\circ}(1 - \varepsilon\Theta), \tag{3.42d}$$

is consistent, with an error of $O(\varepsilon^2)$, when Θ is given by relation (3.43a) and a perturbation of pressure is introduced by its companion relation (3.43b).

Therefore, if we assume that T_{d° is the constant temperature of the free surface in the purely static motionless conduction basic state (which is the plane $z = d$), such that $\Delta T^\circ = T_w - T_{d^\circ}$, in (3.13) and (3.43a), we can introduce the following dimensionless temperature (T_w is the temperature at $z = 0$):

$$\Theta = (T - T_{d^\circ})/\Delta T^\circ, \tag{3.43a}$$

The companion dimensionless perturbation of pressure is:

$$\pi = [1/(\mathrm{Fr}_{d^\circ})^2]\{[(p - p_A)/(g\rho_{d^\circ}d^\circ)] + (z - 1)\}, \tag{3.43b}$$

where p_A is the passive constant atmospheric pressure.

Thanks to the introduction of π by (3.43b), and when the limiting RB process (3.45) is applied to the NS–F system of equations for a weakly expansible liquid, we can easily verify that relation (3.42a) is sufficient (and consistent according to (3.42d)) for the density ρ, for the derivation of the dominant, leading-order, Eqs. 3.44a–c.

Indeed, if now, as the characteristic length scale we choose d°, as characteristic constant velocity $\nu_d{}^\circ/d^\circ$, the characteristic time being $d^{\circ 2}/\nu_{d^\circ}$, where ν_{d° is the constant kinematic viscosity (at $z = d^\circ$), then we can derive, first, the following dimensionless dominant equations (from an NS–F system of equations written for an expansible liquid) for our weakly expansible liquid:

$$\nabla.\mathbf{u}_{\mathrm{liq}} = \varepsilon D\Theta/Dt, \tag{3.44a}$$

$$(1 - \varepsilon\,\Theta)D\mathbf{u}_{\mathrm{liq}}/Dt + \nabla\pi - \mathrm{Gr}\Theta\ \mathbf{k} - \nabla^2\mathbf{u}_{\mathrm{liq}} = O(\varepsilon), \tag{3.44b}$$

$$[1 - \varepsilon\Theta(1 + \Gamma^\circ)]\mathrm{D}\Theta/\mathrm{Dt} - (1/\mathrm{Pr})\nabla^2\Theta = \mathrm{O}(\varepsilon), \tag{3.44c}$$

where

$$\mathrm{C(T)} = \mathrm{C}_{d^\circ}(1 - \varepsilon\Gamma^\circ\Theta).$$

When (limiting O-B case)

$$\mathrm{Lim}^{\mathrm{O-B}} \equiv \varepsilon \to 0, \text{ with } \mathrm{Gr} = \mathrm{O}(1), \mathrm{Pr} = \mathrm{O}(1) \text{ and } \Gamma^\circ = \mathrm{O}(1), \tag{3.45}$$

is performed, from the dominant Eqs. 3.44a–c, we obtain the Oberbeck–Boussinesq model equations for the Rayleigh–Bénard shallow thermal convection in the following form:

$$\nabla.\mathbf{u}_{\mathrm{O-B}} = 0, \tag{3.46a}$$

$$\mathrm{D}\mathbf{u}_{\mathrm{O-B}}/\mathrm{Dt} + \nabla\,\pi_{\mathrm{O-B}} - \mathrm{Gr}\Theta_{\mathrm{O-B}}\,\mathbf{k} = \nabla^2\mathbf{u}_{\mathrm{O-B}}, \tag{3.46b}$$

$$\mathrm{D}\Theta_{\mathrm{O-B}}/\mathrm{Dt} = (1/\mathrm{Pr})\nabla^2\Theta_{\mathrm{O-B}}. \tag{3.46c}$$

In Chap. 8 the reader can find some complementary consequences of the above limiting O-B (3.45) case in the framework of the well-known Bénard heated-from-below thermal convection problem, for a weakly expansible liquid layer with a free surface and temperature-dependent tension.

3.3.4 Stokes–Oseen Model Equations in the Case of a Low Reynolds Number

In the case of an incompressible viscous Navier fluid flow with $\mathrm{S} \equiv 1$, from (3.35b) we obtain:

$$\mathrm{Re}\ \mathrm{D}\mathbf{u}_{\mathrm{N}}/\mathrm{Dt} + \nabla \mathrm{p}_{\mathrm{St}} = \nabla^2\mathbf{u}_{\mathrm{N}}, \tag{3.47a}$$

with

$$\mathrm{p}_{\mathrm{st}} = \mathrm{Re}\ \pi, \tag{3.47b}$$

the Stokes pressure, and for low Reynolds number, when the Stokes limiting process is performed:

$$\mathrm{Lim}^{\mathrm{St}} = \mathrm{Re} \downarrow 0 \text{ with } \mathrm{p}_{\mathrm{st}} \text{ and } \mathrm{t}, \mathbf{x} \text{ fixed}, \tag{3.48}$$

we obtain, for $\mathrm{Lim}^{St}\mathbf{u}_N = \mathbf{u}_{St}(t, \mathbf{x})$ and p_{St}, the so-called Stokes equation:

$$\nabla^2 \mathbf{u}_{St} = \nabla p_{St}. \tag{3.49a}$$

with, from (3.35a),

$$\nabla.\mathbf{u}_{St} = 0. \tag{3.49b}$$

One would expect that the boundary conditions for the steady Stokes equations (3.49a, b) would be same as those for the full Navier equations (3.35a, b); but it was noted by Stokes (1851) [89] that solutions do not exist for stationary 2D flow past a solid which satisfies both conditions:

$$\text{at the solid wall, } \mathbf{u}_{St} = 0 \text{ (no slip)}, \tag{3.49c}$$

as well as

$$\text{at infinity, } \mathbf{u}_{St} \to U_\infty \mathbf{i} (\text{uniform flow}), \tag{3.49d}$$

Indeed, the steady incompressible Stokes flow, governed by the above problem (3.49a–d), in an unbounded domain, exterior to a solid body, is only a "local-proximal" flow valid mainly in the vicinity of the wall of this body (the Stokes paradox).

Far from the wall, near infinity, when $\mathrm{Re}|\mathbf{x}| = |\mathbf{x}_{0s}| = O(1)$, it is necessary to derive another consistent "local-distal" model equation [90] for the Oseen (1910) flow, in place of the above problem (3.49a–d) for the Stokes flow.

Using Oseen time–space variables:

$$t_{0s} (= \mathrm{Re}\ t) \text{ and } \mathbf{x}_{0s}, \tag{3.50a}$$

and also the Oseen pressure,

$$p_{0s} = p_{St}/\mathrm{Re}, \tag{3.50b}$$

we derive, first, in place of the unsteady Navier equation (3.35b), the following dimensionless Navier equation (when $S \equiv 1$) for $\mathbf{u}'_N$ $(t_{0s}, \mathbf{x}_{0s})$:

$$D\mathbf{u}'_N/Dt_{0s} + \nabla_{0s} p_{0s} = (\nabla_{0s})^2 \mathbf{u}'_N. \tag{3.51}$$

But, near infinity, in the Oseen limiting process,

$$\mathrm{Lim}^{0s} = \mathrm{Re} \downarrow 0 \text{ with } p_{0s}, t_{0s} \text{ and } \mathbf{x}_{0s} \text{fixed}, \tag{3.52}$$

a finite body shrinks to a point, which cannot cause a finite disturbance in the viscous fluid flow.

Thus, in the outer, Oseen region, we can assume that for $\mathbf{u}'_N$ the asymptotic expansion

$$\mathbf{u}'_N = \mathbf{i} + \delta(\mathrm{Re})\mathbf{u}_{Os}(t_{Os}, \mathbf{x}_{Os}) + \ldots, \tag{3.53a}$$

is significant, when $\delta(\mathrm{Re}) \ll 1$, assuming that (with dimensionless quantities),

$$\mathbf{u}'_N \to \mathbf{i} \text{ at infinity.} \tag{3.53b}$$

With (3.52) and (3.53a, b), for $\mathbf{u}_{Os}(t_{Os}, \mathbf{x}_{Os})$, we derive, from (3.51), the following unsteady Oseen equation:

$$[\partial/\partial t_{Os} + \nabla_{Os}.\mathbf{i}]\mathbf{u}_{Os} + \nabla_{Os}p_{Os} == (\nabla_{Os})^2\mathbf{u}_{Os}. \tag{3.54}$$

In the above Oseen leading-order equation for $\mathbf{u}_{Os}$, the gauge function, $\delta(\mathrm{Re}) \to 0$, with $\mathrm{Re} \to 0$, is arbitrary, and only via a matching

$$\text{local} - \text{proximal} \Leftrightarrow \text{local} - \text{distal}$$

do we have the possibility of determining this gauge. (See Lagerstrom (1964) [91], pp.163–7, for a discussion of this matching process.)

We observe also that for a compressible fluid flow, when $\mathrm{Re} \to 0$, it is also necessary to specify the role of the Mach number, M. In fact, we must address the problem of the behaviour of full NS–F system of equations when simultaneously,

$$\mathrm{Re} \to 0 \text{ and } \mathrm{M} \to 0, \text{ with t and } \mathbf{x} \text{ fixed.} \tag{3.55a}$$

Obviously, it must no forgotten that for the validity of these full NS–F equations (3.4a–c) with (3.1), for a thermally perfect gas, under the above limiting low Reynolds and Mach numbers (3.55a), it is necessary that the corresponding limit fluid flow remains a low Knudsen, *Kn*, number flow (briefly discussed at the beginning of Chap. 2).

As a consequence of the relation $\mathrm{Kn} = \mathrm{M}/\mathrm{Re}$, the above double limiting process (3.55a) must be made with the following similarity relation:

$$\mathrm{M} = \sigma^\circ \mathrm{Re}^{1+a}, \sigma^\circ = (O(1) \text{ fixed, } a > 0, \text{ when } \mathrm{Re} \to 0. \tag{3.55b}$$

But, with (3.55a, b) it is also necessary to take into account the following asymptotic expansions in the framework of the NS–F equations (3.4a–c):

$$\mathbf{u} = \mathbf{u}_S + \ldots; \tag{3.55c}$$

$$p = 1 + \mathrm{Re}^{1+2a}[p_S + \ldots], \tag{3.55d}$$

$$\mathrm{T} = \mathrm{T}_S + \ldots;\ \rho = \rho_S + \ldots, \tag{3.55e}$$

where the "Stokes" limit functions (with '$_S$' as subscript) depend on **x** (fixed) only, because the above expansions are singular near the time $t = 0$. In fact, the time t (also fixed in (3.55a)), in the compressible Stokes equations (3.56a–c), plays the role of a parameter!

These Stokes compressible (inner) equations are written in two parts:

$$\nabla \cdot [k_S(\mathrm{T}_S)\nabla \mathrm{T}_S] = 0;\ \rho_S = 1/\mathrm{T}_S, \tag{3.56a}$$

and

$$\nabla \cdot \mathbf{u}_S = \mathbf{u}_S \cdot \nabla \log \mathrm{T}_S; \tag{3.56b}$$

$$\nabla \mathrm{p}_S = \gamma\sigma^{o2}\nabla \cdot \{\lambda(\mathrm{T}_S)(\nabla.\mathbf{u}_S)I + 2\mu(\mathrm{T}_S)\mathbf{D}(\mathbf{u}_S)\}. \tag{3.56c}$$

The first equation of (3.56a) with the associated boundary condition (see, for instance, (3.22)) on the boundary $\Gamma = \partial\Sigma$, determines T_S, as soon as the temperature behaviour at infinity, or matching with the outer Oseen compressible equations, is specified (see, for instance, in Lagerstrom (1964) [91], pp. 191–2 and 202–5) for an unbounded fluid flow outside a solid bounded (by Γ) body Σ. The second equation of (3.56a) is a relation (limit form of the equation of state (3.1)) between T_S and ρ_S, and determines ρ_S when T_S is known.

Finally, the system of the two Eqs. 3.56b, c gives a closed system for the determination of the velocity vector $\mathbf{u}_S$ and the perturbation of pressure p_S, when the boundary condition (for $\mathbf{u}_S$) on the wall $\partial\Sigma$ are used (for instance, (3.21)).

The scalar $a > 0$ in (3.55b) and in the second expansion for the pressure in (3.55c) is determined only after the matching with the companion compressible Oseen (outer and valid far at $t = 0$) equations. (Concerning the singular (for $\mathrm{Re} \to 0$) region near $t = 0$, see Sect. 9.3 of our (2002) [26].)

On the other hand, when the rate of temperature fluctuation (the scalar χ in (3.22)) on the wall $\partial\Sigma$ tends to zero with $\mathrm{Re} \to 0$, then we obtain a particular simple solution for Eq. 3.56a: $\rho_S = 1$ and $\mathrm{T}_S = 1$. In such a case, Eqs. 3.56b,c reduce to the classical steady Stokes equation (3.49a) with (3.49b) for an incompressible low Reynolds fluid flow if we assume that:

$$\gamma\sigma^{o2}\mu(1) = 1.$$

Curiously, in his first tentative attempts to derive the leading-order equations for a compressible low Reynolds laminar fluid flow, Lagerstrom (in [91], pp. 190–2), does not take into account our above constraint (3.55b). He assumes that in the NS–F equation (3.4c) for the temperature, the ratio M^2 to Re is unspecified, and obtains a system of equations (where M^2 is present). He also observes that this derived (for Re $\rightarrow$ 0) system is uniformly valid for any value of M, and considers the case where

$$M^2 \text{tend to zero faster than Re, and } M^2 \text{ is of the same order as Re.}$$

There is also the case in which

$$\text{Re tends to zero faster than } M^2,$$

which (Lagerstrom [91], p. 192) "presents some special difficulties and will not be considered here!"

We observe, also, that the steady Stokes equation (3.49a) for an incompressible fluid flow may be obtained either by linearization or by letting Re tend to zero. But it is fortuitous that these two procedures produce the same result in the incompressible case!

For compressible fluids the above, low Reynolds, Stokes equation for T_S (the first of Eq. 3.56a) is a non-linear equation. In Chap. 5 of our (2004) book [47], the reader can find information concerning the unsteady case (the steady case being singular near initial time) and the compressible case (influence of the temperature field).

Here, as a final remark concerning the compressible flow at low Reynolds numbers, it seems me judicious to mention some remarks of Lagerstom ([91], p. 192), related to the *validity* of the NS–F equations:

"It may be objected that the NS–F equations are no longer valid in the limit considered above, especially not in the limit where (M^2/Re) tends to infinity! It must certainly be true that the NS–F equations cease to be valid for extreme values of certain parameters. However, a similar criticism would apply, say, to the classical case of high Reynolds number and zero Mach number! As the Mach number tends to zero, at constant free stream velocity, the velocity of sound, and hence the temperature, tends to infinity! At sufficiently high temperatures any real gas certainly has properties which are not accounted for in the NS–F (and Navier?) equations. The answer to the objection stated above is that letting a parameter tend to zero, say, is a mathematical device for obtaining approximate models for small values of the parameter (more generally one should consider complete expansions of NS–F equations for small values of this parameter!). The method [in fact, our RAM Approach] is physically significant if there are values of the parameter which are sufficiently large for the NS–F equations to be adequate, and at the same time sufficiently small for the mathematical method used to be approximately valid. As yet, it has not been investigated carefully whether or not such values of the parameters exist in the cases considered above."

In order to settle this question it is first necessary to carry out the matching between the various local models (near the time, t = 0, and in the vicinity of the wall, z = 0) with the main approximate model (an example is given in Sect. 9.2, on the framework of low Kibel asymptotics). But the comparison with the experiments, and especially with the results of associated (with derived consistent model) computational simulation (by the numericians), is the more convincing test!

3.3.5 The Case of Non-linear Acoustics

We have already mentioned that in non-linear acoustics, in place of Re (given by (3.5a)), it is necessary to introduce an acoustic Reynolds number Re_{ac}, such that ($Re = L°U°/(\mu°/\rho°)$ and $M = U°/\sqrt{\gamma R T°}$):

$$Re_{ac} = Re/M = \rho°/k°\mu°, \tag{3.57}$$

the characteristic length scale being $L° = 1/k°$, with $k° = \omega_{ac}/\sqrt{\gamma R T°}$ the wave number, ω_{ac} a reference frequency determined by the main frequency $(\omega_{ac}/2\pi)$ of the source-signal, and U° as source velocity.

In fact, the acoustic Reynolds number Re_{ac} given by (3.57) compares the orders of magnitude of the propagation and viscous dissipative terms in non-linear equations (*à la* NS–F) – see Eqs. 3.60a–d below – and $Re_{ac} \gg 1$ (see, for instance, Appendix 1 in Coulouvrat (1992) [92]). On the other hand, in problems of non-linear acoustics it is assumed that $S \gg 1$ and necessarily $M \ll 1$– such that (see [13]):

$$S\,M = 1. \tag{3.58a}$$

The reference acoustic time is then just:

$$t_{ac} = L°/\sqrt{\gamma R T°} \equiv 1/\omega_{ac}. \tag{3.58b}$$

When in place of thermodynamic functions p, ρ, T, we introduce the corresponding perturbations π, ω, θ, such that:

$$p = 1 + M\pi, \quad T = 1 + M\theta, \quad \rho = 1 + M\omega, \tag{3.59}$$

and assume that the dissipative coefficients are constants, then with a constant bulk viscosity μ_v and the help of the relations (3.57) and (3.58a) we can write the following 3D system of unsteady dimensionless (*à la* NS–F-dominant) acoustic equations, for the acoustic velocity vector **u** and thermodynamic perturbations π, θ, and ω:

$$\partial\omega/\partial t + \nabla.\mathbf{u}_{ac} = -M\nabla.(\omega\mathbf{u}_{ac}); \tag{3.60a}$$

$$\begin{aligned}\partial\mathbf{u}_{ac}/\partial t + (1/\gamma)\nabla\pi = {} & (1/Re_{ac})\{\nabla^2\mathbf{u}_{ac} + [(1/3) + (\mu_{v°}/\mu_°)]\nabla(\nabla.\mathbf{u}_{ac})\} \\ & - M[\omega\partial\mathbf{u}_{ac}/\partial t + (\mathbf{u}_{ac}.\nabla)\mathbf{u}_{ac}] + O(M^2);\end{aligned} \tag{3.60b}$$

$$\partial\theta/\partial t + (\gamma-1)\nabla.\mathbf{u}_{ac} = (\gamma/\mathrm{Re}_{ac}\mathrm{Pr})\nabla^2\theta + \mathrm{M}\{\gamma(\gamma-1)/\mathrm{Re}_{ac}[(1/2)[\mathbf{D}(\mathbf{u}_{ac}).\mathbf{D}(\mathbf{u}_{ac})]$$
$$+[(1/3)+(\mu_{v^\circ}/\mu_\circ)](\nabla.\mathbf{u}_{ac})^2]$$
$$-[\omega\partial\theta/\partial t + \mathbf{u}_{ac}.\nabla\theta + (\gamma-1)\pi\nabla.\mathbf{u}_{ac}]\} + \mathrm{O}(\mathrm{M}^2); \quad (3.60c)$$

with

$$\pi - (\omega + \theta) = \mathrm{M}\theta\omega. \quad (3.20d)$$

This above system of basic dimensionless equations (3.60a–d) is the main starting point for the derivation of various consistent and simplified (non-ad hoc) non-linear models of acoustics via the RAM Approach.

Detailed information concerning such acoustic models is included in our (2006) book [13].

In particular, the well-known KZK (Kuznetsov, Zabolotskaya, and Khokhlov) equation (pp. 231–6) was derived from the above system of Eqs. 3.60a–d consistent with a RAM Approach, when it is assumed that the 3D acoustic field is locally plane, such that the non-linear wave propagates in the same way as a linear plane wave over a few wavelengths – the wave profile or amplitude being significantly altered only at large distances away from the source (in the far field).

As a consequence, obviously, the so-called parabolic approximation, which leads to the KZK model equation, may not be valid close to the source (in the near field).

The condition for such an approximation to be valid is that the width of the acoustic source, *d,* should be much larger than the wavelength ($1/k$), so that the ratio

$$\alpha = (1/k)/d \ll 1 \text{ and tends to zero as } \mathrm{M} \downarrow 0. \quad (3.61a)$$

the transverse field variations being slow compared with longitudinal variations along the acoustics axis.

A main hypothesis is also:

$$\mathrm{M}^2 \ll (1/\mathrm{Re}_{ac}) \approx \mathrm{M} \quad (3.61b)$$

and first we introduce two transverse slow coordinates:

$$\eta = \alpha y \text{ and } \zeta = \alpha z, \quad (3.61c)$$

and for $\mathbf{u}_{ac}$ we write:

$$\mathbf{u}_{ac} = \mathrm{u}\,\boldsymbol{i} + \alpha(\mathrm{V}\,\boldsymbol{j} + \mathrm{W}\,\boldsymbol{k}). \quad (3.61d)$$

For low Mach numbers, we consider the following asymptotic expansion:

$$U=(u,\ V,\ W,\ \pi,\ \theta,\ \omega) = U_0 + M\,U_1 + \ldots, \tag{3.61e}$$

with M ↓ 0 fixed and the following two similarity relations between the three small parameters, $1/Re_{ac}$, M and α:

$$1/Re_{ac} = \kappa M,\ \ \alpha^2/M = \beta,\ \ \kappa \text{ and } \beta \text{ are } O(1) \text{ and fixed.} \tag{3.61f}$$

First from the leading-order system for $U_0 = (u_0, V_0, W_0, \pi_0, \theta_0, \omega_0)$ we derive, first, a simple linear acoustic system for u_0 and π_0, θ_0, ω_0, and then for V_0 and W_0 the following two relations, which are valid in far-field, related with the function π_0:

$$\partial V_0/\partial t = -(1/\gamma)\partial \pi_0/\partial \eta \text{ and } \partial W_0/\partial t = -(1/\gamma)\partial \pi_0/\partial \zeta, \tag{3.61g}$$

or

$$\partial/\partial t\,[\partial V_0/\partial \zeta - \partial W_0/\partial \eta] = 0, \tag{3.61h}$$

which is a consistent consequence of our asymptotic approach.

Then from a linear acoustic system we deduce as a solution for u_0:

$$u_0 = F(\tau, \eta, \zeta),\ \text{with}\ \tau = t - x. \tag{3.61i}$$

assuming that the fluid is unbounded and considering a solution for an outgoing wave propagating towards $x > 0$.

However, the acoustic solution (3.61i) is in fact a good solution only for the near-field close to acoustic source!

On the other hand, the KZK equation is a far-field equation, and to avoid various cumulative effects it is necessary to consider the asymptotic expansion (3.61e) as a non-secular two-scale expansion relative to variations along the acoustic axis and to define the slow scale as:

$$\xi = Mx;\ \text{with}\ \partial/\partial x = -\partial/\partial \tau + M\,\partial/\partial \xi \tag{3.61j}$$

and

$$\partial^2/\partial x^2 = \partial^2/\partial \tau^2 + 0(M). \tag{3.61k}$$

In a such case, for u_0 we write

$$u_0 = A(\tau, \xi, \eta, \zeta), \text{ and } \pi_0 = \gamma u_0, \ \omega_0 = u_0, \ \pi_0 = (\gamma - 1)u_0, \quad (3.61l)$$

and as $\partial/\partial t = \partial/\partial\tau$, from (3.61g) we obtain (an unexpected relation?)

$$-\partial/\partial\tau[\partial V_0/\partial\eta + \partial W_0/\partial\zeta] = \nabla_\perp{}^2 A, \quad (3.61m)$$

where

$$\nabla_\perp{}^2 \equiv \partial^2/\partial\eta^2 + \partial^2/\partial\zeta^2.$$

Now, from the linear system of equation for U_1 in (3.61e), taking into account (3.61j–m), we obtain for u_1 a single inhomogeneous acoustic equation in the following rather awkward form:

$$\partial^2 u_1/\partial t^2 - \partial^2 u_1/\partial x^2 = \partial\Theta(A)/\partial\tau + (\beta/\gamma)\nabla_\perp{}^2 A, \quad (3.61n)$$

with

$$\Theta(A) = \Gamma^\circ \partial^2 A/\partial\tau^2 + (\gamma - 1)A\partial A/\partial\tau - 2\partial A/\partial\xi, \quad (3.61o)$$

where

$$\Gamma^\circ = (\kappa/2)[(4/3) + (\mu_{v\circ}/\mu_\circ) + (\gamma - 1)/Pr]. \quad (3.61p)$$

Finally, according to the usual multiple-scale method (see, for instance, Kevorkian and Cole [93]), the non-secularity – validity – condition for the asymptotic expansion (3.61e) leads to the following KZK equation for the amplitude function $A(\tau, \xi, \eta, \zeta)$:

$$\begin{aligned}\partial^2 A/\partial\tau\partial\xi - (\beta/2\gamma)\nabla_\perp{}^2 A - (1/2)(\gamma + 1)\partial/\partial\tau[A\partial A/\partial\tau] \\ = \Gamma^\circ \partial^3 A/\partial\tau^3.\end{aligned} \quad (3.61q)$$

This above KZK equation is a very representative leading-order equation in nonlinear acoutics. The above RAM Approach is a good example of an application of the multiple scales method carried out in a consistent way!

3.3.6 A Sketch of Consistent Models Derived from NS–F Equations Relative to the Considered Reference Parameters

$\mathrm{Re} \uparrow \infty$

⇓

Euler: t and $\vec{\mathbf{x}}$ *fixed*
matching, slip condition
$\downarrow\uparrow$ for Eulerian flow

Prandtl: t, x, y and $\zeta = \dfrac{z}{\mathrm{Re}^{-1/2}}$ fixed
and $w_{Pr} = \dfrac{w_E}{\mathrm{Re}^{-1/2}}$

$\mathrm{Re} \downarrow \mathrm{O}$
(incompressible case)

⇓

Stokes: t, $\vec{\mathbf{x}}$ *fixed → from Navier*
with $p_{st} = \mathrm{Re}\, \pi$
matching → singularity at infinity
$\uparrow\downarrow$

Oseen: $t_{Os} = \mathrm{Re}$ and $\vec{\mathbf{x}}_{Os} = \mathrm{Re}\, \vec{\mathbf{x}}$ fixed
with $p_{os} = \dfrac{p_{st}}{\mathrm{Re}}$

$\varepsilon \downarrow \mathrm{O}$
(expansibility parameter)

⇓

R-B shallow thermal convection:
with: $\mathrm{Gr} = \dfrac{\varepsilon}{\mathrm{Fr}^2_{d^\circ}}$ fixed, $\mathrm{Fr}^2_{d^\circ} \downarrow \mathrm{O}$

$d^\circ >> \left(\nu^2_{d^\circ}/g\right)^{1/3}$

$\mathrm{M} \to \mathrm{O}$

⇓

Navier: t, $\vec{\mathbf{x}}$ *fixed*
for $\vec{\mathbf{u}}_{\mathbf{N}}$ and $\pi = \dfrac{p-1}{\gamma M^2}$
matching, initial condition for $\vec{\mathbf{u}}_{\mathbf{N}}$ at $t = 0$
$\downarrow\uparrow$

Acoustics : $\tau = \dfrac{t}{M}$, $\vec{\mathbf{x}}$ *fixed*

$\mathrm{Re} \downarrow \mathrm{O}$
(compressible case)

⇓

Stokes: t, **x** *fixed → from NS-F*
with $M = \sigma^{o}\, \mathrm{Re}^{1+a}$, $a > 0$
$P = 1 + \mathrm{Re}^{1+2a}\,[P_s + \cdots]$
matching
$\uparrow\downarrow$

Oseen: "compressible"

$\mathrm{Re}_{ac} = \dfrac{\mathrm{Re}}{M} >> 1$

⇓

Non-linear acoustics:
$SM = 1;\ P = 1 + M\pi$
$T = 1 + M\theta,\ \rho = 1 + M\omega$
KZK model eq.

Chapter 4
A Typical RAM Approach: Boussinesq Model Equations

The following derivation-justification of Boussinesq equations via the RAMA for atmospheric flows is a typical test example, in the sense that the various steps of the derivation constitute a logical, well argued, non-contradictory train of thought, which ensures the rational justification of these Boussinesq approximate equations, even if we do not use any rigorous mathematically abstract tools!

An amazing success of our RAMA is the real possibility to derive, also, a second-order consistent, well-balanced, model equation in the framework of the well-known assertion of Boussinesq (1903), which take into account the various non-Boussinesq effects. This, it seems to me, is not really possible if the RAMA below is not applied.

In Monin's book (1969) [94], the reader can find a very pertinent exposition concerning scales of weather processes. The discussion below does not take into account the influence of the Coriolis force, because the sphericity parameter defined by (3.16d) is assumed very small. On the other hand, the parameter ε_s, according to (3.18b), is assumed equal to 1.

The ad hoc approach to justification of simplified Boussinesq equations was developed throughout 1960–1980. Here we mention the papers by: Spiegel and Veronis, Mihaljan, Malkus, Dutton and Fichtl, Perez Gordon and Velarde, and Mahrt, and for precise references see our books [12, 13, 19], and also the survey [25].

4.1 Introductory Remarks Concerning Atmospheric Flow

Below, only the case of non-dissipative and adiabatic atmospheric flows, for simplicity is considered, when the starting "exact" equations are the Euler equations. See Chaps. 1 and 2 of our (1991) book [13] for a detailed derivation directly from NS–F equations. Namely, with the system of Euler equations (1.2a–c) with (1.3) it is necessary, above all, to assume the existence, also, of a hydrostatic reference state.

DOI 10.1007/978-3-642-20746-4_4, © Springer-Verlag Berlin Heidelberg 2012

If $p^*(z^*)$, $\rho^*(z^*)$, $T^*(z^*)$, the three thermodynamic functions are dependent only on z^*, the altitude of the motionless reference state, then we can work with the following three basic equations (for a thermally perfect gas):

$$-\,dT^*/dz^* = \Gamma(z^*); \tag{4.1a}$$

$$dp^*/dz^* + g\,\rho^* = 0; \tag{4.1b}$$

$$p^* = R\,\rho^* T^*, \tag{4.1c}$$

where $\Gamma(z^*)$ is a given function of z^*, in the considered adiabatic case. In the troposphere, a good approximation is obtained when

$$\Gamma(z^*) = \Gamma^\circ = \text{const.}$$

In order to make the governing exact Euler 3D unsteady, compressible, non-viscous and adiabatic equations non-dimensional – which is our *first key step* – we introduce the following characteristic constant quantities:

$$H^\circ,\ U^\circ,\ p^*(0),\ \rho^*(0),\ T^*(0) \equiv p^*(0)/R\,\rho^*(0),$$

where the length scale H° has been chosen to be representative of the vertical motion of the lee-wave regime downstream of the mountain, since it is the effect of gravity with which we are primarily concerned.

The characteristic velocity U° is an average constant value of the velocity profile which is assumed smaller than the speed of sound, defined by

$$c_0 = [g\ p^*(0)/\rho^*(0)]^{1/2}, \tag{4.2a}$$

such that

$$U^o << c_0.$$

Now, if we define the associated dimensionless quantities $\mathbf{u}'$, ρ', p', T', $\mathbf{x}'$, and t', respectively, linked with the above characteristic constant quantities U°, $\rho^*(0)$, $p^*(0)$, $T^*(0)$, H°, and H°/U°, we rewrite the set of Euler equations, for atmosphere in movement, relative to $\mathbf{u}'$, ρ', p', and T', as functions of t' and $\mathbf{x}'$, in the following (exact) dimensionless form (slightly different from (3.24a–d)):

$$D\mathbf{u}'/Dt' + (1/\gamma M^2)(1/\rho')\nabla' p' + (1/Fr_H)^2\mathbf{k} = 0; \tag{4.3a}$$

$$D\rho'/Dt' + \rho'(\nabla'.\mathbf{u}') = 0; \tag{4.3b}$$

$$DT'/Dt' + [(\gamma - 1)/\gamma](1/\rho')(Dp'/Dt') = 0; \tag{4.3c}$$

$$p' = \rho' T'. \tag{4.3d}$$

In (4.3a–d), ∇' is now the gradient operator with respect to $\mathbf{x}'$, and the material derivative (along the trajectories) is:

$$D/Dt' = \partial/\partial t' + \mathbf{u}'.\nabla'.$$

As main parameters, in Eqs. 4.3a–d, we have for the Froude number the following remarkable and very instructive relation:

$$(1/Fr_H)^2 \equiv Bo/\gamma M^2, \tag{4.4}$$

where

$$M^2 = (U^\circ/c_0)^2 \tag{4.5}$$

is the square of the Mach number and

$$Bo = H^\circ/H^* \tag{4.6}$$

our Boussinesq number, where

$$H^* = p^*(0)/g\rho^*(0) \equiv H_S. \tag{4.7}$$

The length scale H* is (see (3.18a)) the altitude of the homogeneous hydrostatic reference state – a judicious characteristic vertical length scale for the altitude z*, such that in dimensionless form we write the relation (Bo is defined by (3.5d)):

$$z^{*\prime} = Boz', \tag{4.8}$$

which plays an important role in the derivation of Boussinesq approximate equations, and gives a deeper and more subtle meaning to the Boussinesq fluid flow case!

Concerning the characteristic horizontal length scale L°, we assume (see (2.18b) and (2.16d)) that:

$$L^\circ \approx H_S \Rightarrow \ \varepsilon_S \approx 1 \text{ and as a consequence } \delta << 1, \tag{4.9}$$

which is a condition for a negligible influence of the Coriolis force, and also for a derivation of Boussinesq model equations, without the hydrostatic approximation, for the lee-waves regime flows downstream from the mountain.

As a consequence, the dimensionless reference hydrostatic state ($p^{*\prime}$, $\rho^{*\prime}$, $T^{*\prime}$) is, in fact, only a function of Boz', and

$$\partial/\partial z^{*\prime} = (1/Bo)\partial/\partial z'. \tag{4.10}$$

Naturally, when for H we choose H*, then Bo $\equiv$ 1 and $z^{*\prime} \equiv z'$.

The above introduction and discussion of various dimensionless quantities and reduced parameters form, in fact, *our first key step* towards the justification of Boussinesq model equations in the framework of a Eulerian fluid flow.

4.2 Asymptotics of the Boussinesq Case

In the case of the Boussinesq approximation, if we want to obtain rationally the corresponding Boussinesq system of simplified equations, a detailed (see below) dimensional analysis shows that when M $< <$ 1, then necessarily that must be also Bo $< <$ 1.

Below, our asymptotic rational approach confirms this observation. The emergence of Bo (thanks to relation (4.8)) was our *second* (significant) *key step*, during our 1969–1974 investigations relative to justification of this Boussinesq approximation, in the framework of our RAMA theory for a consistent derivation of the leading-order model equations, *à la* Boussinesq.

Indeed, with two small parameters, M and Bo, it is necessary to consider, in general, three limiting cases:

$$\text{(I) Bo fixed, } \mathrm{M} \downarrow 0 \text{ and then Bo} \downarrow 0 \tag{4.11a}$$

$$\text{(II) M fixed, Bo} \downarrow 0 \text{ and then } \mathrm{M} \downarrow 0 \tag{4.11b}$$

$$\text{(III) } \mathrm{M} \downarrow 0 \text{ and Bo} \downarrow 0, \text{ such that Bo } = \mathrm{M}^{\mathrm{a}} B^{*} \tag{4.11c}$$

where $B^* = \mathrm{O}(1)$ and $\mathrm{a} > 0$ is a scalar.

The Boussinesq equations are derived (see below) with the particular significant choice $\mathrm{a} = 1$, under the Boussinesq limiting process, corresponding to case (III)–(4.11c), with B^*, γ, t', and $\mathbf{x}'$ fixed during the Boussinesq limiting process (see (4.18) below).

We observe that the relation $\mathrm{Bo} = \mathrm{M}^{\mathrm{a}} B^*$ in (4.11c), when $\mathrm{a} \equiv 1$, is the Boussinesq similarity rule, and B^* is the associated similarity parameter.

From (4.11c) we deduce an important (but unfortunate) feature derived below (see (4.17a–d)) for Boussinesq model equations. Namely, when $\mathrm{a} = 1$, then:

$$B^* \approx 1 \Rightarrow \mathrm{H} \approx (\mathrm{U}^{\mathrm{o}}/\mathrm{g})[\mathrm{RT}^*(0)/\gamma]^{1/2} \equiv \mathrm{H_B} << \mathrm{H}^*, \tag{4.12a}$$

and more precisely, for the usual meteorological values of U^{o} and $\mathrm{T}^*(0)$, we obtain:

$$\mathrm{H_B} \approx 10^3\mathrm{m}, \text{ while } \mathrm{H}^* \approx 10^4\mathrm{m}! \tag{4.12b}$$

This is a very strong restriction for the various applications of the Boussinesq equations in atmospheric motions.

For the prediction of lee waves around and downstream of a mountain in the whole thickness of the troposphere (in troposphere), another set of model equations are necessary – the so-called "deep (Zeytounian) convection equations" and see Chap. 10 in [12] and also the Sect. 6.2.3 of [13]. Here, in Sect. 9.1.4 we also consider four distinguished limiting cases for the 2D steady lee-waves problem.

4.2.1 *Euler Dimensionless Equations for the Thermodynamic Perturbations*

Now a *third key step* is necessary, which is linked with the existence of the dimensionless hydrostatic reference basic state $p^{*\prime}$, $\rho^{*\prime}$, $T^{*\prime}$, function of Boz′, and for this it is necessary to introduce the thermodynamic perturbations, π, ω, θ, such that:

$$p' = p^{*\prime}(\mathrm{Bo}z')[1+\pi], \tag{4.13a}$$

$$\rho' = \rho^{*\prime}(\mathrm{Bo}z')[1+\omega], \tag{4.13b}$$

$$T' = T^{*\prime}(\mathrm{Bo}z')[1+\theta]. \tag{4.13c}$$

With (4.13a–c) the dimensionless Euler equations are rewritten in the following new form, for the velocity vector $\mathbf{u}'$ and thermodynamic perturbations π, ω, and θ:

$$\begin{aligned} &\mathrm{D}\,\omega/\mathrm{D}t' + (1+\omega)(\nabla'.\mathbf{u}') \\ &= (1+\omega)[\mathrm{Bo}/T^*(z^{*\prime})]\{1+\mathrm{d}T^*(z^{*\prime})/\mathrm{d}z^{*\prime}\}(\mathbf{u}'.\mathbf{k}) = 0; \end{aligned} \tag{4.14a}$$

$$\begin{aligned} &(1+\omega)\mathrm{D}\mathbf{u}'/Dt' + [T^*(z^{*\prime})/\gamma M^2]\nabla'\pi \\ &- (1+\omega)[\mathrm{Bo}/\gamma M^2]\theta\mathbf{k} = 0; \end{aligned} \tag{4.14b}$$

$$\mathrm{D}\,\theta/\mathrm{D}t' - [(\gamma-1)/\gamma](\mathrm{D}\pi/\mathrm{D}t') + (1+\pi)\mathrm{Bo}N^2(z^{*\prime})(\mathbf{u}'.\mathbf{k}) = 0, \tag{4.14c}$$

$$\pi = \omega + (1+\omega)\theta, \tag{4.14d}$$

where

$$N^2(z^{*\prime}) = [1/T^*(z^{*\prime})]\{[(\gamma-1)/\gamma] + \mathrm{d}T^*(z^{*\prime})/\mathrm{d}\,z^{*\prime}\}, \tag{4.15}$$

when (4.10) has been used. The relation (4.15) for $N^2(z^{*\prime})$ takes into account the stratification of the hydrostatic reference-motionless state atmosphere (in the

Eulerian case (4.1a–c)), and is related to the dimensionless Väisälä internal frequency.

The consideration of the above full system of dimensionless exact Euler equations, (4.14a–d), for $\mathbf{u}'$ and π, ω, and θ, was indeed our *main third key step* in the quest for the justification of the Boussinesq model equations.

Obviously, for the atmospheric motions, π, ω, and θ are small, by comparison with unity perturbations. As a consequence, we write – in the framework of the *fourth key step* – the following asymptotic expansions:

$$\pi = \mathrm{M}^{\mathrm{b}}\pi_{\mathrm{B}} + \mathrm{M}^{\mathrm{c}}\pi_{*} + \ldots, \tag{4.16a}$$

$$(\omega,\ \theta) = \mathrm{M}^{\mathrm{d}}(\omega_{\mathrm{B}},\ \theta_{\mathrm{B}}) + \mathrm{M}^{\mathrm{d}+1}(\omega_{*},\ \theta_{*}) + \ldots, \tag{4.16b}$$

$$\mathbf{u}' = \mathbf{u}_{\mathrm{B}} + \mathrm{M}^{\mathrm{e}}\mathbf{u}_{*} + \ldots, \tag{4.16c}$$

associated with the limiting process (4.11c). In (4.16a–c), the exponents, b, c, d, and e are real scalars which are determined consistently in Sect. 4.2.2.

4.2.2 Rational Derivation of Boussinesq Equations

The Boussinesq equations, as leading-order approximate equations, are derived from the 'unbalanced' system of Eqs. 4.17a–d, which are obtained when the above asymptotic expansions (4.16a–c) are used in exact Euler dimensionless equations (4.14a–d).

First, we can write (with (4.8)), from Eq. 4.14c, the following 'unbalanced' equation for θ_{B}, when we take into account only the terms with the subscript "B" in expansions (4.16a–c):

$$\begin{aligned} &\mathrm{D}\theta_{\mathrm{B}}/\mathrm{Dt}' + \mathrm{M}^{\mathrm{d}}\omega_{\mathrm{B}}\mathrm{D}\theta_{\mathrm{B}}/\mathrm{Dt}' - \mathrm{M}^{\mathrm{b-d}}[(\gamma-1)/\gamma]\mathrm{D}\pi_{\mathrm{B}}/\mathrm{Dt}' \\ &\quad + (\mathrm{Bo}/\mathrm{M}^{\mathrm{d}})N^2(\mathrm{Bo}z')(\mathbf{u}_{\mathrm{B}}.\mathbf{k}) \\ &\quad + \mathrm{Bo}\ \mathrm{M}^{\mathrm{b-d}}\pi_{\mathrm{B}}N^2(\mathrm{Bo}z')(\mathbf{u}_{\mathrm{B}}.\mathbf{k}) + \ldots = 0, \end{aligned} \tag{4.17a}$$

and it is necessary to take into account also the similarity rule for Bo ($= \mathrm{M}^{\mathrm{a}}B*$) written in (4.11c), when the Boussinesq limiting process is performed:

$$\begin{aligned} &\mathrm{M} \downarrow 0 \text{ with t and } \mathbf{x} \text{ fixed}; B^* = \mathrm{O}(1) \\ &\qquad \text{and } N^2(0) = \mathrm{O}(1) \text{ fixed.} \end{aligned} \tag{4.18}$$

Because $\gamma > 1$, it is necessary, first, with (4.17a), that

$$\mathrm{a} = \mathrm{d}, \text{ and it seems that with}: \ \mathrm{d} = 1,\ \mathrm{a} = 1 \text{ and } \mathrm{b} = 2?, \tag{4.19a}$$

we derive not only a leading-order equation for θ_B:

$$D_B\theta_B/Dt' + B^*N^2(0)(\mathbf{u}_B.\mathbf{k}) = 0. \tag{4.20a}$$

but also we open the way for the derivation of a second-order non-homogeneous equation for θ_*.

For a confirmation of the above (see (4.19a)) rather premature conclusion, it is now necessary to consider the derivation of a leading-order equation for $\mathbf{u}_B$, associated with (4.17a). From (4.14b) with (4.16a–c), we obtain as a second 'unbalanced' equation for $\mathbf{u}_B$:

$$\begin{aligned} &D\mathbf{u}_B/Dt' + M^{b-2}[T^{*\prime}(Boz\prime)/\gamma](\nabla'\pi_B) - (1/\gamma)\ Bo\ M^{d-2}\theta_B\mathbf{k} \\ &\quad = -M^d\omega_B(D\mathbf{u}_B/Dt') - (1/\gamma)BoM^{2(d-1)}\omega_B\theta_B\mathbf{k}. \end{aligned} \tag{4.17b}$$

First the above (4.19a) choice of:

$$b = 2 \text{ is confirmed and then also } d = 1,\ a = 1! \tag{4.19b}$$

Thus, we obtain, as a companion simplified (*à la* Boussinesq) leading-order equation for $\mathbf{u}_B$, associated to equation for θ_B:

$$D_B\mathbf{u}_B/Dt' + [T^{*\prime}(0)/\gamma](\nabla'\pi_B) - (B^*/\gamma)\ \theta_B\mathbf{k} = 0 \tag{4.20b}$$

The two terms in the right-hand side of (4.17b) are linked with the associated, second-order equation for $\mathbf{u}_*$.

Now, from the 'unbalanced' equation of continuity, obtained from (4.14a), with (4.16a–c),

$$\begin{aligned} (\nabla'.\mathbf{u}_B) &= -M^d[D\omega_B/Dt' + \omega_B(\nabla'.\mathbf{u}_B)] \\ &\quad + Bo\ [1/T^{*\prime}(Boz\prime)][1 - \Gamma(Boz\prime)](\mathbf{u}_B.\mathbf{k}) + \ldots, \end{aligned} \tag{4.17c}$$

and obviously our choice of (4.19a) simply produces, at the leading-order, the following divergenceless constraint for the velocity vector $\mathbf{u}_B$:

$$\nabla'.\mathbf{u}_B = 0. \tag{4.20c}$$

Again, in the right-hand side of (4.17c), the two terms are linked with the associated, second-order equation of continuity.

Finally, from (4.14d) we obtain as 'unbalanced' equation:

$$\omega_B + \theta_B = M^e\omega_B\theta_B + M^{b-d}\pi_B, \tag{4.17d}$$

and as Boussinesq equation of state we derive:

$$\omega_B = -\ \theta_B. \tag{4.20d}$$

which determines the perturbation of the density.

In Boussinesq equations (4.20a) and (4.20b) we have:

$$D_B/Dt' = \partial/\partial t' + \mathbf{u}_B.\nabla'$$

and we observe that, obviously, $T^{*\prime}(0) = 1$, but in general $N^2(0)$ is different from zero. The choice $e = 1$, in (4.16c), indeed, follows when we want to derive the second-order *à la* Boussinesq limit model equations for $\mathbf{u}_*$, ω_*, π_* and θ_*, and we have also $d = 1$.

I invite the reader to derive these second-order Boussinesq equations as a profitable and stimulating exercise! In particular, in the 2D steady case, starting from the exact 2D, (x, z) steady $(\partial/\partial t = 0)$ case, it is possible to derive, without any approximations, an analogue to Eq. 4.26b – but significantly complicated – a single equation (see Sect. 9.1).

So, with the following asymptotic expansions:

$$\pi = M^2\pi_B + ..., (\omega, \ \theta) = M(\omega_B, \ \theta_B) + ..., \mathbf{u}' = \mathbf{u}_B + ...$$

when the following "Boussinesq limiting process" is performed in an Euler system of equations (written for the atmospheric flows):

$$M \downarrow 0 \text{ with t and } \mathbf{x} \text{ fixed;}$$

$$B^* = O(1) \text{ fixed}$$

and

$$N^2(0) = O(1) \text{ fixed,}$$

then, at leading-order, we derive the following Boussinesq equations:

$$D_B\theta_B/Dt' + B^*N^2(0)(\mathbf{u}_B.\mathbf{k}) = 0,$$

$$D_B\mathbf{u}_B/Dt' + [T^{*\prime}(0)/\gamma](\nabla'\pi_B) - (B^*/\gamma)\ \theta_B\mathbf{k} = 0.$$

$$\nabla'.\mathbf{u}_B = 0,$$

$$\omega_B = -\theta_B.$$

with

$$N^2(0) = [(\gamma - 1)/\gamma] + [dT^*(z^{*\prime})/dz^{*\prime}]_{z^{*\prime}=0}$$

and

$$B^* = \mathrm{Bo}/\mathrm{M}$$

This rather long, but instructive, derivation of the Boussinesq equations (4.20a–c) is the only RAMA way, in the framework of low Mach (hyposonic, [13]) atmospheric flows, for a consistent rational derivation of not only the Boussinesq system of equations, but also the second-order consistent equations associated with these Boussinesq equations.

4.2.3 Two Particular Cases

It is instructive also to note two particular, less significant limit, cases of the above Boussinesq system of equations. On the one hand, if

$$B^* \uparrow \propto \Leftrightarrow \mathrm{M} << \mathrm{Bo};\ \ \mathrm{Bo\ fixed},$$

or

$$\mathrm{M} \downarrow 0,\ \text{ and then } \mathrm{Bo} \downarrow 0 - \ \text{ case (I)}, \tag{4.21a}$$

then we obtain the "quasi-non-divergent" limit equations for the velocity components, $\mathbf{v}_d = (u_d, v_d)$, w_d, and π_d/γ:

$$\partial \pi_d/\partial z' = 0, w_d = 0,\ \ \theta_d = 0; \tag{4.22a}$$

$$\mathbf{D}.\mathbf{v}_d = 0; \partial \mathbf{v}_d/\partial t' + (\mathbf{v}_d.\mathbf{D})\mathbf{v}_d + \mathbf{D}(\pi_d/\gamma) = 0. \tag{4.22b}$$

where $\mathbf{D} = (\partial/\partial x'; (\partial/\partial y')$.

If on the other hand

$$B^* \downarrow 0 \Leftrightarrow \mathrm{M} >> \mathrm{Bo};\ \ \mathrm{M\ fixed},$$

or

$$\mathrm{Bo} \downarrow 0,\ \text{ and then } \mathrm{M} \downarrow 0 - \text{case (II)}, \tag{4.21b}$$

then we obtain the classical "incompressible" Navier limit equations for u_i, v_i, $w_{i,}$ and π_i/γ:

$$\partial u_i/\partial x' + \partial v_i/\partial y' + \partial w_i/\partial z' = 0. \tag{4.23a}$$

$$D_i u_i/Dt\prime + \nabla'(\pi_i/\gamma) = 0. \tag{4.23b}$$

with

$$D_i \theta_i/Dt' = 0, \tag{4.23c}$$

$$\omega_i = -\theta_i. \tag{4.23d}$$

where

$$D_i/Dt' = \partial/\partial t' + \mathbf{u}_i.\nabla'.$$

In both cases we derive from the above dimensionless dominant full Euler equations a less significant limit system of equations than the inviscid, adiabatic Boussinesq equations.

4.3 The Steady 2D Case

In the steady 2D case, when the Boussinesq fluid flow is considered in the plan (x', z') with, as velocity components, u_B and w_B, we have the possibility of introducing a 2D stream function $\psi_B(,x', z')$ such that:

$$u_B = \partial\psi_B/\partial z' \text{ and } w_B = -\partial\psi_B/\partial x', \tag{4.24}$$

and, because the coefficient $B^*N^2(0) =$ constant, in Eq. 4.20a, we derive a "first integral":

$$\theta_B + B^*N^2(0)z' = F_B(\psi_B). \tag{4.25a}$$

On the other hand, assuming $T^{*\prime}_{(0)} \equiv 1$ if in the two equations obtained from (4.20b):

$$u_B\partial u_B/\partial x' + w_B\partial u_B/\partial x' = -\partial(\pi_B/\gamma)/\partial x',$$

$$u_B\partial w_B/\partial x' + w_B\partial w_B/\partial x' = -\partial(\pi_B/\gamma)/\partial z' + (B^*/\gamma)\,\theta_B = 0,$$

we cancel out the terms with π_B/γ, then we derive the following equation for ψ_B:

$$(\partial\psi_B/\partial x')\partial\Delta_2\psi_B/\partial z' - (\partial\psi_B/\partial z')\partial\Delta_2\psi_B/\partial x' = -(B^*/\gamma)\partial\theta_B/\partial x',$$

where $\Delta_2 = \partial^2/\partial x'^2 + \partial^2/\partial z'^2$.

But

$$\partial\theta_B/\partial x' = (dF_B(\psi_B)/d\psi_B)\partial\psi_B/\partial x',$$

and as a consequence we can write:

$$(\partial\psi_B/\partial z')\partial/\partial x'[\Delta_2\psi_B - (B^*/\gamma)z'(dF_B(\psi_B)/d\psi_B)]$$

$$-(\partial\psi_B/\partial x')\partial/\partial z'[\Delta_2\psi_B - (B^*/\gamma)\,z'(dF_B(\psi_B)/d\psi_B)] = 0.$$

From above, we obtain a second first integral:

$$\Delta_2\psi_B = (B^*/\gamma)z'(\mathrm{d}F_B(\psi_B)/\mathrm{d}\psi_B) + H_B(\psi_B). \tag{4.25b}$$

The functions $F_B(\psi_B)$ and $H_B(\psi_B)$ are two arbitrary functions of ψ_B only.

In the particular case of an airflow over a mountain, if we consider a 2D steady, uniform, constant flow in the direction of $x' > 0$, with π_B and θ_B both $\equiv 0$ at infinity upstream ($x \uparrow -\infty$) of the mountain, then, in place of (4.25b), we obtain for the function

$$\delta_B(x', z') \equiv z' - \psi_B(x', z') \tag{4.26a}$$

the following linear Helmholtz equation:

$$\partial^2\delta_B/\partial x^2 + \partial^2\delta_B/\partial z'^2 + (B^{*2}/\gamma)N^2(0)\,\delta_B = 0. \tag{4.26b}$$

The dominant feature of Eq. 4.26b, for δ_B, from a mathematical point of view, is that its linearity is not related to any particular hypothesis about the small perturbations in the steady 2D Boussinesq equations.

However, from the exact slip-condition on the surface of the mountain, simulated by the dimensionless equation,

$$z' = \mu h'(\lambda x').$$

where μ is an amplitude parameter and $\lambda>0$, we must write the following boundary, slip, non-linear, condition for δ_B on the surface of the mountain:

$$\delta_B(x', z' = \mu h'(\lambda x')) = \mu h'(\lambda x'). \tag{4.26c}$$

At infinity upstream, we have also a "behaviour condition":

$$\delta_B(x' \uparrow -\infty, z') = 0, \tag{4.26d}$$

But, because of the emergence of steady lee-waves at infinity downstream of the mountain, there is only the possibility of assuming one physically realistic boundary condition, namely:

$$|\,\delta_B(x' = +\infty, z')| < \infty. \tag{4.26e}$$

If we add, in particular, to the linear Helmholtz equation (4.26b), with slip condition (4.26c), the following upper condition:

$$\delta_B(x', z' = 1) = 0 \tag{4.26f}$$

then we obtain the well-known Long (1953) problem [7].

However, in reality, the above, (4.26f), condition does not emerge consistently from the exact dimensionless formulation of the lee-waves problem, considered in the framework of the 2D steady Euler equations for the whole troposphere, and where the upper boundary condition is assumed to be slip condition on the tropopause (simulated simply, before the non-dimensionalization, by the equation $z = Hs \approx 10^4$ m).

In reality, in dimensionless form, for the above 2D steady Boussinesq equation (4.26b), the consistent upper condition is:

$$\delta_B(x', z' \uparrow +\infty) = +\infty\,! \tag{4.27a}$$

The Boussinesq equation being a consistent equation only in a layer of the order of 1 km in altitude, as a consequence of (4.27a) it is necessary to solve the lee-waves problem in an unbounded atmosphere with an *à la* Sommerfeld radiation condition, which expresses the condition that

no waves are radiated inwards. (4.27b)

This inner Boussinesq problem, in an unbounded atmosphere, has been considered by Miles (1969) [95], and also by Kozhevnikov (1963) [96].

In a paper by Guiraud and Zeytounian (1979) [97], the associated outer Boussinesq problem is considered, with an upper slip condition on the rigid plane:

$$\zeta = 1/B^* \tag{4.28}$$

with an outer vertical dimensionless coordinate $\zeta = Mz'$.

In Guiraud and Zeytounian's paper (1979) [97] it is shown that the upper and lower boundaries of the troposphere alternately reflect internal short-wavelength gravity waves excited by the lee waves of the inner (Boussinesq) approximate problem, with a wavelength of the order of the Mach number (which is assumed vanishing), on the scale of the outer region.

As a consequence, there is a double scale built into the solution, and we must take care of it. In fact, the important point of our (1979) with Guiraud analysis is that "these short-wavelength gravity waves propagate downstream and that no feedback occurs on the inner Boussinesq flow close to the mountain – to the lowest order at least."

Finally, we would understand the imposed upper boundary at the top of the troposphere as an artificial one, having asymptotically no effect on the inner Boussinesq (lee waves) flow, which is the only really interesting flow.

4.4 The Problem of Initial Conditions

In Boussinesq equations, as a consequence of the Boussinesq limiting process (4.18), in Euler full unsteady Eqs. 4.14a–d with three partial derivatives in time t, we have only two partial derivatives in time t, for $\mathbf{u}_B$ and θ_B! As a consequence, for a Boussinesq system of three equations, (4.20a–c), for $\mathbf{u}_B$, π_B, and θ_B we have only the possibility of assuming the availability of two initial data, namely:

$$t' = 0 : \mathbf{u}_B = \mathbf{u_B}^0(\mathbf{x}) \text{ and } \theta_B = \theta_B{}^0(\mathbf{x}), \tag{4.29a}$$

with

$$\nabla'.\mathbf{u_B}^0 = 0. \tag{4.29b}$$

But, on the other hand, for the Euler exact starting Eqs. 4.14a–d, three initial data are necessary for $\mathbf{u}'$, θ, and ω, at $t' = 0$.

According to Sect. 2.4.2, we know that it is necessary to formulate an associated unsteady adjustment problem to Boussinesq approximate equations by the introduction, in Eqs. 4.14a–c, of a local time,

$$\tau = t'/M, \text{ significant near } t' = 0 \tag{4.30a}$$

and a local (acoustic) limiting process (in place of (4.18)) with local asymptotic expansions. Such a local formulation can be found in Sect. 20 of Chap. 5 in my (1991) book [19].

The initial data (in (4.29a)) at $t' = 0$, for the Boussinesq equations, are derived from a matching between two asymptotic representations:

$$\begin{aligned}&\text{the main one (outer – Boussinesq, relative to time } t'\text{fixed}),\\&\text{at } t' = 0, \text{ and a local one (inner – acoustic, near } t' = 0), \text{ with}\\&\text{local short time } \tau \text{ fixed) when } \tau \uparrow +\infty:\end{aligned} \tag{4.30b}$$

with the matching:

$$\text{Lim}^{\text{Bo}}[t' = 0] \Leftrightarrow \text{Lim}^{\text{Ac}}[\tau \uparrow +\infty]. \tag{4.30c}$$

In reality (!) the result of our (1991) approach is valid only when we assume the following initial conditions for the exact starting Euler system of Eqs. 4.14a–d:

$$t' = 0 : \mathbf{u} = \mathbf{u}^\circ \text{ and } (\pi, \ \omega, \ \theta) = M(\pi^\circ, \ \omega^\circ, \ \theta^\circ), \tag{4.31a}$$

with

$$\mathbf{u}^\circ = \nabla\phi^\circ + \nabla \wedge \psi^\circ. \tag{4.31b}$$

In Sect. 20 of my [19] the reader can find a details derivation of the above initial conditions (4.29a) via matching, at leading-order (4.30b), from the solution of the following initial value problem of the classical acoustics:

$$\partial^2\phi/\partial\tau^2 - \Delta\phi = 0. \tag{4.32a}$$

$$\tau = 0 : \phi = \phi^\circ, \partial\phi/\partial\tau = -\pi^\circ/\gamma. \tag{4.32b}$$

The solution of the above problem (4.32a, b) is straightforward and, from it we obtain at infinity in time:

$$\tau \to \infty : |\nabla \phi| \to 0 \text{ and } |\partial \phi / \partial \tau| \to 0. \tag{4.33}$$

In such a case, for the derived (in Sect. 4.2) Boussinesq approximate equations (4.20a, b) we obtain, for the unknown initial conditions at $t' = 0$, $\theta_B{}^o(\mathbf{x})$ and $\mathbf{u}_B{}^0(\mathbf{x})$, the following relations:

$$\mathbf{u}_B{}^o(\mathbf{x}) = \nabla \wedge \psi^\circ \text{ and } \theta_B{}^\circ(\mathbf{x}) = (\pi^\circ / \gamma) - \omega^\circ. \tag{4.34}$$

4.5 A Sketch of a RAM Approach for a Boussinesq Model

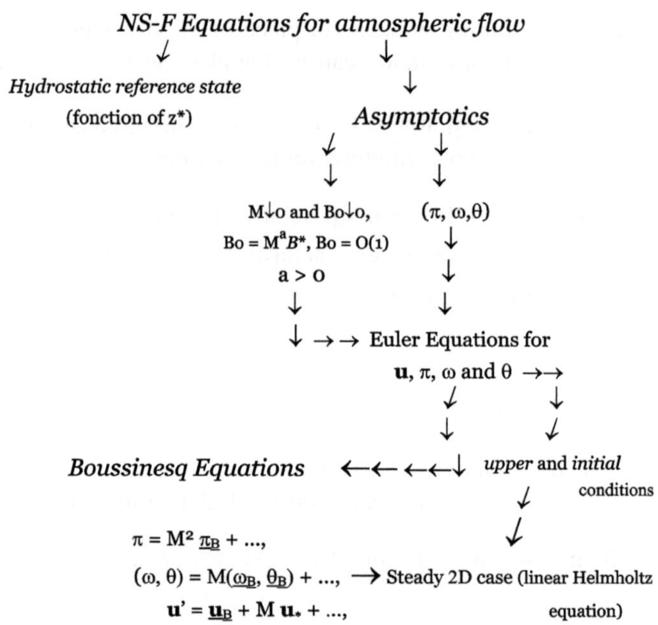

Fig. 4.1 Justification of Boussinesq equations and associated conditions

What happens if the initial values of π, ω, and θ – the solutions of exact starting Euler equations (4.14a–d) – are different from the assumed data M π^o, M ω°, and M θ°, as is the case in (4.31a), are not known? Actually, this is an open problem which deserves further investigation.

In Chap. 9, devoted to applications of our RAMA to some atmospheric flow problems, the reader can find a discussion (in Sect. 9.1) concerning a few companion equations associated with the Boussinesq/Helmholtz equation (4.26b).

Part II
A Sketch of a Mathematical Theory of the RAM Approach

Chapter 5
The Structure of Unsteady NS–F Equations at Large Reynolds Numbers

Numerous papers are devoted to investigations of NS–F equations in the case of large Reynolds numbers, and many references are cited in my *Theory and Applications of Viscous Fluid Flows* [47]. Curiously, the case of unsteady NS–F equations for large Reynolds numbers, when Re tends to infinity, has been very poorly considered. However, in Chap. 4 of [47] (Sect. 4.3) there is a presentation of the asymptotic structure of unsteady state NS–F equations.

Below, in this Chap. 5, I again give a presentation of this structure – the reason being that such unsteady NS–F equations structure, at Re $\uparrow\propto$, very well illustrates the importance of the various limiting processes related to fixed time and space, and forms a guideline in the elaboration of our mathematics for the RAMA.

On the other hand, such a structure illuminates the various consequences of the singular nature of the Prandtl (1904) concept of the boundary layer. A strong singularity for NS–F equations opens (it seems to me) new perspectives for the resolution of various paradoxes encountered in unsteady compressible fluid flow theory. It is surprising that this close initial time singularity of the Prandtl boundary layer, was for so long ignored.

5.1 Introduction

Fluid dynamicians and applied mathematicians have always found fluid dynamics to be a rich and interesting field for investigations, because the basic system of partial derivative equations for a Newtonian fluid – the so-called Navier–Stokes–Fourier equations (NS–F) equations – have a great capacity for producing various particular fluid flow models.

In particular, a large class of such fluid flow models is closely linked with the analysis of a dimensionless form of the NS–F system, and more specifically with the large Reynolds numbers (Re $\uparrow\propto$) for a compressible, weakly viscous, and heat-conducting fluid flow.

R. Kh. Zeytounian, *Navier–Stokes–Fourier Equations*,

DOI 10.1007/978-3-642-20746-4_5,

Indeed, in technologically and geophysically relevant fluid flows, Re is usually quite large. In 1904 Prandtl took into account this fact and derived (in an ad hoc manner) his well-known 'Prandtl Boundary-Layer (BL) equations'. Surprisingly, it seems that the case of an unsteady fluid flow, when initial conditions are prescribed, has not been carefully considered, and this BL Prandtl concept has become default and become singular near $t = 0$! Only in 1980, in a short note [98], was I the (apparently) first to show that:

> The limiting form (at $Re \uparrow \infty$) of the unsteady NS–F equations, near $t = 0$ and in the vicinity of a wall ($z = 0$), bounding below the compressible, viscous and heat-conducting fluid flow, is identified – in place of Prandtl BL equations – rather with the equations of the Rayleigh compressible problem, considered first in 1951 [99] by Howarth.

In Chap. 5 I fully intend to present a more deep and careful investigation of this singular problem, which not only has an obvious theoretical interest, but also seemly a practical one. For the NS–F unsteady full equations, we have considered, as a typical working case, an "emergency situation": namely, a sudden rise in temperature locally on the wall at initial time $t = 0$ – which is a possible application of our new four regions structure for the NS–F unsteady equations!

5.2 The Emergence of the Four Regions as a Consequence of the Singular Nature of BL Equations Near $t = 0$

The main feature of the well-known Prandtl 1904 BL concept in a thin region near the wall ($z = 0$), in an aerodynamics problem for high Reynolds number, is linked with the strong simplification of the equation of motion for the vertical component w of the velocity $\mathbf{u} = (\mathbf{v}; w)$.

In steady and unsteady, in compressible viscous, in heat-conducting, and in incompressible viscous fluid flows, this Prandtl BL concept in all cases produces a very degenerate limit equation, when $Re \uparrow \infty$, for w!

When we work with dimensionless quantities, then, for the variation of the pressure in the direction normal to the (horizontal) wall, $z = 0$, relative to vertical coordinate, z, we obtain (when the gravity force is not taken into account):

$$\partial p/\partial z = 0. \tag{5.1}$$

and in particular, the partial derivative in time for the component w of the velocity disappears in a BL system of equations!

If this failure seems not to have serious consequences in the usual case of steady or unsteady incompressible viscous fluid flows, conversely, this is not the case for an unsteady compressible viscous and heat-conducting fluid flow. In such a case, as an unfortunate consequence of (5.1), we have a new "four regions" structure for NS–F equations governing these fluid flows at high Reynolds numbers.

This new structure of unsteady NS–F equations seems, in particular, very significant for the various applications linked with the heat emergency situations (sudden rise of a thermal source) – explosions, fires, failures of oil and gas pipelines, and so on – in a local domain on the wall in contact with the fluid.

Such a "four regions" structure, at $Re \uparrow \infty$, is linked with four limiting processes in full unsteady NS–F equations and replaces the two classical regions, "Euler–Prandtl" and regular coupling, linked with the following two limiting processes (the horizontal coordinates x and y being fixed):

$$\mathrm{Lim}^{E} = [\varepsilon \downarrow 0, \text{ with t and z fixed}]. \tag{5.2a}$$

and

$$\mathrm{Lim}^{Pr} = [\varepsilon \downarrow 0, \text{ with t and } \zeta = z/\varepsilon \text{ fixed}], \tag{5.2b}$$

As a consequence of the strong degeneracy linked with (5.1), in the unsteady case, the limiting process (5.2b) is singular near $t = 0$. Therefore, because the partial time derivative of the vertical component of the velocity is absent in BL equations, we do not have the possibility of taking into account the corresponding data which is prescribed for full unsteady NS–F equations.

It is necessary to consider a third limiting process, inner in time – a so-called "acoustic" limiting process:

$$\mathrm{Lim}^{Ac} = [\varepsilon \downarrow 0, \text{ with } \tau = t/\varepsilon \text{ and } \zeta = z/\varepsilon \text{ fixed}], \tag{5.3}$$

An acoustic problem must be considered in a third "acoustic region" close to initial time, and then (if possible) a matching with the boundary-layer (BL) region far from the region of the acoustics. Indeed, the consideration, in this acoustic region, of an unsteady adjustment problem, when $\tau \to \infty$, presents the possibility (in principle) of prescribing the correct initial conditions for unsteady BL equations significant only far of the initial time in the "Prandtl BL region".

But a new problem emerges because in this third non-viscous, near the initial time acoustic region, we do not have the possibility of taking into account the thermal condition on the wall $z = 0$, prescribed for NS–F equations in the framework of the heat emergency problem.

As a consequence, it is necessary to consider a fourth limiting process simultaneously near $t = 0$ and $z = 0$:

$$\mathrm{Lim}^{Ra} = [\varepsilon \downarrow 0, \text{ with } \theta = t/\varepsilon^2 \text{ and } \eta = z/\varepsilon^2 \text{ fixed}], \tag{5.4}$$

In such a case, in this corner small fourth region, we derive the unsteady one-dimensional NS–F equations governing the compressible Rayleigh problem, with the corresponding thermal condition on $z = 0$ (Fig. 5.1).

Precisely, the above (related with the limiting process (5.4)) compressible viscous and heat-conducting Rayleigh problem presents the possibility of producing an answer to following question: What is the significant problem, governing the unsteady compressible viscous and heat-conducting fluid flow, emerging as a result of a suddenly rise in temperature locally on the wall, z = 0, at time, t = 0?

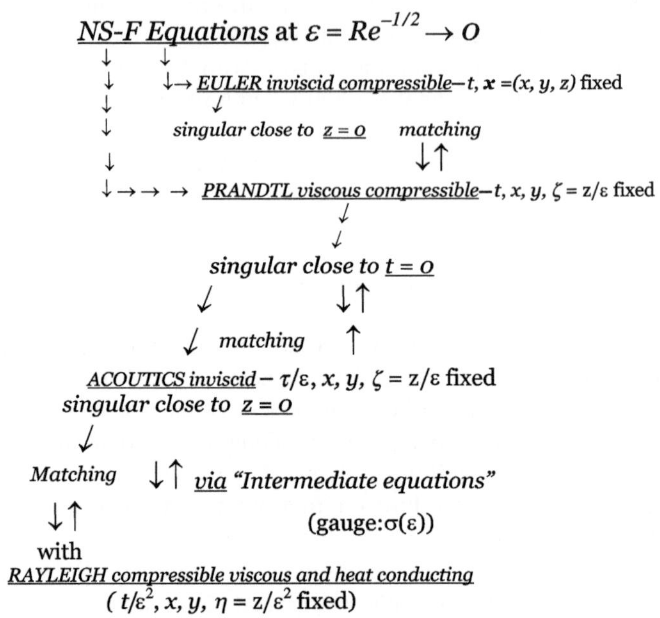

Fig. 5.1 Four regions structure of NS–F unsteady equations at large Reynolds number

In Sect. 5.3 the formulation of the starting unsteady NS–F problem is given, while Sect. 5.4 is devoted to the derivation of the corresponding four problems linked with the above four limiting processes – (5.2a, 5.2b), (5.3), and (5.4). Section 5.5 concerns the problem of the unsteady adjustment (via the Rayleigh problem significant in the fourth region) to Prandtl BL evolution, in time, which is significant in the second region close to wall z = 0, but far from the third and fourth regions near the initial time t = 0. In Sect. 5.6 some conclusions are presented.

5.3 Formulation of the Unsteady NS–F Problem

We consider the atmospheric dry air as a thermally perfect, viscous, and heat-conducting gas, with constant dissipative coefficients. In such a case, the dimensionless NS–F equations are written (for dimensionless functions, **v**, w, p, ρ, and T) in the following form – all functions, variables, and coefficients being non-dimensional:

$$\partial \rho/\partial t + \mathbf{D} \cdot (\rho \mathbf{v}) + \partial(\rho w)/\partial z = 0, \tag{5.5}$$

$$\rho[\partial \mathbf{v}/\partial t + (\mathbf{v}\,.\,\mathbf{D})\mathbf{v} + w\partial \mathbf{v}/\partial z] + (1/\gamma M^2)\mathbf{D}p = \varepsilon^2\{\Delta \mathbf{v} + (1/3)\,\mathbf{D}\,[\mathbf{D}\,.\,\mathbf{v} + \partial w/\partial z]\}, \tag{5.6}$$

$$\rho[\partial w/\partial t + (\mathbf{v}\,.\,\mathbf{D})w + w\,\partial w/\partial z] + (1/\gamma M^2)\partial p/\partial z = \varepsilon^2\{\Delta w + (1/3)\partial/\partial z[\mathbf{D}\,.\,\mathbf{v} + \partial w/\partial z], \tag{5.7}$$

$$\rho[\partial T/\partial t + (\mathbf{v}\,.\,\mathbf{D})T + w\,\partial T/\partial z] + (\gamma - 1)p\,[\mathbf{D}\,.\,\mathbf{v} + \partial w/\partial z] = (\gamma/Pr)\,\varepsilon^2\Delta T + \gamma(\gamma-1)\varepsilon^2 M^2\{\Phi - (2/3)[\mathbf{D}\,.\,\mathbf{v} + \partial w/\partial z]^2\}, \tag{5.8}$$

with

$$p = \rho T, \tag{5.9}$$

where the horizontal (relative to coordinates (x, y)) velocity vector is $\mathbf{v} = (u, v)$, and viscous dissipation is written in the following form:

$$\Phi = [\partial u/\partial z + \partial w/\partial x]^2 + [\partial v/\partial z + \partial w/\partial y]^2 + [\partial u/\partial y + \partial v/\partial x]^2 + 2\left[(\partial u/\partial x)^2 + (\partial v/\partial y)^2 + (\partial w/\partial z)^2\right], \tag{5.10}$$

and $\Delta = \mathbf{D}^2 + \partial^2/\partial z^2$ with $\mathbf{D} = (\partial/\partial x,\ \partial/\partial y)$.

For the above evolution equations (5.5)–(5.8) we write. as initial conditions at initial time $t = 0$:

$$t^- \geq 0 : \mathbf{v} = 0,\ w = 0,\ \rho = 1 \text{ and } T = 1. \tag{5.11}$$

At the horizontal solid wall, $z = 0$, we assume:

$$z = 0 : \mathbf{v} = w = 0 \tag{5.12a}$$

and

$$T = \Theta(t/\beta, P),\ \text{when } t^+ \geq 0,\ \Theta(t/\beta, P) \equiv 0,\ \text{when } t^- \leq 0, \tag{5.12b}$$

where

$$\beta \equiv t^*/t^0 \ll 1, \tag{5.13}$$

is a ratio of two time scales: t^*, a short time scale, in comparison to characteristic evolution time scale t°, which appears in Strouhal number S in (3.3).

In condition (5.12a, 5.12b), the dimensionless function $\Theta(t/\beta, P)$ is used to simulate an emergency of a thermal spot at $t^+ \geq 0$, P being a point on a local domain on the wall, $P \subset D$, for which the reference length scale L° is a diameter for this time-dependent domain $D = D(t)$.

In the above four dimensionless NS–F equations (5.5)–(5.8), we have assumed that (S is the Strouhal number):

$$S \equiv 1 \Rightarrow U^0 = L^0/t^0,$$

and in the fact (t' is the time with dimension):

$$\Theta(t/\beta,\ P) \equiv \Theta(t'/t^*,\ P).$$

The characteristic time $t^\circ = L^\circ/U^\circ$ is a "long" time and characterizes the evolution of fluid flow after the thermal spot emergency, while the characteristic "short" – small – time, $t^* \ll t^\circ$, is linked just with this emerging thermal spot short (time) interval.

The above formulated NS–F initial-boundary value problem, (5.5)–(5.13), is a very complicated mathematical problem, and the rigorous proof concerning its well-posedness is obviously an intractable question!

In fact, our objective below is rather to analyze the specific structure of this problem, when ε tends to zero – for large Reynolds number – and consider the relations, via matching, between the four particular fluid flows regions discussed in Sect. 5.2.

The existence of these four fluid flows regions, at large Reynolds number, shows that we have the possibility of considering this above NS–F system of equations as a "puzzle!" A challenging, but difficult, approach is to "deconstruct"[1] this puzzle, relative to limiting values – vanishing or infinity – of various reference parameters, Re, M, or Pr..., in order to unify – by a RAMA process – the set of various, ill-assorted, partial approximate system of equations, customarily used in classical fluid dynamics.

We observe that from our above formulated mathematical–physical (5.11)–(5.13) problem for the NS–F equations (5.5)–(5.8) with (5.9), it follows that we have the possibility of first investigating not only the initial stage of the motion, emerging as a cause and effect of a sudden rise in temperature locally on the wall $z = 0$ at $t = 0$ in a corner fourth region, but also the evolution of this "thermal accident" in second, Prandtl, and first, Euler, BL viscous and inviscid Eulerian regions, by two matching processes.

[1] In fact, the system of NS–F equations, despite its dimensionless form, do not have a fixed meaning, even if the various reference parameters, in NS–F equations, give a good idea for an investigation in this way! A real meaning is "created", each time, in the act of the RAM Approach, relative to a precise parameter (high or low), via the derivation of a consistent simplified model, this process is just a deconstruction ("à la Derrida") of NS–F system of equations.

The considered physical case is a typical free problem independent of any given external flow. This case can be generalized for the atmosphere, where gravity plays an obvious and important role in emerging convective local motion. For such an atmospheric case, a typical example is a forest fire, which is actually a very bad accident which causes a large amount of damage, and is considered to be an "environmental disaster"!

5.4 Derivation of the Corresponding Four Model Problems

Below we consider the derivation of approximate leading-order equations for the corresponding four limiting processes (5.2a, 5.2b), (5.3), and (5.4), discussed in Sect. 5.2, and linked with the limiting process $\varepsilon \downarrow 0$, from our full NS–F problem: (5.5)–(5.9), with (5.10), and (5.11)–(5.13) – our main objective being a consistent obtention of these four particular systems of equations related to the four limiting processes (5.2a, 5.2b)–(5.4).

5.4.1 Euler–Prandtl Regular Coupling

When we consider the first, Euler, limiting process (5.2a), Lim^{E}, at $\varepsilon \downarrow 0$, with t and z fixed, from NS–F, evolution equations (5.5)–(5.8), with (5.9), we obtain a strong degeneracy (close to the horizontal solid wall $z = 0$), which leads to a Euler system of equations for the leading-order functions, $\mathbf{v}_E$, w_E, p_E, T_E and ρ_E, in the following Euler asymptotic expansion associated with (5.2a):

$$(\mathbf{v}, w) = (\mathbf{v}_E, w_E) + \varepsilon(\mathbf{v}_E{}^1, w_E{}^1) + \ldots, \tag{5.14a}$$

$$(p, T, \rho) = (p_E, T_E, \rho_E) + \varepsilon(p_E{}^1, T_E{}^1, \rho_E{}^1) + \ldots, \tag{5.14b}$$

where the Eulerian terms (with "$_E$" as subscript) are dependent on t, x, y, and z. We then obtain the following system of Euler inviscid compressible, adiabatic, equations:

$$\partial\rho_E/\partial t + \mathbf{D}.(\rho_E \mathbf{v}_E) + \partial(\rho_E w_E)/\partial z = 0; \tag{5.15a}$$

$$\rho_E[\partial \mathbf{v}_E/\partial t + (\mathbf{v}_E.\mathbf{D})\mathbf{v}_E + w_E \partial \mathbf{v}_E/\partial z] + (1/\gamma M^2)\mathbf{D} p_E = 0; \tag{5.15b}$$

$$\rho_E[\partial w_E/\partial t + (\mathbf{v}_E.\mathbf{D}) w_E + w_E \partial w_E/\partial z] + (1/\gamma M^2)\partial p_E/\partial z = 0, \tag{5.15c}$$

$$\rho_E[\partial T_E/\partial t + (\mathbf{v}_E.\mathbf{D})T_E + w_E\partial T_E/\partial z] + (\gamma - 1)p_E[\mathbf{D}.\mathbf{v}_E + \partial w_E/\partial z] = 0, \tag{5.15d}$$

with

$$p_E = \rho_E T_E. \tag{5.15e}$$

Because all the dissipative terms, in the right-hand side of NS–F equations (5.6)–(5.8), are absent in limiting, leading-order, Euler equations (5.15a–5.15d) (see, for instance [100], and [37], Chap. 9), these Euler equations cannot be valid in the vicinity of the wall $z = 0$.

The significant equations valid near $z = 0$, which replace the above Euler equations (5.15a–5.15e), are derived when we introduce, in place of z, an inner vertical coordinate, significant in the vicinity of $z = 0$: namely,

$$\zeta = z/\varepsilon, \tag{5.16a}$$

when we consider the second, Prandtl, limiting process (5.2b), Lim^{Pr} at $\varepsilon \downarrow 0$, with t and ζ fixed.

In such a case, again, from full NS–F equations (5.5)–(5.8), with (5.9), for the leading-order functions $\mathbf{v}_{Pr}, p_{Pr}, T_{Pr}, \rho_{Pr}$, and second–order vertical component of the velocity, ${w_{Pr}}^1$, we consider the following asymptotic, *à la* Prandtl, expansion associated with (5.2b):

$$(\mathbf{v}, w) = (\mathbf{v}_{Pr}, 0) + \varepsilon({\mathbf{v}_{Pr}}^1, {w_{Pr}}^1) + \ldots, \tag{5.17a}$$

$$(p, T, \rho) = (p_{Pr}, T_{Pr}, \rho_{Pr}) + \varepsilon({p_{Pr}}^1, {T_{Pr}}^1, {\rho_{Pr}}^1) + \ldots, \tag{5.17b}$$

where the Prandtl (BL) terms (with "$_{Pr}$" as subscript) are dependent on t, x, y, and ζ. Then, with (5.2b) and (5.17a, 5.17b), we derive the Prandtl BL unsteady equations (see Stewartson [101], and the more recent book by Oleinik and Samokhin [102]):

$$\partial \rho_{Pr}/\partial t + \mathbf{D}.(\rho_{Pr}\mathbf{v}_{Pr}) + \partial(\rho_{Pr}{w_{Pr}}^1)/\partial\zeta = 0; \tag{5.18a}$$

$$\rho_{Pr}[\partial \mathbf{v}_{Pr}/\partial t + (\mathbf{v}_{Pr}.\mathbf{D})\mathbf{v}_{Pr} + {w_{Pr}}^1\partial \mathbf{v}_{Pr}/\partial\zeta] + (1/\gamma M^2)\mathbf{D}p_{Pr} = \partial^2\mathbf{v}_{Pr}/\partial\zeta^2; \tag{5.18b}$$

$$\partial p_{Pr}/\partial\zeta = 0; \tag{5.18c}$$

$$\begin{aligned}\rho_{Pr}[\partial T_{Pr}/\partial t &+ (\mathbf{v}_{Pr}.\mathbf{D})T_{Pr} + {w_{Pr}}^1\partial T_{Pr}/\partial\zeta] \\ &+ (\gamma - 1)p_{Pr}[\mathbf{D}.\mathbf{v}_{Pr} + \partial {w_{Pr}}^1/\partial\zeta] \\ &= (\gamma/Pr)\partial^2 T_{Pr}/\partial\zeta^2 + \gamma(\gamma - 1)M^2|\partial\mathbf{v}_{Pr}/\partial\zeta|^2,\end{aligned} \tag{5.18d}$$

with

$$p_{Pr} = \rho_{Pr} T_{Pr}. \tag{5.18e}$$

It is well known that both the above system of equations – outer, Euler (5.15a–5.15e), and inner, Prandtl (5.18a–5.18e) – are related to the following classical matching relation (discussed in Sect. 6.4.2):

$$\lim_{\zeta\uparrow\infty} \left[\mathrm{Lim}^{Pr}\right] = \lim_{z\downarrow 0} \left[\mathrm{Lim}^{E}\right], \tag{5.19a}$$

and, as a first consequence of (5.19a), we obtain for the Euler outer system of Eqs. 5.15a–5.15e the following single (slip!) condition:

$$w_E = 0 \text{ at } z = 0. \tag{5.19b}$$

Then, as a second consequence of (5.19a), we see that from the strong degenerated equation (5.18c), in the Pranddtl system of Eqs. 5.18a–5.18e, we have the possibility of relating the constant value of p_{Pr}, with respect to vertical BL coordinate, ζ, with the value of p_E at $z = 0$:

$$p_{Pr}(t, x, y) \equiv p_E(t, x, y, 0) = p_{E,0}(t, x, y)$$

and

$$\mathbf{D}p_{Pr} \equiv (\gamma M^2)\rho_{E,0}[\partial w_{E,0}/\partial t + (\mathbf{v}_{E,0}.\mathbf{D})w_{E,0}]. \tag{5.19c}$$

Now, concerning the thermal condition (5.12), with the parameter β given by (5.13), we are obliged to assume that (when t is fixed, in Lim^{Pr}, far of the initial time):

$$\beta = \beta(\varepsilon) \downarrow 0 \text{ with } \varepsilon \downarrow 0, \tag{5.20a}$$

and we have only the possibility of writing, for the BL equations (5.18a–5.18e), the following conditions:

$$\text{on } \zeta = 0 : \mathbf{v}_{Pr} = 0, w_{Pr}{}^1 = 0,$$
$$T_{Pr} = \Theta(\infty, P), \text{ at } t > 0 \text{ fixed.} \tag{5.20b}$$

Obviously, in the framework of the Euler–Prandtl regular coupling we do not have the possibility of taking into account the “thermal accident” linked with the “temperature emergency at initial time on the wall z = 0.”

On the other hand, the above Prandtl system of Eqs. 5.18a–5.18d, with (5.18e), due to the reduced BL equation (5.18c), must be considered as a system of two equations for $\mathbf{v}_{Pr}$ and T_{Pr}, while the Prandtl vertical velocity ${w_{Pr}}^1$ must be computed through the following relation:

$$ {w_{Pr}}^1 = (T_{Pr}/p_{E,0}) \int_0^\zeta \{[\partial/\partial t + (\mathbf{v}_{Pr}.\mathbf{D})]\rho_{Pr} + \rho_{Pr}(\mathbf{D}.\mathbf{v}_{Pr})\} d\zeta, \tag{5.20c} $$

since ${w_{Pr}}^1 = 0$ on $\zeta = 0$, according to the second condition in (5.20b), and the (matching) relation:

$$ \lim_{\zeta\uparrow\infty}\left[{w_{Pr}}^1\right] = {w_{E,0}}^1, \tag{5.20d} $$

is, in fact, a regular coupling condition with the second-order linearized Euler equations for the terms with "${}_E{}^1$" proportional to ε, in Euler asymptotic expansion (5.14a, 5.14b).

However, the problem of two initial conditions for the two unsteady Prandtl equations, (5.18b) and (5.18d), for $\mathbf{v}_{Pr}$ and T_{Pr}, is more subtle, and is a direct consequence of the change of the nature of Prantdl equations relative to the incomplete parabolic character of NS–F equations (see [59, 60, 67]).

In fact, for $\mathbf{v}_{Pr}$ and T_{Pr} we have a system of two hyperbolic–parabolic equations [102]:

$$ \rho_{Pr} D_{Pr}\mathbf{v}_{Pr}/Dt - \partial^2\mathbf{v}_{Pr}/\partial\zeta^2 = \boldsymbol{F}, \tag{5.21a} $$

$$ \rho_{Pr} D_{Pr}T_{Pr}/Dt - (\gamma/Pr)\partial^2 T_{Pr}/\partial\zeta^2 = G, \tag{5.21b} $$

where $D_{Pr}/Dt = \partial/\partial t + \mathbf{v}_{Pr}.\mathbf{D}$, and the right-hand side $\boldsymbol{F}$ and G are a collection of terms with the first-order derivatives relative to $\mathbf{D}$ and ζ.

The continuity equation in the Prandtl system (5.18a–5.18d) is, in fact, an equation determining ${w_{Pr}}^1$, due to (5.20c), and in place of (5.18e) the relation:

$$ \rho_{Pr} = p_{E,0}/T_{Pr}, \tag{5.21c} $$

determines the density ρ_{Pr}.

Without loss, the generality the hyperbolic–parabolic character of the system of Eqs. 5.21a, 5.21b is related to the transmission of the information in the planes $\zeta =$ const, along the trajectories linked with the derivative operator $D_{Pr}/Dt = \partial/\partial t + \mathbf{v}_{Pr}.\mathbf{D}$, supporting the hyperbolicity – this information being instantaneously diffused by vertical coordinate ζ on each normal direction to the wall $\zeta = 0$, at each moment t (which just characterizes the "parabolicity").

We observe also, that the domain of the dependence, for a fixed moment, of the point on the wall has an angular form, but in the unsteady case the precise form of this domain is not easy definable.

The above brief discussion shows explicitly that there is obviously a change in the mathematical character of the fluid dynamics equations, when we pass from a fourth order in time (four partial derivatives in time) unsteady NS–F system of Eqs. 5.5–5.8, to an unsteady Prandtl reduced system of two Eqs. 5.21a, 5.21b. This strong modification leads to a singular nature of the system (5.21a, 5.21b) near the initial time $t = 0$ – this singular nature of the Prandtl boundary-layer concept for the unsteady case being (curiously) ignored up to 1980 (See, for instance, a recent (1994) discussion by Van Dyke: *Nineteenth-century roots of the boundary-layer idea* [103]).

More precisely, it is necessary to prescribe in the framework of Prandtl BL equations (for instance for Eqs. 5.21a, 5.21b) only two initial conditions at $t = 0$. But unfortunately, the initial data for $\mathbf{v}_{Pr}$ and T_{Pr}, at time $t = 0$ (designated by: $\mathbf{v}^0{}_{Pr}$ and $T^0{}_{Pr}$), are certainly different from the initial conditions (in particular, (5.11)) for the full NS–F system of Eqs. 5.5–5.8.

Indeed, the main question is the following. Since two, NS–F, initial conditions are lost during the Prandtl limiting process (5.2b), how are the (unknown?) initial conditions:

$$\mathbf{v}_{Pr} = \mathbf{v}^0{}_{Pr} \text{ and } T_{Pr} = T^0{}_{Pr}, \text{ at } t = 0 \tag{5.22}$$

for the unsteady Prandtl BL equations (5.18b) and (5.18d), and how are the data $\mathbf{v}^\mathbf{0}{}_{Pr}$ and $T^\mathbf{0}{}_{Pr}$ linked with the initial data in conditions prescribed at the start (at $t = 0$) for the unsteady NS–F equations. The answer is strongly related to the obtention, from the full unsteady NS–F equations, of a particular system of equations valid near initial time and written relative to a short time. In fact, we have three short times:

$$\tau = t/\varepsilon, \quad \theta = t/\varepsilon^2, \text{ and } \sigma = t/\beta, \tag{5.23}$$

and the choice of data $\mathbf{v}^0{}_{Pr}$ and $T^\mathbf{0}{}_{Pr}$, is realized via an unsteady adjustment problem, when the adequate(?) short time tends to infinity in unsteady adjustment equations valid near initial time!

5.4.2 *Acoustic and Rayleigh Problems Near the Initial Time* $t = 0$

First, with the limiting process (5.3),

$$\text{Lim}^{Ac}, \text{ when } \varepsilon \downarrow 0, \text{ with } \tau = t/\varepsilon \text{ and } \zeta = z/\varepsilon \text{ fixed,}$$

from the NS–F equations (5.5)–(5.8), with (5.9), for the leading-order functions, $\mathbf{v}_{Ac}, w_{Ac}, p_{Ac}, T_{Ac}, \text{and } \rho_{Ac},$ in the following asymptotic acoustic expansions, associated with (5.3):

$$(\mathbf{v}, w) = (\mathbf{v}_{Ac}, w_{Ac}) + \varepsilon(\mathbf{v}_{Ac}{}^{1}, w_{Ac}{}^{1}) + \ldots, \tag{5.24a}$$

$$(p, T, \rho) = (p_{Ac}, T_{Ac}, \rho_{Ac}) + \varepsilon(p_{Ac}{}^{1}, T_{Ac}{}^{1}, \rho_{Ac}{}^{1}) + \ldots, \tag{5.24b}$$

where the acoustic terms (with "$_{Ac}$" as subscript) are dependent on τ, x, y, and ζ, we derive the following, compressible, non-viscous, adiabatic, unsteady, one-dimensional, (τ, ζ), equations:

$$\partial\rho_{Ac}/\partial\tau + \partial(\rho_{Ac} w_{Ac})/\partial\zeta = 0; \tag{5.25a}$$

$$\rho_{Ac}[\partial w_{Ac}/\partial\tau + w_{Ac}\partial w_{Ac}/\partial\zeta] + (1/\gamma M^2)\partial p_{Ac}/\partial\zeta = 0; \tag{5.25b}$$

$$\rho_{Ac}[\partial T_{Ac}/\partial t + w_{Ac}\partial T_{Ac}/\partial\zeta] + (\gamma - 1)p_{Ac}\partial w_{Ac}/\partial\zeta] = 0, \tag{5.25c}$$

$$p_{Ac} = \rho_{Ac} T_{Ac}. \tag{5.25d}$$

and also the following transport equation for $\mathbf{v}_{Ac}$:

$$\partial\mathbf{v}_{Ac}/\partial\tau + w_{Ac}\partial\mathbf{v}_{Ac}/\partial\zeta = 0. \tag{5.26}$$

The system of Eqs. 5.25a–5.25d – valid simultaneously close to initial time, $\tau = 0$, and in a thin layer in the vicinity of the wall, $\zeta = 0$ – are identical to the usual equations for one-dimensional vertical unsteady motion in (non-viscous, adiabatic) gas dynamics.

Once w_{Ac} has been obtained, through the solution of the system (5.25a–5.25d) with the proper initial conditions (that is, the initial conditions, (5.11), for starting NS–F equations), single (slip) boundary condition:

$$w_{Ac} = 0 \text{ on } \zeta = 0, \quad \tau > 0, \tag{5.27a}$$

and matching condition (in time)

$$\mathrm{Lim}_{\tau\uparrow\infty} w_{Ac} = w_{Pr|t=0} = 0, \tag{5.27b}$$

we may use the transport equation (5.26) in order to compute $\mathbf{v}_{Ac}$.

Unfortunately, with the above non-viscous adiabatic system of Eqs. (5.25a–5.25d), and transport equation (5.26), we do not have the possibility of taking into account our main emergency thermal effect (via the thermal spot $\Theta(t/\beta, P)$ on the wall), since the inviscid (non-viscous, adiabatic) system (5.25a–5.25d) and Eq. 5.26 are not valid close to the wall, where the conditions (5.12a, 5.12b), with (5.13), are prescribed.

The Eq. 5.26 for $\mathbf{v}_{Ac}$ shows that

$$\mathbf{v}^{0}{}_{Pr} = \mathrm{Lim}_{\tau\uparrow\infty}\mathbf{v}_{Ac}, \tag{5.27c}$$

from matching, and it seems (in (5.22)) that we can assume (for our particular case), as a value for $\mathbf{v}^0{}_{Pr}$, zero; but this is certainly not the case for $T^0{}_{Pr}$.

Again, near initial time and close to the wall, where we have the conditions (5.12a, 5.12b), according to Zeyounian [98], it is necessary to consider the Rayleigh limiting process (see (5.4)):

$$\mathrm{Lim}^{Ra},\ \text{when } \varepsilon \downarrow 0,\ \text{with}\ \theta = t/\varepsilon^2 \text{ and } \eta = z/\varepsilon^2 \text{ fixed}, \tag{5.28a}$$

with

$$(\mathbf{v},\ w) = (\mathbf{v}_{Ra}, w_{Ra}) + \varepsilon(\mathbf{v}_{Ra}{}^1, w_{Ra}{}^1) + \ldots, \tag{5.28b}$$

$$(p, T, \rho) = (p_{Ra}, T_{Ra}, \rho_{Ra}) + \varepsilon(p_{Ra}{}^1, T_{Ra}{}^1, \rho_{Ra}{}^1) + \ldots, \tag{5.28c}$$

where the Rayleigh terms, in (5.28a, 5.28b), with "$_{Ra}$" as subscript, are dependent on θ, x, y, and η.

In such a case, from full NS–F unsteady equations (5.5)–(5.8), (5.9), for leading-order functions, $\mathbf{v}_{Ra}, w_{Ra}, p_{Ra}, T_{Ra}$, and ρ_{RA}, we derive below the Rayleigh equations (5.29) and (5.30a–5.30d) used in the compressible Rayleigh problem, which are, in fact, the one-dimensional reduced form of the full NS–F equations valid in a corner region near initial time $\theta = 0$, and close to the wall $\eta = 0$. Namely:

$$\rho_{Ra}[\partial \mathbf{v}_{Ra}/\partial\theta + w_{Ra}\partial \mathbf{v}_{Ra}/\partial\eta] = \partial^2 \mathbf{v}_{Ra}/\partial\eta^2, \tag{5.29}$$

$$\partial\rho_{Ra}/\partial\theta + \partial(\rho_{Ra} w_{Ra})/\partial\eta = 0; \tag{5.30a}$$

$$\begin{aligned}&\rho_{Ra}[\partial w_{Ra}/\partial\theta + w_{Ra}\partial w_{Ra}/\partial\eta] + (1/\gamma M^2)\partial p_{Ra}/\partial\eta\\ &\quad = (4/3)\partial^2 w_{Ra}/\partial\eta^2;\end{aligned} \tag{5.30b}$$

$$\begin{aligned}&\rho_{Ra}[\partial T_{Ra}/\partial\theta + w_{Ra}\partial T_{Ra}/\partial\eta] + (\gamma - 1)p_{Ra}\partial w_{Ra}/\partial\eta]\\ &\quad = (\gamma/\mathrm{Pr})\partial^2 T_{Ra}/\partial\eta^2 + \gamma(\gamma-1)M^2\{|\partial \mathbf{v}_{Ra}/\partial\eta|^2\\ &\quad + (4/3)|\partial w_{Ra}/\partial\eta|^2\},\end{aligned} \tag{5.30c}$$

$$p_{Ra} = \rho_{Ra} T_{Ra}, \tag{5.30d}$$

These above Rayleigh equations, (5.29) and (5.30a–5.30d), are applied in [99] for the Rayleigh compressible problem by Howarth in 1951, but in the case of an infinite flat horizontal plate (submerged in a viscous and heat-conducting and originally quiescent fluid) which is impulsively started moving in its own plane with a constant velocity.

In fact, from Our above RAM Approach I can now affirm that in a corner region (θ, η), which is significant for the small time near initial time and in thin layer close to wall, at the leading-order for large Reynolds number, the above compressible Rayleigh equations, (5.29) and (5.30a–5.30d), for a viscous and heat-conducting fluid flow, consistently replace the full unsteady NS–F equations.

Both these unsteady systems, ((5.29) and (5.30a–5.30d), and (5.25a–5.25d) and (5.26)), both valid near the initial time, are related amongst themselves by the following matching relations:

$$\mathrm{Lim}_{\eta\uparrow\infty} \mathrm{w_{Ra}} = \mathrm{w_{Ac|\zeta=o}}, \tag{5.31a}$$

$$\mathrm{Lim}_{\theta\uparrow\infty}\left[\mathbf{v}_{\mathrm{Ra}}, \mathrm{w_{Ra}}, \rho_{\mathrm{Ra}}, \mathrm{T_{Ra}}\right] = \left[\mathbf{v}_{\mathrm{Ac}}, \mathrm{w_{Ac}}, \rho_{\mathrm{Ac}}, \mathrm{T_{Ac}}\right]_{|\tau=0}. \tag{5.31b}$$

The reader can find in Antontsev et al. [104], Chap. 2, some mathematically rigorous results concerning the above, *à la* Rayleigh, equations (5.29) and (5.30a–5.30d); and see also the review paper by Solonnikov and Kazhykhov [105].

Finally, if we assume that in a thermal spot

$$\Theta(\mathrm{t}/\beta, P), \text{ with } \beta \ll 1,$$

β defined by (5.13), is equal to ε^2, then, for the emergency of the "temperature accident" we have the possibility of taking into account all starting initial (5.11) and wall (5.12a, b) conditions, in the framework of an initial-boundary values Rayleigh problem.

We therefore write, for Eqs. 5.29−5.30d, the following initial conditions:

$$\theta^- \leq 0 : \mathbf{v}_{\mathrm{Ra}} = 0,\ \mathrm{w_{Ra}} = 0,\ \rho_{\mathrm{Ra}} = 1 \text{ and } \mathrm{T_{Ra}} = 1, \tag{5.32a}$$

and, at the horizontal solid wall, $\eta = 0$, we assume:

$$\eta = 0 : \mathbf{v}_{\mathrm{Ra}} = \mathrm{w_{Ra}} = 0 \text{ and } \mathrm{T_{Ra}} = \Theta(\theta, P),\ \theta^+ \geq 0. \tag{5.32b}$$

The above "starting problem", (5.29)−(5.30a–5.30d) with (5.32a, 5.32b), is a typical problem for various "emergency−temperature–accident phenomena" which develop when $\theta^+ \geq 0$.

5.5 Adjustment Processes Towards the Prandtl BL Evolution Problem

If we want to take into account the sudden heat emergency, at the time $\theta^+ \geq 0$, in a local domain, $P \subset D$, on the wall $\eta = 0$, then it now seems justifiable that the main working problem is just the above compressible, viscous, and heat-conducting Rayleigh problem ((5.29), (5.30a–5.30d), (5.32a, 5.32b)).

This Rayleigh problem is valid, simultaneously, near the initial time and close to wall – in a small corner (fourth) region – with a "physical size" of order $(\nu^{\circ}/U^{\circ 2})$ relative to time and (ν°/U°) relative to vertical coordinate – these time and length scales being exactly those used by Howarth [99].

5.5.1 Adjustment Process Via the Acoustics/Gas Dynamics Equations

Once the above Rayleigh problem – (5.29), (5.30a–5.30d), and (5.32a, 5.32b) – is solved (numerically), then we have the possibility (first) of prescribing, by matching relations (5.31a, 5.31b) to Eqs. 5.25a–5.25d, with (5.26), of gas dynamics – significant in the third inviscid region near the time, $\tau = 0$, and characterized by τ and ζ – the consistent conditions at $\tau = 0$ and $\zeta = 0$.

As a consequence of this above matching, it seems that we can expect that in conditions (at $\tau = 0$ and $\zeta = 0$) for Eqs. 5.25a–5.25d with (5.26) of gas dynamics, the influence of the wall condition for the temperature is taken into account, but only via the limit value $\Theta(\propto, P)$.

We observe that in the wall, $\zeta = 0$ at $t > 0$, condition (5.20b) for the Prandtl BL equation (5.18a–5.18e), this same function (independent of time?) $\Theta(\propto, P)$ is also present.

The acoustic/gas dynamics equations (5.25a–5.25d) with (5.26), with these initial conditions (5.31b) and single boundary condition (5.31a), for w_{AC} at $\zeta = 0$, which take into account the (partial?) influence of thermal spot (but independent of time function $\Theta(\propto, P)$), present the possibility of considering, for $\tau \to \infty$, an unsteady adjustment inviscid problem for the initialization of the Prandtl BL equations. As a typical example, see, for instance, our paper co-authored with Guiraud [106], which determines, in particular, the initial data $T^{0}{}_{Pr}$.[2]

When both $\mathbf{v}_{Pr}{}^{0}$ and $T_{Pr}{}^{0}$ are known, as a result of the above unsteady adjustment inviscid problem, then later, via the initial-boundary value BL problem, significant in the second Prandtl, (t, ζ) BL region, we have the opportunity to investigate the quasi-steady evolution of the "temperature accident" arising from the Rayleigh corner fourth region.

[2] In [106], with Guiraud, we have formulated, for the "primitive Kibel equations" (see Sect. 9.2) – which are derived from the hydrostatic approximation to the Euler equations for non-viscous and adiabatic motion – a problem analogous to the one that was considered by Rossby (1938) concerning the quasigeostrophic approximation (a problem which is now well known as the adjustment to geostrophy). The major conclusion of our "adjustment to hydrostatic balance" is that the initial conditions for the primitive equations may be derived from a full set of initial conditions, for the full Euler equations, where in these Eulerian initial conditions the initial data need not fit the hydrostatic balance. This obtention of initial conditions for primitive equations is realized by solving the associated one-dimensional unsteady adjustment problem of vertical motion to hydrostatic balance.

Curiously, however, this unsteady inviscid adjustment scenario, from gas dynamics to BL, does not seem to be the only one possible. Indeed, a more detailed analysis (see Sect. 5.5.2 below) shows the existence of a fifth matching region between Rayleigh (fourth) and Prandtl (second) regions – the existence of such an intermediate fifth region ensuring matching between the Rayleigh corner and Prandtl BL regions.

5.5.2 *Adjustment Process Via the Rayleigh Equations*

This fifth intermediate matching region appears when we investigate, first, the far behaviour of the Rayleigh equations (5.29) and (5.30a–5.30d), for large values of θ and η. In the Rayleigh corner, fourth region, with Prandtl variables,

$$t = \varepsilon^2 \theta \text{ and } \zeta = \varepsilon\eta \text{ as } \varepsilon \downarrow 0, \tag{5.33a}$$

we can introduce new intermediate variables, t* and z*:

$$\theta = t^*/\kappa(\varepsilon) \text{ and } \eta = z^*/\sqrt{\kappa(\varepsilon)},\ 0 < \kappa(\varepsilon) \downarrow 0 \text{ with } \varepsilon \downarrow 0, \tag{5.33b}$$

where $\kappa(\varepsilon)$ is an arbitrary gauge, and t* and z*, the new intermediate variables, are fixed when $\varepsilon \downarrow 0$, in such a way that both Rayleigh variables θ and η tend to infinity.

In this intermediate matching fifth region we have, as leading-order functions:

$$[\mathbf{v}_{Int}, w^*_{Int}, \rho_{Int}, T_{Int}] = \mathrm{Lim}_{\kappa(\varepsilon)\downarrow 0}[\mathbf{v}_{Ra}, w_{Ra}/\sqrt{\kappa(\varepsilon)}, \rho_{Ra}, T_{Ra}], \tag{5.33c}$$

where all intermediate functions (with subscript "$_{Int}$") are dependent on time–space variables, t*, z*, and x, y.

When we take into account that the (unknown) gauge $\kappa(\varepsilon)$ is certainly in an order between ε^2 and ε, such that we derive from the Rayleigh equations, (5.29) and (5.30a–5.30d), due to (5.33b, 5.33c), the following intermediate-matching model equations for $\mathbf{v}_{Int}$, w^*_{Int}, ρ_{Int}, p_{Int}, and T_{Int}:

$$\partial\rho_{Int}/\partial t^* + \partial(\rho_{Int} w^*_{Int})/\partial z^* = 0; \tag{5.34a}$$

$$\partial p_{Int}/\partial z^* = 0; \tag{5.34b}$$

$$\begin{aligned}\rho_{Int}[\partial T_{Int}/\partial t^* + w^*_{Int}\partial T_{Int}/\partial z^*] + (\gamma-1)p_{Int}\partial w^*_{Int}/\partial z^* \\ = (\gamma/Pr)\partial^2 T_{Int}/\partial z^{*2} + \gamma(\gamma-1)M^2|\partial\mathbf{v}_{Int}/\partial z^*|^2,\end{aligned} \tag{5.34c}$$

$$\rho_{Int}[\partial\mathbf{v}_{Int}/\partial t^* + w_{*Int}\partial\mathbf{v}_{Int}/\partial z^*] = \partial^2\mathbf{v}_{Int}/\partial z^{*2}; \tag{5.34d}$$

with

$$p_{Int} = \rho_{Int} T_{Int}. \tag{5.34e}$$

Obviously, the above intermediate matching model equations, (5.34a–5.34e), first pointed out in our short note [98], are those derived when we carry out, on the Rayleigh equations (5.29) and (5.30a–5.30d), the usual approximations of the classical Prandtl boundary-layer theory.

In particular, as in BL region, the pressure p_{Int} is independent of z^* and is determined by matching:

$$\lim_{z^* \uparrow \infty}[\mathrm{Lim}^{Int}] = \lim_{\eta \downarrow 0}[\mathrm{Lim}^{Ra}], \tag{5.35a}$$

where

$$\mathrm{Lim}^{Int} = [\varepsilon \downarrow 0, \text{ with } t^* = t/\sigma(\varepsilon),\ z^* = z/\varepsilon\sqrt{\sigma(\varepsilon)},\ \text{fixed}], \tag{5.35b}$$

the intermediate variables (t^*, z^*) being directly related to the starting (in NS–F equations) variables (t, z).

The intermediate gauge $\sigma(\varepsilon)$, in Lim^{Int}, (5.35b), is linked with the above gauge $\kappa(\varepsilon)$ by the following relation:

$$\kappa(\varepsilon)\,\sigma(\varepsilon) = \varepsilon^2. \tag{5.36}$$

More precisely, this compatibility relation (5.36) is a direct consequence of the investigation of the behaviour of the Prandtl BL equations (5.18a–5.18e), when t and ζ both tend to zero, towards the intermediate fifth region.

Indeed, if we write (again with t^* and z^* fixed):

$$t = \sigma(\varepsilon)t^*,\quad \zeta = \sqrt{\sigma(\varepsilon)}z^*,\ 0 < \sigma(\varepsilon) \downarrow 0 \text{ with } \varepsilon \downarrow 0, \tag{5.37a}$$

and if

$$[\mathbf{v}_{Int}, w^*_{Int}, \rho_{Int}, T_{Int}] = \mathrm{Lim}_{\sigma(\varepsilon)\downarrow 0}[\mathbf{v}_{Pr}, \sqrt{\sigma(\varepsilon)}w_{Pr}, \rho_{Pr}, T_{Pr}], \tag{5.37b}$$

then again we derive the same above intermediate matching model equations (5.34a–5.34d) with (5.34e), but from (5.18a–5.18e). The relation (5.36) is, in fact, a consequence of the compatibility between (5.33b, 5.33c) and (5.37a, 5.37b).

Unfortunately, the precise localization of this intermediate matching region (characterized by the gauge $\sigma(\varepsilon)$), between the Rayleigh and Prandtl regions, does not seem possible at this stage of asymptotic analysis, and more careful (second-order?) investigations are obviously necessary.

Finally, we observe that if on the one hand, when $\sigma(\varepsilon) = \varepsilon^2$, then $t^* = \theta$ and $z^* = \zeta/\varepsilon = z/\varepsilon^2 = \eta$, then we recover the Rayleigh region; and if on the other hand, when $\sigma(\varepsilon) = \varepsilon^0 \equiv 1$, then $t^* = t$ and $z^* = \zeta$, and we recover the Prandtl region.

The existence of such an intermediate region is a striking indication that it seems possible (as a conjecture) to directly match the Rayleigh and Prandtl equations by an adjustment problem, via the intermediate equations (5.34a–5.34e), when the intermediate time $t^* \downarrow \propto$, using the matching condition:

$$\lim_{t^* \downarrow \propto}[\mathbf{v}_{\mathrm{Int}}, \mathrm{T}_{\mathrm{Int}}] = \left[\mathbf{v}^0{}_{\mathrm{Pr}}, \mathrm{T}^0{}_{\mathrm{Pr}}\right]. \tag{5.38}$$

This presents the possibility of obtaining, consistently, the associated initial data, ${\mathbf{v}_{\mathrm{Pr}}}^0$ and ${\mathrm{T}_{\mathrm{Pr}}}^0$, in (5.22), for Prandtl unsteady equations (5.18b) and (5.18d) for $\mathbf{v}_{\mathrm{Pr}}$ and T_{Pr}.

5.6 Some Conclusions

First, it is clear that the above problem of matching, (5.38), deserves careful consideration and may be an interesting numerical/computational problem.

We then observe that it is possible to considerably simplify the matching problem between the Rayleigh and Prandtl equations, if we assume that the Mach number M, in Rayleigh equations, is a small parameter, and assume, for this, that in the wall the thermal condition (5.32b) can be written in the following form:

$$\Theta(\theta, P) = 1 + \Lambda_0 \mathrm{M}^2 \Sigma(\theta, P), \tag{5.39}$$

where $\Lambda_0 = \mathrm{O}(1)$, and $\Sigma(\theta, P)$ replace thermal spot $\Theta(\theta, P)$.

In such a case, the solution of the Rayleigh problem is also expanded relative to a low Mach number, $\mathrm{M} \ll 1$ (as in Howarth's paper [99]). But here we do not proceed further.

A third remark concerns the fact that further investigations are necessary for a complete understanding of the above intriguing five-regions structure, which is very interesting, because it is unusual and does not have an obvious clear interpretation! But the above new five-regions (four regions plus the intermediate region) structure of NS–F equations, at large Reynolds number, as a consequence of the singular nature of the unsteady Prandtl BL equations near the initial time, do not restrict investigations to emergency phenomena, and have fundamental importance in the RAM Approach of NS–F equations.

I think that from this detailed further re-examination of boundary-layer Prandtl theory, it is now possible to resolve some singularities arising in various unsteady boundary-layer problems (see, for instance, Stewartson [101]).

A final remark concerns the pedagogical interest of such partition of NS–F equations, in five regions, for large Reynolds number fluid flows, and this RAM Approach presents the possibility of deriving a new logical interpretation of Euler,

Prandtl, Rayleigh, acoustic/gas dynamics, and intermediate equations, as five significant and particular models of full NS–F unsteady equations for Newtonian fluid flow at large Reynolds number.

Again we observe that not only Prandtl in 1904, but also (it seems) Blasius and Schlichting (in Germany), Lagerstrom, Cole, and Kaplun (at Caltech), Van Dyke (at Stanford), Stewartson and Smith (in England), and Germain (in France), did not realize that indeed the concept of boundary-layer, which is an extension to long-waves approximation in the case of a viscous fluid flow, is singular, in the case of an unsteady fluid flow, near initial time, where initial data are prescribed in a well-posed initial-boundary value problem.

Chapter 6
The Mathematics of the RAM Approach

6.1 Our Basic Postulate for the Realization of the RAM Approach

As was determined at the outset, as a basis of our mathematics (below) for realization of our RAM Approach, it is laid down that:

"*If a leading order, an approximate, model is derived from an NS–F fluid flow problem, then it is necessary that a RAMA be adopted to make sure that terms neglected, in such a full unsteady NS–F stiff problem, really are much smaller than those retained in derived approximate simple (but consistent), no-stiff leading-order, model problem.*"

This above "basic postulate" – a statement that is accepted as true, despite its simplicity – is in fact a very good and suitable general rule – a statement of what the reader is advised to do in various modelling situations. (See, in particular, Chaps. 7–9 below, on applications of the RAM Approach.)

Until this postulate is applied to the derivation of approximate/rational and consistent models for technologically or geophysically/atmospheric stiff problems, it will be difficult to convince the (possibly sceptical) reader (as noted in Germain's *The "New" Mechanics of Fluids of Ludwig Prandtl* (2000) [107], p. 34) of its value as a guide-line for our RAMA, on behalf of numerical/computational simulation.

Massive computations[1] are actually capable of bringing so much to our understanding, but there seems to be no indication that they are in competition with our RAMA, Both are useful and complementary, and, what is more, I think that the RAMA, which is a new way of looking at the derivation (in place of a doubtful ad hoc approach) of the approximate rational model problem on behalf of numerical

[1] Recently, a new Intensive Computation Centre was inaugurated at Bruyères-le-Châtel (Essone), France, and in 2011 the new super-high-speed computer "Curie" – capable of making 1.6 million of billion operations per second – should be fully operational!

R. Kh. Zeytounian, *Navier–Stokes–Fourier Equations*,

DOI 10.1007/978-3-642-20746-4_6,

simulation, is the only one rational way to obtain a non-ad hoc consistent, simplified, non-stiff, well balanced model problem from the full unsteady NS–F stiff problem.

Obviously: the ultimate goal (especially for the numericians in the framework of their simulation processes) is to find the mathematical key which explains the success of the modelling used in industrial and technological applications.

In theoretical fluid dynamics, perhaps the most significance advance in research and understanding during the last 50 years has been the use of asymptotic techniques in order to settle, on a rational basis, a number of approximate models which were often, much earlier, derived by the various and often very questionable ad hoc procedures.

Actually, it might have been clear that asymptotic techniques were well suited for deriving mathematical models amenable to numerical treatment, rather than for obtaining approximate closed forms of solutions.

It is now also evident that asymptotic techniques are very powerful tools in the process of constructing well argued, non-contradictory, consistent, and simplified mathematical models for problems which are stiff, from the point of view of numerical analysis.

In the framework of our RAMA, the main goal is the modelling, and not the finding of solutions, via the derivation of rational consistent models from an initial-boundary value problem formulated for a real fluid flow.

This RAMA has, as its main objective, the generation of simpler adequate models amenable to numerical simulation. But in most cases an efficient method of achieving this goal will be to retain more the spirit (the real meaning and purpose) of asymptotic techniques rather than their complete and formal structure – nothing being said of the rigorous mathematics.

Concerning "our mathematics" for the realization of our RAMA, we have in mind, first, the above formulated postulate and also several key steps which play a decisive role during the rational derivation of consistent models (as clearly indicated in Chap. 4, with the justification of Boussinesq equations, and also in Chap. 5, in the framework of the five-regions structure of unsteady NS–F equations at large Reynolds numbers).

Concerning, precisely, the role of our "basic postulate", we observe also that the knowledge and skill that I acquired, with J.-P. Guiraud, during 1970–1980, working within the framework of the RAMA, presents the possibility, thanks to this postulate, of deriving, in various cases, the dominant concepts and ingredients of a full solution of the starting fluid flow stiff problem.

In fact, fluid dynamics inspired by this "basic postulate" is a new way of approaching classic fluid flow problems – via NS–F full unsteady 3D equations – which is very useful in gas dynamics (from hyposonic to hypersonic), compressible aerodynamics, hydrodynamic instability, technologically and geophysically (atmospheric) interesting (but very stiff) flows, and various thermal and thermocapillary convection problems.

The necessity of such a basic postulate is obvious. Computerized numerical simulation – using a mathematical model created artificially in order to study what could exist in a real and very complicated fluid flow – is a very expensive activity (costing a lot of money). Therefore, as a consequence, confidence is necessary concerning the consistency of the model used and the abilities of this model as a valuable substitute for reality!

Unfortunately, the models derived by numericians and applied mathematicians using ad hoc procedures obviously do not have the essential qualities mentioned above! The cause and effects of such a situation are linked with the appearance of internal inconsistencies in most of the relevant intensive engineering computations based on relatively ad hoc models.[2]

On the other hand, applied mathematicians, who are interested in the application of their rigorous results, based on abstract (modern) non-linear functional analysis, have always found fluid mechanics to be a rich and interesting field, because the basic equations (NS–F equations) have an almost unlimited capacity for producing complex solutions that exhibit unbelievably interesting properties.

Unfortunately, these rigorous investigations (proofs of the existence and uniqueness of solutions) are based on relatively ad hoc and very simplified approximate models which are rife with various internal inconsistencies. This is the case, as far I know, in France especially, in works relative to the incompressible limit of compressible fluid flows and in various papers devoted to models for oceanic and atmospheric motions. Again, the question is: "What is the scientific value of such rigorous results based on inconsistent ad hoc models"?

On the other hand, as observed by Germain in his Anniversary Volume (2000) [108], p. 13:

> Brilliant young physicists who began to be somewhat attracted by fluid dynamics and not simply by hard physics, knew the usefulness of approximations and of non-dimensional scaling. But they did not know that a systematic technique was available for building approximate mathematical models and trying to measure quantitatively their validity ... I showed that the approximation is very often tied to the existence of a small parameter, coming out from the non-dimensional form of the equations [concerning some facets of this non-dimensionalization, see Sect. 6.3], and I intended to show that the process is sustained by asymptotic singular expansions, and insisted on the methodology, in particular the matching conditions and the concept of significant degeneracy [see Sect. 6.4.3] recently created by Eckhaus (see [109]). I liked very much this last one, because it gives a systematic

[2] I took the opportunity of expressing my (very critical) opinion during the conference debate, organized by the Académie des Sciences, which took place in Paris on 29 June 2010. In his presentation entitled "Simulation by super-high-power computers: today and tomorrow", Olivier Pironneau spoke mainly about the significance of numerical simulation and the increasing capacity of super-computers, but did not mention, in any way, the role of fluid dynamics modelling for simulation with the help of a consistent and rational model. The next three invited papers, concerning numerical simulations of very interesting and difficult real problems ("Earthquakes at the planetary scale", "Molecular structure of the living", and "Advanced computations in meteorology and climatology"), are indeed of practical importance, and the results of simulations, which were very attractively presented, gave a favourable impression – but again without any reference to the nature of the selected model and its relation with the starting real simulated problem! It seems to me that Pironneau was not entirely satisfied by the lack of information concerning the process of fluid dynamics modelling in the above-mentioned simulations. Eventually, he was asked the question – one which requires serious thought – "How sure are you of your results? All these numerical simulations will cost money."

way to find out what should be the various stretchings. I thought that this might be attractive to physicists, because it is a quasi-systematic way of comparing the respective weights of various terms in the equations, which measure the physical importance of phenomena they are likely to describe.

A final comment concerns a quite curious recent aspect of the modelling for hydro-aerodynamical problems conceived by applied mathematicians.

The latter to take into consideration – initially a very simple single equation or two coupled simple equations – shows various numerical results (elegant and illuminating) corresponding to a fluid flow over, for example, a simple hollow or a rectangular obstacle. Then, in these single or coupled equations the numerician adds, step by step, various terms which simulate (very naively) the influence of the viscosity, compressibility, gravity or non-linearity, and possibly other physical effects, numerically computing the corresponding fluid flows. As the final result we have a family of fluid flows which possibly mimic a series of realistic flows.

Again, the question is: "What is the scientific interest of such numerical investigations in relation to the simulation of realistic fluid flows?"

6.2 The Mathematical Nature of NS–F Equations: A Fluid-Dynamical Point of View

Rigorous proof of the existence, uniqueness, smoothness, and stability of solutions of problems in fluid dynamics are needed to give meaning to the equations (NS–F equations) and corresponding initial and boundary conditions that govern these problems.

Their formulation (see Sects. 2.3 and 2.4) shows conclusively that the NS–F system of equations is a closed system, and that with the assigned initial and boundary conditions they produce certainty, relative to the well-posedness of problems in fluid dynamics, that these equations are a numerical solution. Therefore, simulation via the RAM Approach is required – indeed, rigorously, as for any arbitrary reasonable choice of a class of admissible initial data, a problem in fluid dynamics must be well-posed (in the Hadamard sense). This means ([47], Chap. 8) that:

(a) The problem has a solution for any initial data in this class.
(b) The solution is unique for any initial conditions.
(c) The solution depends continuously on the initial data.

Obviously, this "well-posedness" is strongly linked with the mathematical nature of NS–F equations as a partial differential system of equations:

Roughly speaking, we can expect that, in particular, the equations for a viscous fluid flow (in a simple, incompressible, case of the Navier equations) are parabolic, and that the equations for an inviscid, non-viscous, adiabatic fluid flow (the Euler equations) are hyperbolic.

However, a more detailed analysis of the structure of full NS–F unsteady equations shows that this above conclusion is not quite correct (at least for a compressible, viscous, and heat-conducting (NS–F) fluid flow). Indeed, the equation of continuity with respect to density ρ is always, for a compressible fluid flow, hyperbolic even for non-trivial viscosity or heat conduction!

Thus, to be more exact, we can say that a system of full unsteady Navier–Stokes–Fourier equations is hyperbolic–parabolic or incompletely parabolic, according to the definitions suggested in [59] and [62] in the study of the mathematical properties of these equations. Concerning, more precisely, the linear incompletely parabolic systems see the paper by Gustafsson and Sundström [67]. In the linear case (when the coefficients of viscosity and heat conduction are constant) the NS–F system becomes:

$$\partial U/\partial \mathrm{t} + \mathrm{A_k}\partial U/\partial \mathrm{x_k} = L_{\mathrm{jk}}\ \partial^2 U/\partial \mathrm{x_j}\partial \mathrm{x_k}, \tag{6.1}$$

where $\mathrm{A_k}$ et L_{jk} are constant square matrices, and $U = (\mathrm{u_1, u_2, u_3, T}, \rho)$.

If we consider the following decomposition for U:

$$U = (\mathrm{u_1, u_2, u_3, T}, 0) + (0, 0, 0, 0, \rho) = U_I + U_{II} \tag{6.2a}$$

then in a such case we write, in place of (6.1), a system of two coupled equations:

$$\begin{aligned} \partial U_I/\partial \mathrm{t} + \mathrm{A_{k;I,I}}\ \partial U_I/\partial \mathrm{x_{k+}}\ \mathrm{A_{k;I,II}}\ \partial U_{II}/\partial \mathrm{xk} \\ = L_{\mathrm{jk;I,I}}\ \partial^2 U_I/\partial \mathrm{x_j}\partial \mathrm{x_k}, \end{aligned} \tag{6.2b}$$

$$\partial U_{II}/\partial \mathrm{t}\ + \mathrm{A_{k;II,I}}\partial U_I/\partial \mathrm{x_{k+}}\mathrm{A_{k;II,II}}\ \partial U_{II}/\partial \mathrm{x_k} = 0. \tag{6.2c}$$

That is just the system (6.2b, c) of two coupled equations for U_I and U_{II}, which is called incompletely parabolic if, on the one hand,

$$\partial U_I/\partial \mathrm{t} = L_{\mathrm{jk\,;I,I}}\ \partial^2 U_I/\partial \mathrm{x_j}\partial \mathrm{x_k} \tag{6.3a}$$

is parabolic, and if, on the other hand,

$$\partial U_{II}/\partial \mathrm{t}\ + \mathrm{A_{k;II,II}}\ \partial U_{II}/\partial \mathrm{x_k} = 0 \tag{6.3b}$$

is hyperbolic.

In two of my survey papers – (1999) [79] and (2001) [29] – the reader can find various theoretical results concerning the NS–F system of equations. In addition, in two books referring to non-viscous (2002 [37], Chap. 9) and viscous (2004 [47], Chaps. 8–10) fluid flows, the reader can also find a fluid-dynamical point of view relative to the well-posedness, existence, uniqueness, stability, turbulence, and strange attractors for NS–F equations.

Concerning, first, the rigorous mathematical results for the Navier incompressible system of equations, see, in Temam (2000) [83], pp. 1049–1106, a discussion of the development of Navier–Stokes (incompressible, in fact, Navier) equations in the second half of the twentieth century, initiated by Leray in 1933 [84]. Leray's first rigorous work, published in 1931, was only in fluid dynamics, and concerns only the Navier equations. He proved the basic existence and uniqueness results [84] for the Navier, viscous, and incompressible equations, and Temam, in [110], has expressed the view that: "No further significant rigorous work on Navier equations was done until that of Hopf in 1951."

The interested reader can find in "No. spécial de la Gazette des Mathématiciens" (2000) – a tribute to Jean Leray (1906–1998) – various papers by some well-known mathematicians and, in particular, the paper by Chemin [111], pp. 70–82, which presents a remarkable summary via an illuminating analysis of chapters of the well-known Leray paper of 1934 [84]: "Sur le mouvement d'un liquide visqueux emplissant l'espace", published in *Acta Mathematika*.

This magnificent paper by Leray was the principal stimulation (with two companion papers published in *Journ. de math. pures et appl.*, vol. 12, 1–82, 1933, and vol. 13, 331–418, 1934) for several of his own papers on the uniqueness, stability, and regularity of viscous and incompressible fluid motions, and indeed was the inspiration for a vast modern literature on the subject.

A particular place in the rigorous mathematical theory of viscous imcompressible fluid flows is also undoubtedly deserved by Ol'ga Aleksandrovna Ladyzhenskaya (see, for instance, the paper devoted to her on her eightieth birthday, in *Russian Math. Survey*, vol. 58(2), 395–425, 2003) who, with the numerous papers and books published on the Navier equations problem during 1957–2002, has played a particularly important role, having a considerable influence on the development of the mathematical theory of Navier problems.

A typical "*à la* Ladyzhenskaya" result is the following major theorem: "Suppose that V is a generalized solution of the Navier unsteady problem:

$$V_t - \nu\Delta V + V.\nabla V + \nabla p = f,\ \mathrm{div}V = 0; V|_{t=0}, V|_{\partial\Omega\times(O,T)=0} \tag{6.4}$$

in $Q_T = \Omega \times [O, T]$ that belongs to the class $L_{m,n}(Q_T)$ with *m*, s satisfying one of the conditions:

$$(n/m) + (2/s) = 1, m \in (n, \propto], s \in [2, \propto) \tag{6.5a}$$

and

$$m = n + \varepsilon,\ \varepsilon>0,\ s = \propto . \tag{6.5b}$$

If $f \in L_2(Q_T), V^\circ \in H(\Omega) \subset (W^12)^\circ$ and Ω is a bounded domain in $\Re^3$ with $\partial\Omega \subset C^2$, then V belongs to the space $W^22(Q_T)$, the corresponding pressure p is such that $\nabla p \in L_2(Q_T)$, and (V, p) satisfy all the conditions of problem (6.4)."

Thus, Ladyzhenskaya's above theorem guarantees that the solution V has generalized derivatives V_x, V_{xx}, and V_t in $L_2(Q_T)$, and that V and the corresponding pressure, *p*, satisfy the Navier system for almost all $(x,\ t) \in Q_T$.

It is clear that under the given conditions on f and V° the solution cannot be smoother. However, if the smoothness of f and V° increases, then V and p also become smoother, as already follows from the results of the linear theory described by Ladyzhenskaya (1961) [112] (English translation). Several mathematicians have essentially repeated this theorem – but unfortunately without any reference to Ladyzhenskaya!

The significance of Ladyzhenskaya's theorem is indisputable. It provides a good guideline for those who wish to try themselves in solving the "sixth problem of the millennium", announced by the Clay Mathematics Institute of Cambridge.[3]

However that may be today, despite tremendous progress in many mathematical aspects of fluid flow theory since Leray's pioneering thesis/paper of 1933, it is necessary to be just a little more modest if we have in mind the (large) number of problems that still remain open! For instance, we have not determined, until now (in the framework of Navier problem), whether a solution that is initially smooth can develop a singularity at some (finite) later time, or whether singularities are a fundamental feature of turbulence.

In spite of the fact that Navier viscous and incompressible equations are a very simplified form of full unsteady NS–F equations, they are useful because they describe the physics of many things of academic and economic interest. They may be used to model the weather, ocean currents, water-flow in a pipe, air-flow around a wing, the motions of stars in a galaxy, and so on. Despite their simplified (relative to NS–F equations for real compressible and heat-conducting fluid flow) forms, they help with the design of aircraft and cars, the study of blood flow, the design of power stations, the analysis of pollution, and many other things. Coupled with Maxwell's equations, they can be used to model and study magnetohydrodynamics.

The Navier equations are also of great interest in a purely mathematical sense. Somewhat surprisingly, given their wide range of practical uses, mathematicians have not yet proven that (what are called Navier existence and smoothness problems) "in three dimensions solutions always exist (existence), or that if they do exist, then they do not contain any singularity (smoothness)."

The Clay Mathematics Institute has called this one of the seven most important open problems in mathematics, and has offered a prize of $1 million for a solution or a counter-example. The official problem description is provided by Ch. L. Fefferman on the website at http://www.Claymath.org/millennium/.

As far as the numerical solution of the unsteady incompressible Navier–Stokes (Navier) equations are concerned, Quartapelle [113] presents a unitary view of the

[3] In *Russian Mathematical Surveys*, vol. 58(2) (2003), pp. 251–86, there is a paper by O. A. Ladyzhenskaya – "The sixth problem of the millennium: Navier–Stokes equations, existence, and smoothness" – which presents the main results concerning solubility of the basic initial-boundary value problem and the Cauchy problem for the three-dimensional non-stationary Navier (incompressible) equations, together with a list of what to prove in order to solve the sixth problem of the "seven problems of the millennium" proposed on the Internet, at http://www.Claymath.org/.

methods which reduce the equations for viscous incompressible flows to a system of second-order equations of parabolic and elliptic type. Concerning the rigorous mathematical results for Navier–Stokes compressible equations for viscous and heat-conducting fluid flows (denoted, in this book, as "Navier–Stokes–Fourier", NS–F, equations), the mathematically rigorous theory is actually still in its infancy. In our (2004) [47], Chap. 8, the reader can find more recent references concerning the published papers and books devoted to NS–F equations. In 2004, two books were published by Feireisl [114] (dynamics of viscous compressible fluids) and Novotny and Straskraba [115] (introduction to the mathematical theory of compressible flow).

Actually, it is firmly established that we are able to prove an existence theorem for compressible viscous (but baroptropic N–S equations) fluid flows, which is global in time, without assuming smallness of the data, although uniqueness is an open problem (the solution is "weak"). On the other hand, global existence of a "strong" (each term in N–S equations exists as an element of Hilbert space, at least) and regular solution is proven, but assuming that the data are small enough and that in such a case the uniqueness hold.

6.3 Formulation of Dimensionless Equations for Applications of the RAM Approach

The non-dimensional approach of NS–F equations, realized in Chap. 3, is a first key step in our Mathematics for a possible application of the RAM Approach, thanks to the presence – in these dimensionless equations and conditions of the starting problem, at least – of a small (or high) parameter.

If this non-dimensionalization does not often an easy approach, the judicious formulation of dimensionless equations is a very decisive step for the application of the RAM Approach. Below I present a few examples of such distinctive formulations for some basic fluid flow problems.

6.3.1 Turbomachinery Fluid Flows

First, one cannot escape the fact that in a turbomachine the blades of a row are usually very closely spaced. As a consequence, as basic small parameter ε, we choose the reciprocal of the number,

$$N = (2\pi/\varepsilon) >> 1, \tag{6.6}$$

of blades encountered along the periphery of a row. This small parameter is a geometrical one, and does not appear in fluid dynamics equations. In a very simple case, for a 3D steady incompressible non-viscous fluid flow through a row in the turbomachine (an axial compressor, assuming that the row is localized between two infinitely long coaxial cylinders).

Obviously, when far upstream of the row, velocity is uniform and constant and we have irrotationality, and the basic equations are very simple. For the velocity vector $\mathbf{u}(u, v, w)$, we write:

$$\nabla.\mathbf{u} = 0, \tag{6.7}$$

and

$$\nabla \wedge \mathbf{u} = 0, \tag{6.8}$$

where

$$\nabla = (\partial/\partial r)\mathbf{e}_r + (1/r)(\partial/\partial\theta)\mathbf{e}_\theta + (\partial/\partial z)\mathbf{e}_z, \tag{6.9}$$

in a system of (r, θ, z) coordinates, such that the axis z is the axis of the turbomachine. The row is between $r = R°$ and $r = r°$. If we want to use the smallness of the $\varepsilon = 2\pi/N$, in an asymptotic theory, obviously it is necessary to make a change of variables, such that $\varepsilon \ll 1$ appears in the above two Eqs 6.7 and 6.8. But, at once, we observe that Eq. 6.7 is integrated if we write for the velocity vector $\mathbf{u}$:

$$\mathbf{u} = \nabla(\varepsilon\chi) \wedge \nabla\psi \;\; \text{with} \;\; \mathbf{u}.\nabla\chi = 0 \;\text{and}\; \mathbf{u}.\nabla\psi = 0, \tag{6.10}$$

and the two stream functions χ and ψ are of order one (the output in a channel between two consecutives blades being of the order ε).

We assume that the surfaces

$$\theta = \Theta(t, r, z) \tag{6.11}$$

are the blade skeletons in a row, when $\varepsilon \downarrow 0$, and outside of the row are material surfaces that are the extensions of blade skeletons.

By

$$\Delta(r, z), \tag{6.12}$$

we denote the breadth of the channel from blade to blade, with $\Delta(r, z) \equiv 1$ outside the row. In the row, between two consecutives blades, we can write:

$$\theta = \Theta(r, z) + (\varepsilon/2)\Delta(r, z), \tag{6.13a}$$

and

$$\theta = \Theta(r, z) - (\varepsilon/2)\Delta(r, z). \tag{6.13b}$$

As a consequence of the above, with an error (at most) $O(\varepsilon^2)$, we have the possibility, for $\varepsilon<<1$, to write in the blade-to-blade channel the following change for θ:

$$\theta \approx \Theta(t, r, z) + \varepsilon\Delta(r, z)\chi \Rightarrow \varepsilon\chi \approx [\theta - \Theta(t, r, z)]/\Delta(r, z) \tag{6.14}$$

This relation allows us to write, in place of (6.10), for the velocity vector $\mathbf{u}(u, v, w)$:

$$\mathbf{u} \approx (\nabla[\theta - \Theta(t,r,z)]/\Delta(r,z)) \wedge \nabla\, \psi(r, \Theta, z), \tag{6.15}$$

the equation $\nabla.\mathbf{u} = 0$, and the slip condition, for a non-viscous fluid on the blade, being both automatically satisfied with (6.15).

Now, if we make the change from (r, θ, z) to $(r, \Theta(t, r, z) + \varepsilon\Delta(r, z)\chi, z)$, then:

$$\mathbf{u}(r, \theta, z) \Rightarrow \mathbf{u}^*(r, \chi, z; \varepsilon), \quad \psi(r, \theta, z) \Rightarrow \psi^*(r, \chi, z; \varepsilon), \tag{6.16a}$$

and

$$\partial\psi/\partial r = \partial\psi^*/\partial r - (1/\Delta)(\partial\psi^*/\partial\chi)\ [(1/\varepsilon)\partial\Theta/\partial r + \chi\partial\Delta/\partial r], \tag{6.16b}$$

$$\partial\psi/\partial z = \partial\psi^*/\partial z - (1/\Delta)(\partial\psi^*/\partial\chi)[(1/\varepsilon)\partial\Theta/\partial z + \chi\partial\Delta/\partial z], \tag{6.16c}$$

$$\partial\psi/\partial\theta = (1/\varepsilon\Delta)(\partial\psi^*/\partial\chi) \tag{6.16d}$$

and also, from (6.15):

$$u^* = (1/r\Delta)\partial\psi^*/\partial z, \quad w^* = -(1/r\Delta)\partial\psi^*/\partial r \tag{6.17a,b}$$

$$v^* = ru^*[\partial\Theta/\partial r + \varepsilon\chi\partial\Delta/\partial r] + rw*[\partial\Theta/\partial z + \varepsilon\chi\partial\Delta/\partial z]. \tag{6.17c}$$

Finally, it is necessary to use the equation (6.8), $\nabla \wedge \mathbf{u}^* = 0$, for a derivation of well-adapted non-dimensional starting equations for our RAM Approach to turbo-machinery flow.

Taking into account the above relations (we observe that $r\, v^* = \Gamma^*$ is the circulation), we derive, in place of (6.8), as a starting non-dimensional system of equations (unless any simplification, for an application, in Sect. 7.1), for u^*, Γ^* and w^*, the following system of non-dimensional equations (where the small parameter ε is present):

$$\begin{aligned}(1/r\Delta)[\partial w^*/\partial\chi + (\partial\Theta/\partial z)\partial\Gamma^*/\partial\chi] \\ = (\varepsilon/r)\partial\Gamma^*/\partial z - (\varepsilon/r)(\chi/\Delta)(\partial\Delta/\partial z)\partial\Gamma^*/\partial\chi\end{aligned} \tag{6.18a}$$

$$(1/r\Delta)[\partial u^*/\partial\chi + (\partial\Theta/\partial r)\partial\Gamma^*/\partial\chi] \\ = (\varepsilon/r)\partial\Gamma^*/\partial r - (\varepsilon/r)(\chi/\Delta)(\partial\Delta/\partial r)\partial\Gamma^*/\partial\chi \tag{6.18b}$$

$$(1/\Delta)[(\partial\Theta/\partial z)\partial u^*/\partial\chi - (\partial\Theta/\partial r)\partial w^*/\partial\chi] \\ = \varepsilon(\partial u^*/\partial z - \partial w^*/\partial r) \\ + \varepsilon(\chi/\Delta)[(\partial\Delta/\partial r)\partial w^*/\partial\chi - (\partial\Delta/\partial z)\partial u^*/\partial\chi] \tag{6.18c}$$

(Concerning my work with Guiraud, related to various facets of an asymptotic theory for the turbomachinery fluid flow, see some commentaries in [11], pp. 16 and 17.)

In Sect. 7.1 the above system of Eqs. 6.18a–c is analyzed in the case when ε tends to zero, and a model problem is derived for the through-flow model.

6.3.2 *The G–Z "Rolled-up Vortex Sheet" Theory: Vortex Sheets and Concentrated Vorticity*

Vortex sheets are one of the basic ingredients of non-viscous fluid flows, which can be viewed as vorticity concentrated on a core of small but finite diameter. They can be considered as two-dimensional surfaces of zero thickness carrying a truly concentrated vorticity.

We clearly see that as far as a compressible fluid is concerned, concentrated vorticity is not the entire matter. We should add a concept of concentrated baroclinity vector $\boldsymbol{B} = \nabla S$, where S is entropy, by considering that along any surface orthogonal to $\boldsymbol{B}$, variation of p and ρ, at constant time, are related by a barotropic relation. Then, by writing

$$B = B^{\mathrm{S}} + \delta_\Sigma|[S]|\mathbf{n} = B^{\mathrm{S}} + B^{\mathrm{C}} \tag{6.19a}$$

we exhibit a concentrated baroclinicity $\boldsymbol{B}^{\mathrm{C}}$ ($\delta_\Sigma$is the Dirac distribution uniformly spread over Σ and $|[S]|$ the jump in S when Σ is crossed in the sense of unit normal **n**) which is orthogonal to the concentrated vorticity, and we may state that

$$|[1/\rho]| = (\partial\check{T}/\partial p)|[S]| \tag{6.19b}$$

where $\partial\check{T}/\partial p$ is calculated at p, and $\check{S} = \alpha\, S^+ + (1 - \alpha)\, S^-$ with $0< \alpha<1$.

The situation that we want to describe is one in which there are very many vortex sheets closely spaced. We may mention two flow configurations in which this occurs. The first one is the core of a highly rolled vortex sheet, while the second concerns the set of trailing vortex sheets which are formed at the trailing edges of the blades in a row of an axial turbomachine.

We assume that there is a small parameter built into the flow, which is the ratio of the spacing between two consecutive sheets to the width of the region covered by the sheets. We set C for this parameter, and following our G-Z paper (1977) [116] we call it a "closeness parameter".

The purpose of the RAM Approach, in this situation, is to derive a model which avoids the stiffness of the problem, for a computation of the flow with a numerical code capable of capturing many sheets.

The problem is one of multiple scaling, in the terminology of asymptotics. A version of the multiple-scale technique (see Sect. 6.4.4) especially suited for this kind of problem was devised with Guiraud with the purpose of describing rolled vortex sheets, but its scope is more general, as may be seen from the analysis which follows.

We start from the assumption that the velocity **u**, pressure p, density ρ, and entropy S – all suitably non-dimensional – are functions of time t, position **x**, and of one fast variable $C^{-1}\ \chi(\mathrm{t},\ \mathrm{x})$, with $\nabla\chi$ approximately orthogonal to the sheets. We use the notation:

$$U = (\mathbf{u},\ \mathrm{p},\ \rho, S)^{\mathrm{T}} \tag{6.20a}$$

and set

$$U(\mathrm{t},\mathbf{x}) = U^*(\mathrm{t},\mathbf{x}, C^{-1}\chi(\mathrm{t},\mathbf{x})) \tag{6.20b}$$

There are two ingredients in the technique used. The first one is a formal expansion,

$$U^* = U_0^* + CU_1^* + \ldots \tag{6.20c}$$

while the second one is the obvious observation that setting

$$C^{-1} = \partial\chi/\partial\mathrm{t} \ = \ \theta_0 + C\ \theta_1 + \ldots \tag{6.21a}$$

$$\boldsymbol{k} = \nabla\chi = \mathbf{k}_0 + C\mathbf{k}_1 + \ldots \tag{6.21b}$$

we have

$$\partial U/\partial\mathrm{t} \ = C^{-1}\ \theta(\partial U/\partial\chi) + \partial U^*/\partial\mathrm{t} \tag{6.22a}$$

$$\nabla\mathrm{U} \ = \mathrm{C}^{-1}(\partial\mathrm{U}/\partial\chi)\mathbf{k} + \nabla\mathrm{U}^* \tag{6.22b}$$

Substituting the above relations (6.20c)–(6.22b) into the equations of motion for a compressible non-viscous fluid, we obtain, at zeroth order, a set of equations from which we conclude that, provided $\chi =$ const is not a Mach wave, we must have that

$$\theta_0 + \mathbf{k}_0.\mathbf{u}_0^* \text{ does not depend on } \ \chi.$$

On the other hand, by specifying that χ is constant on the vortex sheets we obtain

$$\theta_0 + \mathbf{k}_0.\mathbf{u}_0^* = 0 \tag{6.23a}$$

and as a consequence

$$\mathbf{k}_0.\partial\mathbf{u}_0^*/\partial\chi = \partial p_0^*/\partial\chi \tag{6.23b}$$

Then, for the vorticity we obtain

$$\boldsymbol{\omega}^* = C^{-1}\mathbf{k} \wedge (\partial\mathbf{u}^*/\partial\chi) + \nabla \wedge \mathbf{u}^*$$

so that if we ensure that $\boldsymbol{\omega}^*$ is O(1), then, as was the case in the study of rolled-up vortex sheets, we have the relation:

$$\mathbf{k}_0 \wedge \left(\partial\mathbf{u}_0^*/\partial\chi\right) = 0$$

which in turn enforces that (for density and specific entropy):

$$\partial\rho{*}_0/\partial\chi = 0 \text{ and } \partial S_0^*/\partial\chi = 0 \tag{6.23c}$$

Now, the model of rolled vortex sheets is recovered by assuming that U^* is 2π-periodic with respect to χ. Then, by writing the equations of motion and for vorticity at order one, it may be concluded that $\mathbf{V} = \mathbf{k}_0 \wedge \mathbf{u}^*_1$ and $\Sigma = S^*_1$ are solutions of the following two equations:

$$\begin{aligned}\partial\mathbf{V}/\partial t &+ \mathbf{u}_0^*.\nabla\mathbf{V} - \mathbf{V}.\nabla\mathbf{u}_0^* + (\nabla.\mathbf{u}_0^*)\mathbf{V}\\ &+ \left\{\left[\mathbf{k}_0.(\nabla \wedge \mathbf{u}_0^*)\right]/|\mathbf{k}_0|^2\right\}(\mathbf{k}_0 \wedge \mathbf{V})\\ &\quad + (1/\gamma M^2)\{[\mathbf{k}_0 \wedge \nabla p_0^*]/\gamma\, \rho_0^*\}\,\Sigma = 0\end{aligned} \tag{6.24a}$$

$$\partial\Sigma/\partial t + \mathbf{u}_0^*.\nabla\Sigma + \{[\mathbf{k}_0 \wedge \nabla S_0^*]/|\mathbf{k}_0|^2\}.\mathbf{V} = 0 \tag{6.24b}$$

The important point is that χ does not occur in (6.24a, b). They are a set of ordinary differential equations along the trajectories of the velocity field at zeroth order approximation $\mathbf{u}^*_0$. Via (6.24a, b), a number of configurations are allowed with this description.

In our paper with Guiraud (1982) [117], two of them are presented. The first is an extension of the Kaden problem, and the second concerns Mangler and Weber's solution.

We observe that the closeness parameter C is related – often, but not always – to by $C = \sigma^2$, to a 'slenderness' parameter (σ) with respect to which $\mathbf{u}^*_0$ may be expanded. However, we emphasize here that obtaining a closed form solution should not be considered as the ultimate goal of the analysis. In our opinion, the analysis should rather be considered as a model allowing us to generate a flow with closely spaced vortex sheets, rather than another one which does not involve such a feature!

Again concerning my work with Guiraud relating to various facets of an asymptotics of rolled vortex sheets, see some commentaries in [11], pp. 18–19.

6.3.3 The G–Z Asymptotic Approach to Non-linear Hydrodynamic Stability

The problem of hydrodynamic stability may be posed in the following terms.

Let u_0 be a particular steady (independent of time t) solution of Navier incompressible and viscous equations, written as:

$$S\mathrm{du}/\mathrm{dt} \;=\; \mathrm{F(u))} \text{ and such that } \mathrm{F}(u_0) = 0 \tag{6.25a}$$

If $\mathrm{u} = u_0 + \mathrm{v}$, in another solution, then the problem of the stability of the solution u_0 is strongly related with: whether or not v remains small all the time, provided that it is small initially?

Obviously, the stability has a direct relation with the third condition, (c), in Hadamard well-posedness (mentioned at the beginning of Sect. 6.2). For the investigation of such a stability, we consider the following evolution equation for v:

$$\mathrm{dv/dt} \;=\; \mathrm{L}(u_0)\mathrm{v} \,+\, \mathrm{Q(v,\ v)},\ \text{with } \mathrm{u} = u_0 + \mathrm{v} \tag{6.25b}$$

where $\mathrm{L}(u_0)$ is a linear operator, depending on u_0, and Q(v, v) is a quadratic operator, such that $\mathrm{Q}(\alpha v,\ \alpha\mathrm{v}) = \alpha^2\mathrm{Q(v,\ v)},\ \alpha =$ const. But for any v and w:

$$\mathrm{Q(v\,+\,w,\ v\,+\,w)} = \mathrm{Q(v,\ v)} + \mathrm{M(v,\ w)} + \mathrm{Q(w,\ w)}$$

where M(v, w) is bilinear and is a function of two variables.

We observe that if we take the inner product $<\cdot;\cdot>$ of (6.25b) with v, we derive from (6.25b):

$$\partial(|\mathrm{v}|^2/2)/\partial\mathrm{t} \;= <\mathrm{L}(u_0)\mathrm{v};\mathrm{v}> \tag{6.26a}$$

because $<\mathrm{Q(v,\ v)};\mathrm{v}> \equiv 0$, in the case of the Navier problem.

If we now let

$$\gamma = \inf\{<\mathrm{L}(u_0)\mathrm{v};\mathrm{v}>/\left[|\mathrm{v}|^2/2)\right]\},\ \text{subject to the } \textit{constraint},\ \mathrm{div\ v} \;= 0 \text{ and } \mathrm{v} \;= 0 \text{ on } \partial\mathrm{D} \tag{6.26b}$$

This above variational problem (6.26b) is similar to the classical variational characterization of the first eigenvalue of the Laplacian. Consequently, from (6.26a), we can write:

$$\partial(|v|^2/2)/\partial t + \gamma\, |v|^2 \leq 0$$

Integrating this inequality, we obtain:

$$|v(t)| \leq |v(0)| \exp(-2\gamma t), \text{ for } \gamma > 0$$

u_0 is unconditionally stable

6.3.3.1 Confined Perturbations and the Landau–Stuart Equation

We assume that u is defined on a bounded spacial domain and, as a consequence, $L(u_0)$ has (one may prove) a discrete spectrum; that is, there exists an infinite sequence of eigenvalues, in fact complex: σ_n, $n = 1, 2, \ldots$ with associated eigenfunctions v_n, such that

$$L(u_0)v_n = \sigma_n, \text{ and the equation } L(u_0)v - \sigma = f \tag{6.27}$$

with a given f, is uniquely solvable provided σ is distinct from all σ_n.

Let $\sigma_n = \lambda_n + i\mu_n$, and we are interested here mainly by the unstable case but restrict ourselves to "weak instability". For this, we partition the indexes n into two parts N and A (according to Guiraud (1980) [118]):

$$n \in N \Rightarrow \lambda_n = O(\varepsilon^q) \text{ and } n \in A \Rightarrow \lambda_n < 0 \tag{6.28}$$

where ε is a small parameter and q an exponent that is introduced here for later convenience.

We must comment on the partition (6.28), which we have introduced as working hypothesis to be checked in each particular case.

The flow u_0 depends on parameters (for example, a Reynolds number Re), and when $Re < Re_c$, the flow configuration is stable, whereas it is unstable for $Re > Re_c$.

By continuity, this corresponds to the fact that when Re crosses Re_c increasing, some eigenvalues σ_n, those corresponding to $n \in N$, cross the imaginary axis from left to right.

When Re is close to Re_c on either side, the

$$\lambda_n \text{ corresponding to indexes } n \in N \text{ are small}$$

and this is expressed through the first relation of (6.28). The most frequent are those when N has just one element with σ_n real or two elements with a complex conjugate of σ_n, $\sigma_n{}^*$.

We now start from a normal-mode decomposition, and according to above partition (6.28),

$$v = \Sigma_{n \in N} C_n v_n + \Sigma_{n \in A} C_n v_n = X + Y \qquad (6.29a)$$

and we set, $X = P^* v$, $Y = P^{**} v$, with P^* and P^{**} two projection operators. Then, from (6.28), we see that

$$n \in N \Rightarrow \sigma_n = \varepsilon^q \lambda_{n^0} + i\eta_n \qquad (6.29b)$$

and taking (6.29a) into account, we obtain the following system of two "bifurcation-type" equations for X and Y in place of non-linear evolution equation (6.25b):

$$dX/dt = \varepsilon^q L_1 X + L_2 X + P^* Q(X + Y, X + Y) \qquad (6.30a)$$

$$dY/dt = L_A Y + P^{**} Q(X + Y, X + Y). \qquad (6.30b)$$

It may be stated that

$\varepsilon^q L_1$ corresponds to $\varepsilon^q \lambda_{n^0}$

L_2 to $i\eta_n$

both for $n \in N$, and

L_A corresponds to σ_n

for $n \in A$.

Below, we shall concentrate only on the case when

$$P^* Q(X, X) = 0$$

and in such a case, when we ignore the transient unsteady-state phase $t = O(1)$, to concentrate on the non-linear phase, for long time, and use a multiple-scale technique with two times, t and

$$t = \varepsilon^q t, \; X = \varepsilon^r X^* \text{ and } Y = \varepsilon^{2r} Y^*. \qquad (6.31)$$

The adequate choice being $q = 2$ and $r = 1$, from (6.30a, b) we obtain for X* and Y* the following system (with two times):

$$\partial X^*/\partial t - L_2 X^* = -\varepsilon^2 [\partial X^*/\partial \tau - L_1 X^* - P^* M(X^*, Y^*)] + O(\varepsilon^3) \qquad (6.32a)$$

$$\partial Y^*/\partial t - L_A Y^* = P^{**} Q(X^*, X^*) + O(\varepsilon) \qquad (6.32b)$$

with

$$X^* = X^{*0} + \varepsilon X_1^* + \varepsilon^2 X_2^* + \ldots;\ Y^* = Y_0^* + \varepsilon Y_1^* + \ldots \tag{6.33}$$

we obtain from (6.32a, b) the following leading-order equations:

$$\partial X_0^*/\partial t - L_2 X_0^* = 0; \partial X_1^*/\partial t - L_2 X_1^* = 0, \tag{6.34a}$$

$$\partial Y_0^*/\partial t - L_A Y_0^* = P^{**} Q(X_0^*, X_0^*) \tag{6.34b}$$

$$\partial X_2^*/\partial t - L_2 X_2^* = -\partial X_0^*/\partial t + L_1 X_0^* - P^* M(X_0^*,\ Y_0^*) \tag{6.34c}$$

The solution for X^*_0 is written in the following form :

$$X_0^* = \Sigma_{n\in N} X_{0,n}^*(\tau)\exp(i\eta_n t)$$

and for Y^*_0 we obtain as a solution :

$$Y_0^* = \Sigma_{p\in N}\Sigma_{q\in N} Y_{0,pq}^*(\tau)\exp[i(\eta_p + \eta_q)t]$$

where $Y^*_{0,\ pq}(\tau)$ is a function of $X^*_{0,\ p}(\tau)$ and $X^*_{0,\ q}(\tau)$ and L_A.

It is now necessary to consider Eq. 6.34c for X^*_2, seeing that we want to determine the dependence of $X^*_{0,\ n}(\tau)$ relative to the long time τ. With the above relations, in place of (6.34c), we have the following equation for X^*_2:

$$\begin{aligned}\partial X_2^*/\partial t - L_2 X_2^* &= -\Sigma_{n\in N}[\partial X_{0,n}^*/\partial t - L_1 X_{0,n}^*]\exp(i\eta_n t)\\ &- \Sigma_{p\in N}\Sigma_{q\in N}\Sigma_{r\in N} P^* M\left(X_{0,p}^*, Y_{0,qr}^*\right)\exp[i(\eta_p + \eta_q + \eta_r)t]\\ &= 0\end{aligned} \tag{6.35}$$

In the rather complicated Eq. 6.35, according to the multiple-scale technique (see, for instance, Sect. 6.4), it is necessary to eliminate secular terms in solution for X^*_2, if we want the expansion (6.33) to be uniformly valid with respect to t and also τ, ignoring the transition phase. The result of this elimination is the system of ordinary differential Eq. 6.36, which determines the coefficients $X^*_{0,n}(\tau)$:

$$dX_{0,n}^*/d\tau - L_1 X_{0,n}^* = \Sigma_{(R_n)} P^* M\left(X_{0,n}^*, Y_{0,qr}^*\right) \tag{6.36}$$

In (6.36), $\Sigma_{(Rn)}$ represents the sum over resonant (p, q, r) triplets, where R_n is a resonance condition:

$$R_n \Rightarrow \eta_p + \eta_q + \eta_r = \eta_n$$

The above derived Eq. 6.36 is, in fact, very similar to an *à la* Landau equation.

When N contains only two elements, the corresponding σ_n being complex conjugates, X* is well determined by a single complex amplitude $S(\tau)$, which must be a solution of the Landau–Stuart equation:

$$dS(\tau)/d\tau - \sigma S - KS|S|^3 = 0. \tag{6.37}$$

6.3.3.2 Unconfined Perturbations and the DH–S Evolutionary Amplitude Equation

We observe that when dealing with modes spread over a continuum, the extension of the Stewartson and Stuart (1971) [119] technique to the whole of temporal evolution is not at all obvious, and the difficulty has its root in the fact that, to our knowledge, no standard mathematically rigorous technique exits for dealing with bifurcation from a continuous spectrum!

In (1978) [120], Guiraud and Zeytounian succeeded in filling this gap, in a purely formal way, for continuously spread Tollmien–Schlichting (T–S) waves that exhibit close similarity to the technique used above for discrete modes. For example, in [120] we have, with Guiraud, elucidated that: "The process by which, from a fairly arbitrary initial perturbation, the wave packet is first organized and then evolves, is related to four time scales of evolution."

The first one, of O(1) duration, is devoted to the decay of all but the amplified modes, and the second phase, of much longer duration, $O(1/\varepsilon)$, is a passive one with respect to the organization of the amplitude of the perturbation. The wave packet is dominated by the most amplified of the T-S wave, and is convected with the group velocity associated with the packet and the amplitude which is $O(\varepsilon^4)$, practically unchanged. During the third phase of duration $O(1/\varepsilon^2)$ the amplitude is modulated according to an exponential law which predicted by linear theory. Finally, it is only during the last period of duration $O(|\mathrm{Ln}\varepsilon|/\varepsilon^2)$ that non-linear effects come into play, leading eventually to bursting, and to the well-known envelope, evolutionary, equation of Davey, Hocking, and Stewarston (DH–S), discovered in 1974. Thus we derive, for a leading-order amplitude $A(\tau)$, the following DH–S equation:

$$\partial A/\partial t - \gamma A - (1/2)[\alpha\partial^2 A/\partial\xi^2 + \beta\partial^2 A/\partial\eta^2] = F(A) \tag{6.38a}$$

with

$$F(A) = k_1 A\,|A|^2 + k_2 AB, \tag{6.38b}$$

where k_1 and k_2 are constants, and B is a function solution of the equation:

$$\partial^2 B/\partial\xi^2 + \partial^2 B/\partial\eta^2 = \partial^2(|A|^2)/\partial\eta^2. \tag{6.38c}$$

In our (2004) [47], Chap. 9, the reader can find some aspects of the stability theory of viscous fluid flow and, in particular, a detailed presentation of our (with Guiraud) weakly asymptotic non-linear stability theory of Navier incompressible fluid flow.

In De Coninck, Guiraud, and Zeytounian (1983) [121], a kind of unified theory is presented in which the purely discrete and the continuous cases are treated by two facets of a somewhat unique technique. Our main goal is the derivation of the equations that rule the evolution of the amplitude of the most rapidly amplified modes of linear theory.

A slightly different asymptotic modelling is applied, in [121], to convective Rayleigh–Bénard instability. The distinguishing feature of this type of instability, with respect to asymptotic modelling, is that the area of wave vectors for amplified modes is within a circular annulus of small thickness. Our results show that instead of finding a unique packet of waves one finds six of them organized around the vertices of a hexagon which may interact quadratically with each other. As a matter of fact, one may superpose an infinity of such hexagonal figures which may evolve, to lowest order, independently.

6.3.4 A Local Atmospheric Thermal Problem: A Triple-Deck Viewpoint

In the framework of an atmospheric thermal problem, on a flat ground surface, $z' = 0$, we have (in a 2D steady case) as a typical boundary condition:

$$T'/T'^*(0) = 1 + \tau^\circ\Theta(x'/l^\circ), |x'/l^\circ| \leq 1 \tag{6.39a}$$

with a temperature parameter

$$\tau^\circ = (\Delta T')^\circ/T'^*(0) \tag{6.39b}$$

which is assumed a small ($<<1$) parameter, where $(\Delta T')^\circ$ is a temperature rate for the given function $\Theta(x'/l^\circ)$.

In (6.39a), $T^*(0)$ is the temperature (at $z'^* = 0$) in the hydrostatic reference state (dependent only on the vertical coordinate z'^*), and l° is the local horizontal length scale.

Far upstream, when $x' \to \infty$ and $\Theta \equiv 0$, we assume that we have a basic undisturbed flow which is characterized by an Ekman layer profile[4]:

$$U_{Ek}(X, z/k^\circ) = U'_G(X)\{1 - \exp(-z/\kappa^\circ)\cos(-z/\kappa^\circ)\}, \tag{6.40a}$$

---

[4] In Section 9.2, in the framework of the "quasi-hydrostatic dissipative model", the reader can find the derivation of the "geostrophic relation" and also "Ackerblom's model problem".

where $X = x'/L^\circ$ and $z = z'/l^\circ$, with

$$\kappa^\circ = (1/l^\circ)[\Omega^\circ \sin\phi^\circ/\nu^\circ]^{-1/2} \equiv [Re_{l^\circ}/2Ro_{l^\circ}] \tag{6.40b}$$

the local Reynolds and Rossby numbers being based on the local horizontal length scale: $Re_{l^\circ} = U^\circ\, l^\circ/\nu^\circ$ and $Ro_{l^\circ} = (U^\circ/l^\circ)/\Omega^\circ \sin\phi^\circ$.

If in the local thermal problem we non-dimensionalize the horizontal and vertical coordinates with l° – for example $\Theta(x'/l^\circ) = \Theta(x)$ – then in the dimensionless local problem there also appears a local Boussinesq number, $Bo_{l^\circ} = g\, l^\circ/RT'^*(0)$, and if $l^\circ \approx 10^3$m, then $Bo_{l^\circ} << 1$, and in such a case we also have $Re_{l^\circ} >> 2Ro_{l^\circ} >> 1$; κ° being, below, our main small parameter (see the local Eqs. 6.42a–e).

Therefore, in this case we can assume:

$$2Ro_{l^\circ} = [Re_{l^\circ}]^{-1/a} \Rightarrow \kappa^\circ = [Re_{l^\circ}]^{-1/m} \tag{6.41a}$$

with $(0 < a < 1)$

$$m = [(2 - a)/(1 - a)] > 2 \tag{6.41b}$$

As an example, if $U^\circ \approx 10\text{m/sec}$, $\nu^\circ \approx 5\text{m}^2/\text{sec}$, and $f^\circ = 2\,\Omega^\circ \sin\phi^\circ \approx 10^{-4} 1/\text{sec}$, the considered case, $l^\circ \approx 10^3$m, leads to $m = 5$. For such a case, we have the possibility of using (M is the mach number):

$$l^\circ \approx (U^\circ/g)[RT'^*(0)/\gamma\,]^{1/2} \Rightarrow Bo_{l^\circ}/M = B^* \equiv 1 \tag{6.41c}$$

and the Boussinesq approximation is correct.

Curiously, the value $m = 5$ is the same as the one used by Smith and co-workers for the flow over an isolated 2D short hump in the boundary layer.

The Boussinesq stratified fluid flow is also used by Sykes (1978) as a starting equations for the application of a triple-deck theory. When $m = 5$ we can show that a typical triple-deck case exists:

$$l^\circ/L^\circ \approx (Re_{L^\circ})^{-3/8} \text{ where } Re_{L^\circ} = L^\circ U^\circ/\nu^\circ \tag{6.41d}$$

but in our LNP (1987) [17], pp. 211–20, the reader can find a more general approach. Also in [19], Sect. 31, there are various references concerning the application of the triple-deck theory in various meteo and environmental problems.

Now, according to the Boussinesq approximation, taking into account the above relations, we have the possibility of formulating the following dimensionless local steady 2D thermal problem, for the local velocity components (u, w), and thermodynamic perturbations θ and π:

$$u\partial u/\partial x + w\partial u/\partial z + (1/\gamma)\partial\pi/\partial x = k^{o5}(\partial^2 u/\partial x^2 + \partial^2 u/\partial z^2) \tag{6.42a}$$

$$\begin{aligned} u\partial w/\partial x + w\partial w/\partial z + (1/\gamma)\partial\pi/\partial z - (1/\gamma)B^*\,\theta \\ = k^{o5}(\partial^2 u/\partial x^2 + \partial^2 u/\partial z^2) \end{aligned} \tag{6.42b}$$

$$\partial u/\partial x + \partial w/\partial z = 0 \tag{6.42c}$$

$$u\partial\theta/\partial x + w\partial\theta/\partial z \; + \; B^{*}\Lambda(0)\; w = (1/\mathrm{Pr})k^{o5}(\partial^{2}\theta/\partial x^{2} + \partial^{2}\theta/\partial z^{2}) \tag{6.42d}$$

$$\omega = -\; \theta, \tag{6.42e}$$

when we perform the following limiting process:

$$\tau^{\circ} \to 0 \text{ with } M \to 0 \text{ such that } \; \tau^{\circ}/M \; = \tau^{*} \approx 1 \tag{6.43a}$$

which gives the possibility of writing a relation for the reference velocity U°: namely,

$$U^{\circ} \approx (\Delta T)^{\circ}[\gamma R/T'^{*}(0)]^{1/2}. \tag{6.43b}$$

For our above local system of Eqs. 6.42a–e as boundary conditions, we first write the usual non-slip and temperature conditions:

$$z \; = 0: \; u = w = 0, \quad \theta = \; \tau^{*}\Theta(x), \quad 0 < x < 1 \tag{6.44a}$$

because, in dimensionless Eqs. 6.42a–c, according to Boussinesq approximation, we have :

$$T' = T'(z'^{*})[1 + \; M\theta] \text{ and } p' = p'(z'^{*})\left[1 + M^{2}\,\pi\right].$$

A second condition for the local problem is an interaction condition between the local thermal spot and the Ekman atmospheric layer (see (6.40a, b)) far upstream:

$$x \; \to -\; \propto: \; u \; \to 1 - \exp(-z/\kappa^{\circ})\cos(-z/\kappa^{\circ}) = U^{\propto}(z/\kappa^{\circ}) \tag{6.44b}$$

$$w \; = \; \pi = \; \theta \to 0 \tag{6.44c}$$

and we observe also that:

$$\text{if } z/\kappa^{\circ} \to \propto, \text{ then } u \; \to 1, \text{ for } x \; \to -\propto \tag{6.44d}$$

and

$$\text{if } z/\kappa^{o} \to 0, \text{ then } u \; \approx \; z/\kappa^{\circ}, \text{ for } x \; \to -\propto \tag{6.44e}$$

A detailed account the triple-deck theory is included in our (2002) [26], Chap. 12.

In Sect. 6.4.6 we return to the above formulated problem –(6.42a)–(6.42e), with (6.44a)–(6.44e) – in the framework of the triple deck theory.

6.3.5 Miscellanea

We now enumerate a few particular fluid flow problems showing the ubiquitous nature of the small parameter. A first example, illustrating the occurrence of the small parameter in the boundary conditions, concerns high-aspect-ratio wings. Here the small parameter is the inverse of this aspect ratio, and it occurs in the model when one enters into the details of the no-slip condition on the wing.

A more subtle occurrence concerns the nature of the domain of flow. The best-known example of such a situation is provided by flow in thin films, and one may find a quite large variety of applications. Here we mention just one of them: the coating of thin wires with a film of very viscous melt which freezes before having time to pour away.

The second type of situation corresponds to the fact that the small parameter is built into the particular solution one looks for without being directly apparent in the formulation of the problem.

The third type of situation occurs when the model suited for the physical setting is substituted by another one. Then, in general, a small parameter occurs in the relation between the two models (for example, in the theory of fluid–fluid interfaces with material properties, and also in the so-called "moving contact line" problem).

The last type of situation that we consider briefly here is when two models are considered for the same physical phenomenon, and when the coupling between them involves a small parameter. A very broad field of applications of the idea of asymptotic modelling may be included under this heading.

An interesting example, it seems to me, has to do with asymptotic and numerical simulation, when what is really done, in numerical simulation, is the substitution by a mathematical model involving, say, partial differential equations – one of them involving only algebraic ones!

The coupling between the two above models is characterized by a small parameter which is the ratio of the mesh, in numerical simulations, to the same characterized length (or time) in the continuous model.

Whenever a numerical method is chosen for some problem, it more or less involves asymptotic modelling.

6.4 Some Key Steps for the Application of the Basic Postulate

Several key steps which are often used – and are of great value in the RAM Approach – are linked with the discovery of similarity rules – these being a necessary task because, in many cases (including in various places in the preceding chapters) one encounters a double or multiple limiting process, in which two or more parameters approach their limits simultaneously. Therefore, one must frequently specify the relative rates of such an approach, since the order of carrying out these several limits cannot in general be interchanged.

In unsteady fluid flow problems, when it is necessary to take into account given initial conditions at initial time ($t = 0$), often the passage from the full (assumed exact) starting dimensionless equations with given initial conditions to a limit system of simplified dimensionless model equations is, in general, singular near $t = 0$! This singular nature is mainly expressed by the fact that, often via the limit passage, some partial derivatives in time (present in the full starting dimensionless equations) disappear in the derived system of model equations, and as a consequence, it is not possible to apply all the initial data at $t = 0$ – this derived system of equations being invalid (inconsistent) in the vicinity of $t = 0$.

We know that in such a case a short-time-scale, local time, rational analysis is necessary. In the framework of our RAM Approach, the logical and rational way for solving the associated local/short-time problem is the consideration of an unsteady adjustment problem, in an initial time layer near $t = 0$, which allows us to take into account the (long-time) effect of the transient behaviour.

This unsteady adjustment problem presents the possibility of obtaining an answer to following question: "What initial conditions can be imposed to a derived approximate model unsteady problem, and how are these initial conditions related to the starting given data for full (exact) starting equations?"

The possibility mentioned above, however, is obviously strongly linked with the validity of the Matched Asymptotic Expansions (MAE) technique, and mainly with the matching process and the concept of significant degeneracy (created by Eckhaus (1979) [109]).

This last concept, we know, provides a systematic way of determining what should be the various stretchings, and is a quasi-systematic way of comparing the respective weights of various terms in the equations, which measure the physical importance of phenomena that they are likely to describe – a very efficient tool for a constructive realization of our above postulate.

In the paper by Feuillebois and Lasek [122], this concept is applied to some problems in fluid mechanics, and we observe that significant degeneracies are, in fact, a mathematical formalism for the principle of least degeneracy defined by Van Dyke (see [14], Sect. 5.5).

It is here worth including some remarks inspired by the Conclusion in Paul Germain's (2000) [107]: "Quite often, the modelling of stiff fluid flow problems may be found by various empirical procedures or by an ad hoc approach. But it seems obvious to me that the ultimate goal is to find the mathematical key which explains not only the success, but the validity and consistency of these procedures in practice during the numerical simulation."

Actually, it seems me that our RAM Approach is an adequate procedure for such a full realization. In [26] the reader can find a preliminary, tentative account of our RAM Approach to various fluid flow phenomena – the present book being a new accomplishment of this major approach.

6.4.1 *Similarity Rules: Small Mach and Large Reynolds Numbers Flow*

We now illustrate the significance of similarity rules in the RAM Approach when at least two dimensionless parameters, Mach and Reynolds, tend respectively to zero and infinity:

$$\mathrm{Re} \to \infty, \ \mathrm{M} \to 0 \tag{6.45}$$

In such a case, as in Sect. 4.2, the combined effect of vanishing viscosity and very low compressibility is related with three possible limiting processes:

$$\mathrm{Re}\ \text{fixed},\ \mathrm{M} \to 0,\ \text{and then}\ \mathrm{Re} \to \propto \tag{6.46a}$$

$$\mathrm{M}\ \text{fixed},\ \mathrm{Re} \to \propto,\ \text{and then}\ \mathrm{M} \to 0 \tag{6.46b}$$

$$\mathrm{M} \to 0,\ \text{and}\ \mathrm{Re} \to \propto,\ \text{simultaenously} \tag{6.46c}$$

such that:

$$1/\mathrm{Re} = \lambda^{\circ}\mathrm{M}^{\beta},\ \text{with}\ \beta > 0\ \text{and}\ \lambda^{\circ} = \mathrm{O}(1) \tag{6.47}$$

where λ° is the similarity parameter (fixed, of the order one), and β a positive scalar (constant) which must be determined during the RAM Approach.

As in the case of the justification of the Boussinesq equations (4.17a–d), we have that the third (6.46c) limiting process, with the similarity rule (6.47), is the more significant, and presents the possibility of taking into account, simultaneously, two small but important physical effects which, via coupling, lead to a finite effect, of the order O(1), when

$\beta > 0$ is judiciously selected during the matching.

On the other hand, limiting process (6.46c) is the more significant because, first, the limiting process (6.46a) is rediscovered when $\lambda^{\circ} \to \propto$, and secondly, the limiting process (6.46b) is rediscovered when $\lambda^{\circ} \to o$.

For instance, if we consider the classical Blasius problem for a slightly compressible flow, with a vanishing viscosity, then a consistent RAM Approach leads to

$$\beta = 4,\ \text{and with}\ \lambda^{\circ} \approx 1 \Rightarrow (1/\mathrm{Re})/\mathrm{M}^4 \approx 1, \tag{6.48}$$

and for details, see Godts and Zeytounian (1990) [123].

Here we observe only that the above similarity rule, (6.48), is obtained via matching, which is linked with a least-degeneracy principle, and presents the possibility of obtaining a consistent, leading-order solution for the incompressible Eulerian outer flow.

6.4.2 The Matching Principle

The essentials of the matching principle are discussed by W. Eckhaus in his "Matching Principles and Composite Expansions" – a short (31-pages!) preprint (NR.42) issued by the Department of Mathematics at the University of Utrecht. Here, however, we do not enter into detailed analysis of this paper – which is undoubtedly very interesting as a rigorous reasoning of asymptotics for applied mathematicians (see also Eckhaus (1979) [109]) – and take into account only a few extracted remarks.

In the more recent paper by Eckhaus (1994) [124], the reader can find a review of the foundations of the two mainstream ideas in matching.

The intermediate matching and the asymptotic matching principles and the interrelations between the two procedures are discussed. It appears, first, that Van Dyke's intuition was correct. But the necessity of overlap seems doubtful!

Any intelligent practitioner of applied analysis will find a way to correct matching in a given problem – no matter what his convictions about the intrinsic value of the overlap hypothesis!

Matching principles are the backbone of asymptotic analysis by MAE. In applications of our RAM Approach, by means of these rules one determines, via the above-mentioned similarity rules, various undetermined constants.

There are, indeed, two schools of matching. One of them (originated by Kaplun and Lagerstrom) employs intermediate variables, and the other (originated by Van Dyke) postulates the validity of an "asymptotic matching principle". Matching in intermediate variables can be deduced from the so-called "overlap hypothesis"; that is, the assumption that the (extended) domains of validity of regular (outer) and local (inner) expansions have a non-empty interaction.

In fact, what is most important is that for a large class of fluid flow problems, overlap implies the validity of a generalized asymptotic matching. But overlap is a sufficient but not necessary condition for the validity of an asymptotic principle; and an asymptotic matching principle holds while there is no overlap at all!

In Van Dyke (1975) [14], Chap. V, the reader can find a detailed consideration of various facets of matching, which is the crucial feature of the method of MAE – involving loss of boundary conditions.

An outer expansion cannot be expected to satisfy conditions that are imposed in the inner region (near the initial time or in the vicinity of the wall of a body). Conversely, the inner expansion will not in general satisfy distant (at infinity) conditions.

On the other hand, the possibility of matching rests on the existence of an overlap domain where both the inner and outer expansions are valid. The existence of an overlap domain implies that the inner expansion of the outer expansion should, to appropriate orders, agree with the outer expansion of the inner expansion (see, for instance, Lagerstrom (1988) book [125]).

A more usual, *à la* Van Dyke, matching principle is:

$$\begin{aligned}&\text{The m} - \text{term inner expansion of (the n} - \text{term outer expansion)}\\&= \text{the n} - \text{term outer expansion of (the m} - \text{term inner expansion)},\end{aligned}$$

where m and n are any two integers. In practice, m is usually chosen as either n or n + 1. This matching principle appears to suffice for any problem to which the MAE can be successfully applied.

In 1957, Kaplun introduced the concept of a continuum of intermediate limits, lying between the inner and outer limits – and in fact, such a case was considered in Sect. 5.5.2 above.

But although the gap between inner and outer limits has been bridged by the intermediate problem, it is not yet apparent that there exists an overlap domain! However, this is assured by Kaplun's well-known "extension theorem", which asserts: "The range of validity of the inner and outer limit extends at least slightly into intermediate range."

Thus we can match the intermediate expansion with the outer expansion at one end of the range and with the inner expansion at the other end.

Finally, we can formulate an "intermediate matching principle": "In some overlap domain the intermediate expansion of the difference between the outer (or inner) expansion and the intermediate expansion must vanish to the appropriate order."

As a conclusion of this above discussion concerning the matching principle in MAE, we now present a simple example of our practical matching rule.

We assume that near y = 0, the considered model is singular when the small parameter $\varepsilon \to o$, with y fixed (outer limit). As a consequence, according to MAE we introduce a local y, significant near y = 0, namely:

$$\hat{y} = y/\varepsilon, \text{ with (inner limit) } \varepsilon \to 0, \text{ with } \hat{y} \text{ fixed} \tag{6.49}$$

In such a case, for the function $U(y;\varepsilon)$, according to the MAE method, we consider two asymptotic expansions:

$$U(y;\varepsilon) = \bar{U}_o(y) + \varepsilon\bar{U}_1(y) + \varepsilon^2\bar{U}_2(y) + O(\varepsilon^3), \text{ 'outer'} \tag{6.50a}$$

$$U(y;\varepsilon) = \hat{U}_0(\hat{y}) + \varepsilon\hat{U}_1(\hat{y}) + \varepsilon^2\hat{U}_2(\hat{y}) + O(\varepsilon^3), \text{ 'inner',} \tag{6.50b}$$

when $\varepsilon \to o$, respectively at y fixed and $\hat{y}$ fixed.

Here we assume that, obviously, in the framework of the method of the MAE, the limiting values in inner expansion (6.50b) of:

$$\hat{U}_0(\propto), \hat{U}_1(\propto) \text{ and } \hat{U}_2(\propto) \text{ exist and are well defined.} \tag{6.51a}$$

On the other hand, it is also assumed that the outer expansion (6.50a) remains valid close to y = 0, according to Kaplun's extension theorem, such that we can write:

$$\begin{aligned} U(y;\varepsilon) = {} & \bar{U}_0(0) + y\,(d\bar{U}_0/dy)_{y=0} + (y^2/2)(d^2\bar{U}_0/dy^2)_{y=0} + \dots \\ & + \varepsilon\bar{U}_1(0) + \varepsilon y\,(d\bar{U}_1/dy)_{y=0} + \dots + \varepsilon^2\bar{U}_2(0) + \dots \end{aligned} \tag{6.51b}$$

If (6.51a) is really verified, and if this is also the case for (6.51b), the matching between the (6.50a, b) outer and inner expansions is expressed by the following equality:

$$\begin{aligned}\bar{U}_0(0) + &\ \varepsilon\{\hat{y}(d\bar{U}_0/dy)_{y=0} + \bar{U}_1(0)\} \\ &+ \varepsilon^2\{(\hat{y}^2/2)(d^2\bar{U}_0/dy^2)_{y=0} + \hat{y}(d\bar{U}_1/dy)_{y=0} + \bar{U}_2(0)\} \\ &+ O(\varepsilon^3) = \hat{U}_0(\infty) + \varepsilon\hat{U}_1(\infty) + \varepsilon^2\hat{U}_2(\infty) + O(\varepsilon^3)\end{aligned} \tag{6.52}$$

where, in place of y, we have written $\varepsilon\hat{y}$.

As a consequence of (6.52) we obtain the following three matching relations, up to $O(\varepsilon^2)$:

$$\bar{U}_0(0) = \hat{U}_0(\infty) \tag{6.53a}$$

$$\hat{y}(d\bar{U}_0/dy)_{y=0} + \bar{U}_1(0) = \hat{U}_1(\infty) \tag{6.53b}$$

$$(\hat{y}^2/2)(d^2\bar{U}_0/dy^2)_{y=0} + \hat{y}(d\bar{U}_1/dy)_{y=0} + \bar{U}_2(0) = \hat{U}_2(\infty) \tag{6.53c}$$

The relation (6.53a) is the most classical and well-known of all matching relations, and Eckhaus (1994) [124], pp. 435–8, writes:

> At the first confrontation it may baffle serious students because it says [that] the regular (outer) approximation when extended to values where it is no longer valid equals the local (inner) approximation when extended to values where it is no longer valid ... Intermediate matching is based on extension theorems and on the assumption of overlap of the extended domains of validity. Asymptotic matching principles are based on assumptions of the structure of uniform expansions; they contain no direct reference to extension theorems and overlap hypotheses.

Eckhaus also asks a natural question: "What are the interrelations between the two procedures?" The interrelations were studied in some detail in his (1979) [109], Chap. 3, where the following facts were established:

> The existence of an overlap domain assures the validity of an asymptotic matching principle, provided that certain explicit conditions on the structure of the regular and the local expansions are satisfied [Theorem 3.7.1 in [109]] ... [but] the non-existence of an overlap domain does not preclude the validity of an asymptotic matching principle.

Finally, it appears that Van Dyke's intuition was correct. As to Lagerstrom's conviction in the necessity of overlap, Eckhaus remarks that counter-examples had already been presented by Fraenkel in 1969 and in his (1979) [109], and some discussion is included in Lagerstrom (1988) [125] – but in the latter there is no reference to Eckhaus's monograph of 1979.

6.4.3 The Least-Degeneracy Principle and the Significant Degeneracies

As mentioned in Van Dyke [14], Sect. 5.5, a crucial step in the MAE method is the choice of inner variables.

The guiding principles are that the inner problem should have the least possible degeneracy, that it must include in the first approximation any essential elements omitted in the first outer solution, and that the inner and outer solutions should match.

As an example, consider the case in Sect. 6.4.1, linked with the choice of β. Indeed, the significant degeneracies, according to Eckhaus, are a mathematical formalism for the principle of least degeneracy, by Van Dyke. In several papers of Feuillebois, and Feuillebois and Lasek (see [122]), the reader can find various applications. The method makes it possible to obtain, with the use of a computer, the boundary layer equations for a singular perturbation problem.

The straightforward but tedious hand-calculation of the significant degeneracies may be avoided by the use of just such a computer! However, this method is not concerned with the full solution to some problems (in fluid mechanics), but only with the process required for obtaining the significant degeneracies.

Usually, the class of problems considered typically involves many parameters, for which multiple boundary layers may exist.

The application essentially provides some boundary-layer equations to solve, together with the order of magnitude of the region in which they are valid. But it seems that there is no proof that the solution of such equations is the required significant approximation to the solution of the original problem.

However, there is a useful heuristic "principle of correspondence" (Eckhaus): "If there exists a significant approximation, then the degeneracy in the boundary layer variable is significant."

Practically, this principle limits the field of research of the significant approximation to the solutions of the boundary-layer equations obtained as significant degeneracies, and this principle holds for many physical problems in fluid mechanics.

In 1984, Feuillebois (see [122]) considered an interesting example which is relative to the full Navier (Navier–Stokes incompressible) steady equations for a large Reynolds number. As a results, he demonstrated the existence of the following seven significant degeneracies:

1. The ordinary boundary layer equations.
2. Equations for an ideal fluid with high thermal conductivity.
3. Boundary layer equations with a low outer pressure gradient.
4. Thermal boundary layer equations of an ideal fluid, with a low outer pressure gradient.
5. Equations for an ideal fluid, with a complete energy equation.
6. The momentum equation as for an ideal fluid, but the energy equation contains a viscous term.
7. Somewhat similar to (3), but with viscous dissipation.

We observe that the cases when Pr (the Prandtl number) is low (high thermal conductivity) may be of practical interest for liquid metals flow. An interesting exercise is the application of the Feuillebois approach to the unsteady Navier equations case, and also (but certainly more complicated) to NS–F, large a Reynolds number case, considered in Chap. 5.

Germain (1977) [126], pp. 89–96, includes a report on his views after reading Eckhaus's papers relative to the concept of significant degeneracy.

6.4.3.1 A Last Comment Concerning the "Non-standard Analysis"

In "Going on with asymptotics", in Guiraud (1994) [127], pp. 257–307 (esp. pp. 299–302), there is a short account of this topic in which the author considers that "although it deals with the foundation of mathematics, in quite an abstract way, yet it seems to me that it has some impact on asymptotics, as I shall try to show at least superficially."

On the other hand, Eckhaus, in his remarks concerning the non-standard analysis (in his (1994) [124], p. 434), writes:

> In the last decade a group of French mathematicians attracted attention by the use of non-standard analysis in problems of singular perturbations. Non-standard analysis is an invention in mathematical logic due to Robinson (see his (1966) [128]) book, in which infinitesimally small numbers are introduced back into the analysis. Robinson attempted to link the non-standard analysis to a concept of asymptotic expansions in a book (edited in 1975) with Lightstone [129], but without much response. For an introduction to the new developments one can consult [130], edited in 1981, and the literature quoted there, and more recently the Proceedings of ICIAM' 91, p. 342. There is an interesting link between these developments and the lemma given above (and concerning the extension theorems of Kaplun introduced in 1957, and formulated in upper part of the page 432 in [124]). In the new applications of non-standard analysis a very important role is played by the so-called 'Robinson's Lemma', which is the subject of much veneration.

Eckhaus also writes:

> In discussions which I had with non-standard analysts (in particular Marc and Francine Diener at Oberwolffrach in August' 1981) it became clear that the lemma quoted in Sect. 3 of [124] is a standard counterpart of Robinson's lemma. The extension theorems are very seldom (almost never) used as an explicit tool of analysis in specific problems. An exception is contained in my analysis of the 'Canard' problem [131], first treated by non-standard analysts (see [130] for references). More recent use of extension theorems, in a related problem, is given in [132].

6.4.4 *The Method of Multiple Scales (MMS)*

Various problems in fluid dynamics are characterized by the presence of a small disturbance which, because of being active over a long time, has a non-negligible cumulation effect.

In solving such a given problem, the main approach is to combine appropriate techniques to construct an (asymptotic) expansion which is uniformly valid over long time intervals. A central feature of such a (MMS) method is the impossibility of the matching (which is just the central feature of the MAE method).

Often, the inner asymptotic expansion does not tend to a defined limiting state when the short (distorted) variable (time or space) tends to infinity. In particular, in the case of unsteady fluid flow, the adjustment to initial state is not possible (when the short time tends to infinity) for the outer model equations.

The fundamental characteristics of the MMS can be summarized as follows (according to Germain [126]):

> When the data of a fluid flow problem show (after careful physical inspection) that the small parameter ε is, in fact, the ratio of two scales (of time or space), the MMS consists of first introducing two variables constructed with these scales (one of them possibly being 'distorted'). Next, the formal expansion of the solution in ε is considered, each coefficient of the equations being a function of two variables introduced (e.g., of time) which are considered as independent during the entire calculation. In order to completely determine a coefficient of this expansion, it is not enough to solve the equation where it appears for the first time! The indeterminants which necessarily remain are chosen by making sure that the equation in which the next term appears will lead to a solution which does not destroy, but on the contrary, best guarantees, the validity of the approximation which is sought. From a certain a priori knowledge of the solution, we generally assume that U depends on the variables t and $\mathbf{x} = (x, y, z)$ so that a rapid variation having a repetitive character analogous to an oscillatory phenomenon is made to appear. This variation is itself modulated from one moment to another and from one point of space to another.

Such a structure is described by the following dependence of the U:

$$U = U^*(\chi(\mathrm{t}, \mathbf{x});\ \mathrm{t}, \mathbf{x}) \tag{6.54}$$

where χ is a so-called fast intermediary variable because this function χ varies rapidly as a function of t and $\mathbf{x}$. If only dimensionless variables are used (which is the first main operation for a realization of the RAM Approach), then this property can be characterized by writing:

$$\partial\chi/\partial\mathrm{t} \;=\; -\omega/\varepsilon \text{ and } \nabla\chi = \lambda/\varepsilon \tag{6.55}$$

where ω and $|\lambda|$ are of the order unity. From (6.55) we deduce two relations:

$$\partial U/\partial\mathrm{t} \;=\; -\,(\omega/\varepsilon)\partial U^*/\partial\chi + \partial U^*/\partial\mathrm{t} \tag{6.56a}$$

$$\nabla U = \;-\,(1/\varepsilon)(\partial U^*/\partial\chi)\ \lambda + \nabla U^* \tag{6.56b}$$

with analogous formulae for the higher-order derivatives.

By substituting these expressions for the derivatives into the equations governing the considered fluid flow, the small parameter ε is introduced into these equations, even if it does not appear initially.

In a second step, we write:

$$U^* = U_0^* + \varepsilon U_1^* + \ldots \text{ and } \chi = \chi_0 + \varepsilon\chi_1 + \ldots \tag{6.57}$$

and then, according to considered starting equations, when both of the above relations (6.56a, b) are taken into account, and by setting at zero the terms proportional to the successive powers of ε, a hierarchy of systems of equations is derived for $U_0^*, U_1^*, \ldots$

The first system in this hierarchy determines U_0^* at best in its dependence with respect to χ_0, but not with respect to t and **x**. It is usually while seeking to determine U_1^* – or even other higher-order terms – that the dependence of U_0^* with respect to t and **x** is prescribed by the cancelling of the secular terms; that is, of terms which in U_1^* do not remain bounded when χ_0 increases indefinitely. Indeed, if we want (6.57) to cover an interval 0(1) in variation of t and **x**, then because of (6.55), this corresponds to a variation of χ_0 which is $O(1/\varepsilon)$.

A typical example is strongly related to the description of the progressive waves (and see Germain (1971) [133]). As noted by Germain [108], pp. 12–13:

> Progressive waves occur generally when a physical phenomenon is thought to be represented by the occurrence of steep gradients in one variable only, across three-dimensional manifolds, in four-dimensional spacetime, with much smoother gradients in other directions. The mathematical structure of the representation looks like a phenomenon in five-dimensional spacetime. In such a case the phenomenon is quantified by an n-dimensional vector U. Assume that the manifold across which the gradients are steep is $F(t, \mathbf{x}) =$ const. Then, $U(t, \mathbf{x}, F/\varepsilon;\ \varepsilon)$ so that $\chi = F/\varepsilon$, is considered as a fifth variable. There is apparently nothing in the equations which allows us to single out the dependency of U on χ. But all is changed when we add the ansatz that the proper physical solution may be obtained as an expansion with respect to χ, t and **x** being fixed when proceeding to the limit of vanishing ε. As a matter of fact, the method of multiple scales (MMS), through the requirement of vanishing of secular terms, provides the way by means of which that dependency can be figured out. A key to the existence of progressive waves is that U, supposed to be dependent on χ only, at leading order, exists as a planar wave solution ruled by a linear system with $\partial F/\partial t$, $\partial F/\partial \mathbf{x}$ obeying a dispersion relation. That relation defines a wave speed, and may be considered as an eikonal of the Halmilton–Jakobi equation, the solution of which is built by means of rays. The planar wave, being a solution to a linear system, is determined only up to a scalar amplitude factor ... This amplitude obeys an equation which is obtained when going to higher order in the asymptotic expansion and eliminating secular terms – this amplitude/transport equation being the main result of the application of MMS. The details depend on the particular phenomenon considered, and there exist quite a variety of situations that may be described mathematically by such a procedure. The small parameter ε characterizes the steepness of the transversal gradient, and if the fluid dynamics process is non-linear, and non-linearity is measured by the order of magnitude of the amplitude, and if, furthermore, the starting equations are first-order quasi-linear, as is the case with inviscid gas dynamics, then the amplitude obeys a partial differential equation which is generally an inviscid Burgers' equation along each ray. If there are second-order derivatives present in the starting equations, with a small coefficient, then the amplitude obeys a partial-differential equation which is generally Burgers'. The role of time is played by the distance along each ray, while the role of space is played by χ. One may deal with third-order derivatives, and another small parameter, yielding then the *KdV* equation; and both phenomena may occur simultaneously.

Thus, when both dispersive and dissipative effects are present, the transport equation is a Burgers–Korteweg–de Vries (*KdV*) equation of the following form:

$$\partial A/\partial \tau + A\partial A/\partial \xi + (\nu/2)^2 \partial^3 A/\partial \xi^3 = \mu \partial^2 A/\partial \xi^2 \qquad (6.58)$$

Even if the Germain presentation is new and tries to be systematic, the essential ideas may be found, in particular, in Guiraud (1969) [134], with a method which is essentially equivalent.

For another application of the two-scales technique, see Sect. 3.3.5, where the derivation of the well-known non-linear model Eq. 3.61q of acoustics – the KZK (Kuznetsov, Zabolotskaya, and Khokhlov) equation – was examined.

In LNP (1987) [16], pp. 150–7, the reader can also find an application related to high Strouhal (S), high Reynolds (Re), and small Mach (M) numbers, where we assume that

$$\mathrm{SM} = \mathrm{O}(1) \text{ ans } \mathrm{Re/S} = \mathrm{O}(1) \tag{6.59}$$

the Mach number M <<1 being the main small parameter.

In such a case, the leading term $\mathbf{u}_0$ in $\mathbf{u} = \mathbf{u}_0 + \mathrm{M}\,\mathbf{u}_1 + \ldots$ is the solution of the equation for the spherical waves (which is derived from the full NS–F unsteady system)

$$\partial^2\mathbf{u}_0/\partial t^2 = \Delta\mathbf{u}_0 \tag{6.60a}$$

and appears a wave front

$$F(\mathbf{x}) - \mathrm{t} = 0, \text{ and } \chi = [F(\mathbf{x}) - \mathrm{t}]/\mathrm{M} \tag{6.60b}$$

As a consequence, a two-scales $(\chi, \mathbf{x})$ expansion, in M, is necessary for: $\mathbf{u}$, ρ, and T. At the leading order we obtain (from full NS–F unsteady system) a linear and homogeneous system relative to first-order derivative, $\partial/\partial\chi$, of $\mathbf{u}_0(\chi, \mathbf{x})$, $\rho_1(\chi, \mathbf{x})$ and $\mathrm{T}_1(\chi, \mathbf{x})$, with $\mathrm{t} = F(\mathbf{x}) + \mathrm{M}\,\chi$. As a consequence we derive the following eikonal equation:

$$\mathbf{k}^2 \equiv [\nabla F(\mathbf{x})]^2 = 1 \tag{6.61a}$$

and, for instance, we have as a solution for $\mathbf{u}_0(\chi, \mathbf{x})$,

$$\mathbf{u}_0(\chi, \mathbf{x}) = A(\chi, \mathbf{x})[\mathbf{k}/(\gamma - 1)] + \mathbf{U}_0(\mathbf{x}) \tag{6.61b}$$

where the function $\mathbf{U}_0(\mathbf{x})$ is arbitrary and is determined from the compatibility conditions linked with the second-order system of equations, which is a linear and non-homogeneous system of equations for

$$\partial\mathbf{u}_1(\chi, \mathbf{x})/\partial\chi,\ \partial\rho_2(\chi, \mathbf{x})/\partial\chi \text{ and } \partial\mathrm{T}_2(\chi, \mathbf{x})/\partial\chi$$

and also with the matching, initial, and boundary conditions of the considered problem. Finally, the amplitude function $A(\chi, \mathbf{x})$ in solution (6.61b) is solution of following transport equation:

$$\partial\mathrm{A}/\partial\tau + \alpha A\partial\mathrm{A}/\partial\chi + \gamma\partial\mathrm{A}/\partial\chi + \delta A = \beta\, \partial^2\mathrm{A}/\partial\chi^2 \tag{6.62}$$

where the "time" τ (increasing with F) is a parameter which varies along the characteristics – the characteristic form of eikonal (6.61a) being

$$dF/d\tau = 1 \text{ and } d\mathbf{k}/d\tau = 0 \text{ along } d\mathbf{x}/d\tau = \mathbf{k}$$

Finally, it is interesting to mention the opinion of Van Dyke [127], p. 7 concerning our (with Guiraud) the multiple scales approach:

> In fluid mechanics, however, Guiraud, working with Zeytounian, has applied multiple scales to two quite different problems, where it is clearly the appropriate method that is mentioned in their titles. First, in the theory of axial-flow machinery we have *Application du concept d'échelles multiples à l'écoulement dans une turbomachine axiale* (G-Z, 1974). Clearly, the long scale is that of the mean flow, and the short scale that of the individual blades, the small parameter being the reciprocal of the number of blades or of stages. Second (G–Z, 1977) we have *A double-scale investigation of the asymptotic structure of rolled-up vortex sheets*. This applies to the rolled-up vortex sheets from the trailing edge of a high-aspect-ratio wing or from the leading edges of a delta wing, the small parameter being the number of turns.

In [11], the reader can find some comments concerning these two G–Z papers. Van Dyke [127], p. 8, also writes, concerning asymptotics allied with numerics:

> Eight years ago, Guiraud and Zeytounian (1986) argued persuasively that although asymptotic techniques will in special circumstances continue to be used to derive closed-form solutions, their future role will be primarily as an adjunct to numerical simulation. They illustrated this view with a number of examples from fluid dynamics, concluding with several drawn from their own research. From casual observation I suggest that the number of papers appearing in technical journals with the word 'asymptotic' in their titles is actually increasing steadily. On the other hand, the cooperation between experts on asymptotics and numerics – which I agree is desirable and inevitable-seems not yet to have gained momentum. Meanwhile, I have for some years been experimenting with a quite different alliance between asymptotics and the computer.

I emphasize here that in the first published G–Z paper (1970), relative to the multiple scales approach for axial-flow machinery, a decisive moment is the appearance of a source term, in the derived leading-order model, which is proportional to the jump in pressure (the difference between the two sides of one and the same blade), orthogonal to the material blade surface in the row. This equivalent homogeneous force occurs due to the redistribution (homogenization) of forces exerted on the flow by the blades of the row. In fact, in 1970 we had already conceived an homogenization technique (without our knowing it!) that presents the possibility of replacing the effect of the blades of the row on the flow in a channel between two consecutives blades, by an equivalent source force in model equations.

As a result of the RAM Approach, an equivalent model – the through-flow – is derived, in a simple cylindrical domain without any row with blades, but with the source force in governing model equations.

6.4.5 The Homogenization Analysis

The homogenization analysis is in fact a more sophisticated form of the MMS approach discussed in the previous section. Seepage through a porous media – for instance: water gradually escaping by seepage through the ground – is considered one of the first examples (if we do not take into account our G–Z (1971) asymptotic analysis of the flow in an axial turbomachine) to which the method of homogenization was applied (Sanchez-Palencia 1974 [135]). It is a good example for explaining the role of physical scales and the mathematical procedure, and it can be used to illustrate the derivation of many properties of the constitutive coefficients.

Let a rigid porous medium be satured by an incompressible Newtonian fluid of constant density ρ_0 (see, for instance, *Advances in Applied Mechanics*, vol. 32 (1996) [136], pp. 288–92). Driven by a steady ambient pressure gradient, the steady flow velocity u_i and pressure p in the pores are governed by the Navier equations:

$$\partial u_i/\partial x_i = 0 \text{ and } \rho_0 u_j \partial u_i/\partial x_j = -\partial p/\partial x_i + \mu \nabla^2 u_i \tag{6.63a,b}$$

On the wetted surface of the solid matrix Γ, there is no slip condition,

$$u_i = 0 \text{ when } x_i \in \Gamma \tag{6.63c}$$

For slow flows (low Reynolds number), the two terms on the right-hand side of (6.63b), representing the pressure and the viscous force, must be dominant. Then, both the pore pressure and the flow velocity vary according to two very different scales: the local or microscale l characteristic of the size of pores and grains, and the global or macroscale L imposed by the global pressure gradient. During the homogenization analysis the main working hypothesis is the ratio of the above two length scales:

$$\varepsilon = l/L << 1 \tag{6.64a}$$

defining the velocity scale U. A moment of reflexion leads to the following dimensionless (primed) quantities:

$$x_i' = x_i/l,\ p' = p/P, u_i' = (\mu L/l^2 P)u_i' \tag{6.64b}$$

where p in (6.64b) denotes the change in pressure (because it appears only in differential form), P/L being the magnitude of the global pressure gradient. Equation (6.63b) becomes, formally in dimensionless form:

$$\text{Re}\ u_j'\partial u_i'/\partial x_j' = -(1/\varepsilon)\partial p'/\partial x_i' + \mu \nabla^2 u_i' \tag{6.64c}$$

where Re $= \varepsilon(\rho_0 l^2 P/\mu^2)$ is just the Reynolds number (micro $\equiv Ul/(\mu/\rho_0)$).

With P $= \mu^2/\rho_0 l^2$, and returning to dimensional variables, but retaining the ordering symbol ε, we obtain, as a starting equation for our homogenization analysis, the following equation:

$$\varepsilon^2\ \rho_0 u_j \partial u_i/\partial x_j = -\partial p/\partial x_i + \varepsilon\ \mu \nabla^2 u_i \tag{6.65}$$

Let us assume that the geometry of the porous matrix is periodic on the micro-scale l, although the structure may still change slowly over the macroscale L. Each periodic cell Ω is a rectangular box of dimension $O(l)$, u_i and p being spatially periodic from cell to cell. We introduce, in addition to x_i also $\xi_i = \varepsilon x_i$, and the perturbation expansions

$$u_i = u_i^{(0)} + \varepsilon u_i^{(1)} + \varepsilon^2 u_i^{(2)} + \ldots \tag{6.66a}$$

$$p = p^{(0)} + \varepsilon p^{(1)} + \varepsilon^2 p^{(2)} + \ldots \tag{6.66b}$$

where $u_i^{(k)}$ and $p^{(k)}$, $k = 0, 1, 2, \ldots$, are functions of x_i and ξ_i.

First, from $\partial u_i/\partial x_i = 0$ we obtain

$$\partial u_i^{(0)}/\partial x_i = 0 \tag{6.67a}$$

$$\partial u_i^{(1)}/\partial x_i + \partial u_i^{(0)}/\partial \xi_i = 0 \tag{6.67b}$$

$$\partial u_i^{(2)}/\partial x_i + \partial u_i^{(1)}/\partial \xi_i = 0 \tag{6.67c}$$

and similarly, we obtain from (6.65),

$$0 = -\partial p^{(0)}/\partial x_i \tag{6.68a}$$

$$0 = -\partial p^{(0)}/\partial \xi_i - \partial p^{(1)}/\partial x_i + \mu \nabla^2 u_i^{(0)} \tag{6.68b}$$

$$\rho_0 u_j^{(0)} \partial u_i^{(0)}/\partial x_j = -\partial p^{(1)}/\partial \xi_i - \partial p^{(2)}/\partial x_i + \mu \nabla^2 u_i^{(1)}. \tag{6.68c}$$

On the wetted interfaces Γ, the velocity vanishes; hence,

$$u_i^{(0)} = 0, u_i^{(1)} = 0, u_i^{(2)} = 0, \ldots \text{ when } x_i \in \Gamma. \tag{6.69a}$$

In a typical Ω cell the flow must be periodic:

$$u_i^{(0)}, u_i^{(1)}, u_i^{(2)}, \ldots p^{(0)}, p^{(1)}, p^{(2)}, \ldots \text{ are } \Omega - \text{periodic} \tag{6.69b}$$

First, from (6.68a) it is clear that:

$$p^{(0)} = p^{(0)}(\xi_i). \tag{6.70a}$$

and because of the linearity of (6.68b), $u_i^{(0)}$ and $p^{(1)}$ can be formally represented by

$$u_i^{(0)} = -K_{ij} \partial p^{(0)}/\partial \xi_j, p^{(1)} = -A_j \partial p^{(0)}/\partial \xi_j + p*^{(1)}(\xi_i). \tag{6.70b}$$

It then follows from (6.67a), (6.68b), (6.69a), and (6.69b) that

$$\partial K_{ij}/\partial x_i = 0 \tag{6.70c}$$

$$-\partial A_j/\partial x_i + \mu\nabla^2 K_{ij} = -\delta_{ij} \tag{6.70d}$$

where

$$K_{ij} = 0 \text{ on } \Gamma, \text{ and } K_{ij}, A_j \text{ are } \Omega - \text{periodic} \tag{6.70e,f}$$

and (Ω_f is the fluid volume inside the Ω cell)

$$<A_j> = [1/|\Omega|] \int_{\Omega f} A_j \, d\Omega \tag{6.71}$$

can be set to zero.

Equations 6.70c and 6.70d with 6.70e,f define a canonical Stokes' flow boundary-value problem in Ω cell, which can, in principle, be solved numerically for any prescribed microstructure.

From (6.70b) we obtain, at the leading order, a well-known equation:

$$<u_i^{(0)}> = -<K_{ij}>\partial p^{(0)}/\partial \xi_j \tag{6.72a}$$

and also:

$$p^{(1)} = n\ p*^{(1)}(\xi_i) \tag{6.72b}$$

where n is the porosity – the ratio of fluid volume in the cell to the total cell volume:

$$n = |\Omega_f|/|\Omega|. \tag{6.72c}$$

Eq. 6.72a being just the celebrated law of Darcy with $<K_{ij}>$ the hydraulic conductivity.

On the other hand, the Ω-average of (6.67b) gives

$$\partial<u_i^{(0)}>/\partial\xi_i + [1/|\Omega|] \int_{\Omega f} (\partial u_i^{(1)}/\partial x_i) d\Omega = 0$$

and the interchange of volume integration with respect to x_i and differentiation with respect to ξ_i is allowable when n = constant in the above relation. Otherwise the same is justifiable by virtue of the spatial averaging theorem (see, for instance, [136], Sect. IIIF, pp. 300–2).

By using the Gauss theorem and the boundary conditions, we see that the volume integral vanishes; hence: $\partial<u_i^{(0)}>/\partial\xi_i = 0$, and this result implies, in turn, that:

$$\partial[<K_{ij}>\partial p^{(0)}/\partial\xi_j]/\partial\xi_i = 0. \tag{6.73}$$

This equation (6.73) governs the seepage flow in a rigid porous medium on the macroscale. Theoretical derivation in the present manner was first presented by Snachez-Palencia in 1974.

If, now, the medium is isotropic and homogeneous on the L scale, we have

$$<K_{ij}> = K\,\delta_{ij}$$

where K is a scalar constant, and it follows from (6.73) that

$$\partial^2 p^{(o)}/\partial\xi_k\partial\xi_k = 0$$

With proper boundary conditions on the macroscale, $p^{(0)}$can then be found. In fact, when the Reynolds number is assumed to be finite (Re = O(1)), then the convective inertia would be on the order of (ε), and (6.68b) would be replaced by

$$\rho_o u_j{}^{(0)}\partial u_i{}^{(0}/\partial x_j = -\partial p^{(0)}/\partial\xi_i - \partial p^{(1)}/\partial x_i + \mu\nabla^2 u_i{}^{(0)} \tag{6.74a}$$

with

$$\partial u_i{}^{(0)}/\partial x_i = 0. \tag{6.74b}$$

The above cell problem, (6.74a, b), is fully non-linear, and (6.73) and Darcy's law (6.72a) no longer hold.

Finally, concerning this "seepage flow in rigid porous media", the reader can find in [136] various complementary results concerning the uniqueness of the cell boundary-value problem ((6.70c)–(6.70e,f)) with zero on the right of (6.70d), properties of hydraulic conductivity, numerical solution of the cell problem, the effects of weak inertia, a spatial averaging theorem, and porous media with three or more scales.

We observe that the typical steps of the homogenization theory are:

1. Identify the microscales and macroscales. This identification of scales is a consideration of physics and is crucial to the success of the mathematical theory.
2. Introduce multiple-scale variables, in non-dimensional form of the starting fluid flow problem, and expansions, and deduce cell boundary-value problems at successive orders. The leading-order cell problem is homogeneous; either the solution itself or the coefficient of the homogeneous solution are indeterminate and independent of the microscale coordinates.
3. Use linearity and express the next-order solution in terms of the leading-order solution and deduce an inhomogeneous cell problem.
4. Determine the solvability of the inhomogeneous cell problem – mathematically, the solvability condition for the inhomogeneous problem – given that the homogeneous problem has a non-trivial solution!
5. Derive the equation governing the evolution of the leading-order solution (or the coefficient of the homogeneous solution) and calculate the constitutive coefficients from the solution of a canonical cell problem.

We have previously mentioned (for instance, in Chap. 3) that the case of low Mach number in gas dynamics is a very difficult problem when the gas is contained in a bounded container Ω, with an impermeable but eventually deformable (with time t) wall, so that the volume occupied by the gas is a given function of time: namely, $V_0(t)$. In such a stiff case, it is necessary to use an homogenization technique with an infinity of microscales. Such a problem is considered in Sect. 7.2.

6.4.6 Asymptotics of the Triple-Deck Theory

Here we do not discuss in depth this triple-deck theory, as a detailed and thorough discussion is presented in our FMIA 64 (2002) [26], Chap. 12. Some of the features of this theory were mentioned in Sect. 6.3.4, and we now discuss the various steps required for the derivation of model equations for viscous lower motion.

6.4.6.1 The Local Atmospheric Thermal Problem

Here I consider the formulated problem of Eqs. 6.42a–6.42e with conditions (6.44a)–(6.44e).

First, if we take into account the boundary conditions on the ground z = 0, it is necessary to introduce an inner variable ž such that

$$\check{z} = z/\kappa^{o\alpha}, \quad \alpha>1$$

and in this case

$$u \approx \kappa^{o\alpha-1}\check{z}, \text{ for } x \to -\propto. \tag{6.75}$$

Then, from the first Eq. 6.42a, for u, we verify that

$$\text{if } \check{z} = z/\kappa^{o\alpha} \text{ and } u \approx \kappa^{o\alpha-1}\check{u}(x, \check{z}),$$

and from

$$\kappa^{o\alpha-1}\partial\check{u}/\partial\check{z} + \ldots = \kappa^{o5-2\alpha}\partial^2\check{u}/\partial\check{z}^2$$
$$+ \ldots \Rightarrow \alpha - 1 = 5 - 2\,\alpha \Rightarrow \quad \alpha = 2. \tag{6.76}$$

As a consequence, we establish that three vertical coordinates are necessary (z, ź, ž), for a rational asymptotic (triple-deck) analysis of the system of Eqs. 6.42a–6.42e: namely,

1. z, for the upper non-viscous region, where u $\approx u_{up} \to 1$, whenx $\to -\propto$;
2. ź = z/κ°, for the middle intermediate region, where $u \approx u_{mid} \to 1 - \exp(-\acute{z})\cos(-\acute{z}) = U^{\propto}(\acute{z})$, when x $\to -\propto$;

3. $\check{z} = z/\kappa^{o2}$, for the lower wall viscous region, where $u \approx \kappa^{\circ}\, u_{low}$, and $u_{low} \rightarrow \check{z}$, when $x \rightarrow -\propto$.

Note that here we consider only the case when (m = 5):

$$l^{\circ}/L^{\circ} \approx \kappa^{o3}$$

For the other case,

$$l^{\circ} < \kappa^{o3}\, L^{\circ} \text{ and } l^{\circ} > \kappa^{o3}\, L^{\circ}$$

it is necessary to apply different asymptotic analysis. The case m = 6 and m = 4 can be analyzed from the equations analogous to (6.42a)–(6.42e). For the case m = 3 it is necessary to start from other equations, where the Boussinesq approximation does not emerge.

In particular, for m = 3, we have $l^{\circ} = 10^4$ m, and we may neglect the terms related to the Coriolis force in the local non-Boussinesq equations.

Beginning our triple-deck analysis with the middle deck where x and $\acute{z} = z/\kappa^{\circ}$ are the order one coordinates, we expand the flow variables as:

$$\begin{aligned} u &= U^{\propto}(\acute{z}) + \kappa^{o\phi} u_{mid} + \ldots \\ w &= \kappa^{o\psi} w_{mid} + \ldots \\ \pi &= \kappa^{o2}\, \pi_{mid} + \ldots \\ \theta &= \kappa^{o\sigma}\, \theta_{mid} + \ldots \end{aligned} \tag{6.77a}$$

and from system of Eqs. 6.42a–6.42e to find for the lowest order:

$$\begin{aligned} &U^{\propto}(\acute{z})\partial u_{mid}/\partial x + (d\, U^{\propto}/\acute{z}) w_{mid} = 0 \\ &\partial u_{mid}/\partial x + \partial w_{mid}/\partial \acute{z} = 0 \\ &\partial \pi_{mid}/\partial \acute{z} = B^{*}\theta_{mid} \\ &\partial \theta_{mid}/\partial x = 0. \end{aligned} \tag{6.77b}$$

The choice

$$\phi = 1, \quad \psi = 1 + \phi \text{ and } \sigma = 1 \tag{6.77c}$$

is necessary if we want to obtain a significant degeneracy of the Eqs. 6.42a–6.42e in the lower viscous deck region in the vicinity of $\check{z} = 0$ near the wall.

We observe that the effects of the expansion of the boundary layer are $O(\kappa^{\circ})$ in u and $O(\kappa^{o2})$ in w. Furthermore, in the boundary layer (lower deck), if we take into account that $\check{z} = z/\kappa^{o2}$, we necessarily have

$$\pi = \kappa^{o2}\pi_{low} + \ldots$$

and, by continuity, we obtain the form of the expansion for π in (6.77a).

Solutions for u_{mid} and w_{mid} satisfying the upstream boundary conditions are:

$$u_{mid} = A(x)\ dU^{\propto}/d\acute{z} \text{ and } w_{mid} = -\ (dA(x)/dx)U^{\propto}(\acute{z}) \quad (6.77d)$$

which represent simply a vertical displacement of the streamlines through a distant – κ°A(x).

On the other hand, the flow in the upper non-viscous deck is driven by an outflow from the middle deck. Far from (6.77d) we have:

$$\mathrm{Lim}_{z'\to\propto}[w_{mid}(x, \acute{z})] = -(dA(x)/dx) \quad (6.77e)$$

In the upper deck, with z = κ° ź, as flow expansion we write:

$$u = 1 + \kappa^{o2}u_{up} + \ldots, (w,\ \pi,\ \theta) = \kappa^{o2}(w_{up},\ \pi_{up},\ \theta_{up}) + \ldots \quad (6.78a)$$

and gain substitution in Eqs. 6.42a–6.42e to find for the lowest order the following equations for an inviscid motion:

$$\begin{aligned} &\partial u_{up}/\partial x + (1/\gamma)\partial\pi_{up}/\partial x = 0,\\ &\partial w_{up}/\partial x + (1/\gamma)\partial\pi_{up}/\partial z = 0\\ &\partial u_{up}/\partial x + \partial w_{up}/\partial z = 0,\\ &\partial\theta_{up}/\partial x + B^*\Lambda(0)w_{up} = 0, \end{aligned} \quad (6.78b)$$

and, as a consequence we derive, for π_{up}, a Helmholtz equation:

$$[\partial^2/\partial x^2 + + {K_0}^2]\partial\pi_{up}/\partial x = 0 \quad (6.78c)$$

where

$${K_0}^2 = (B^*/\gamma)\Lambda(0).$$

In fact, according to upstream conditions we have:

$$\mathrm{Lim}_{z\to o}[\pi_{up}(x,\ z)] \equiv \pi_{up}(x, 0) = P(x) \equiv \mathrm{Lim}_{z'\to\propto}[\pi_{mid}(x, \acute{z})] \quad (6.78d)$$

but

$$w_{up}(x, 0) = -\ dA(x)/dx \quad (6.78e)$$

and, as consequence of (6.77e) and matching between upper and middle decks, we obtain, for Eq. 6.78c, the following rather strange condition for π_{up}:

$$\partial/\partial x[(\partial\pi_{up}/\partial z)_{z=0}] = \gamma\ [{K_0}^2(dA/dx) + d^3A/dx^3]. \quad (6.78f)$$

It is obvious that the middle deck solution (6.77d) does not satisfy the no-slip condition on $\check{z} = 0$ – a situation which is remedied by the analysis of the lower viscous deck, near the wall, where we have the following relation:

$$\check{z} = z/\kappa^{o2} \equiv \acute{z}/\kappa^{o}$$

Now, matching with the middle (6.77a) expansion, when in $U^{\infty}(\acute{z})$, with $\acute{z} = \kappa^{o}\,\check{z}$ and $\check{z}$ is fixed, implies the following lower deck expansions:

$$(u,\ w,\ \pi,\ \theta) = (\ \kappa^{o}u_{low},\ \kappa^{o3}w_{low},\ \kappa^{o2}\ \pi_{low},\ \theta_{low}) + \ldots \tag{6.79a}$$

and from Eqs. 6.42a–6.42e to find for the lowest order the following model equations for viscous lower motion:

$$u_{low}\partial u_{low}/\partial x\ + w_{low}\partial u_{low}/\partial\check{z} + (B^*/\gamma)\int_{\infty}^{\check{z}}(\partial\theta_{low}/\partial x)d\check{z}$$
$$+(1/\gamma)dP(x)/dx = \partial^2 u_{low}/\partial\check{z}^2, \tag{6.79b}$$

$$\partial u_{low}/\partial x\ + \partial w_{low}/\partial\check{z} = 0, \tag{6.79c}$$

$$u_{low}\partial\,\theta_{low}/\partial x\ + w_{low}\partial\theta_{low}/\partial\check{z} = (1/Pr)\partial^2\theta_{low}/\partial z^2, \tag{6.79d}$$

with the boundary conditions:

$$\check{z} = 0 : u_{low} = w_{low} = 0,\ \theta_{low} = \ \tau^*\Theta(x),\ 0< x <1 \tag{6.80a}$$

$$\check{z} \to +\infty\text{:}\ u_{low} \to \check{z}, w_{low} \to 0,\ \theta_{low} \to 0, P(x)\ \text{and}\ dA/dx \to 0 \tag{6.80b}$$

$$\check{z} \to -\infty\text{:}\ u_{low} \to \check{z} +\ A(x), w_{low} \to -\check{z}(dA/dx),\ \theta_{low} \to 0 \tag{6.80d}$$

after matching with the middle deck.

The specification of the above viscous problem in the lower deck is completed by the relation (6.78f) linked with P(x) (if we take into account that according to (6.78d) $\pi_{up}(x, 0) = P(x)$) and the function A(x).

The well-known interpretation of the “strange” condition (6.78f) is as follows:

The pressure P(x) driving the flow in the lower deck is itself induced in the main stream, i.e. by the upper deck solution, as a consequence of the displacement thickness of the lower deck transmitted through the middle deck by the passive effect of displacement of the streamlines.

One the other hand:

This so-called “strong singular self-induced coupling” (*à la* Stewartson) arises because the above problem to be solved in the lower viscous layer does not accept the function P(x) as data known prior to the resolution – as is the case in the classical Prandtl boundary layer problem.

Conversely, this pressure perturbation $P(\mathrm{x})$ must be calculated at the same time as the velocity components u_{low} and w_{low} as well as the temperature perturbation θ_{low}. Nevertheless, it must be emphasized that the function $P(\mathrm{x})$ is not completely arbitrary, and that it is connected to the function A(x) through a relation which is derived via the analysis of inviscid fluid motion in the upper deck via (6.78f).

The reader can find in Guiraud's paper "Going on with asymptotics" [127], pp. 262–71, a very pertinent and personal point of view concerning the asymptotics of the triple-deck theory. In particular, Guiraud writes:

> Triple-deck theory deserves a special mention among the numerous applications of high Reynolds numbers laminar flows. It is somewhat misleading to associate high Reynolds numbers with laminar flow, although such an association is not completely false at the outset. What can be said here is that, provided some caution is exercised when applying the theory, it has proven to give very powerful results. What is unfortunate with this kind of asymptotics is that it seems to be very esoteric, on the one hand, while it looks suspect, on the other hand, for at least two reasons. The first is that in its simplest version – the only one reported here – it is based on $(\mathrm{Re}_{\mathrm{L}^\circ})^{-1/8}$ being much smaller that unity; the second is that the higher $\mathrm{Re}_{\mathrm{L}^\circ}$, the less plausible the laminar conditions seem to be. It is outside the scope of this paper to discuss this last issue, and we ask the reader to believe that laminar conditions can exist (obviously within a limited range). I only intend to debunk, if possible, the esoteric blurring which appears to have for long discouraged many scientists from investing in it. Let me explain the main goal, assuming the reader is familiar with basic boundary layer theory and knows the corresponding hierarchy: computation of the inviscid flow first, followed by boundary-layer computation as such, finally (eventually) computations of corrections, generally of order $(\mathrm{Re}_{\mathrm{L}^\circ})^{-1/2}$, to both. The classical theory is based on what might called an asymptotic ansatz: namely, that a thin (of order $(\mathrm{Re}_{\mathrm{L}^\circ})^{-1/2}$ thick) layer of fluid surrounds the surface of an undeformable smooth body on a O(1) scale. Each time this asymptotic ansatz is broken, one should expect a failure of classical boundary-layer theory.

The triple-deck structure is now seen as a useful and, indeed, valuable element in aerodynamics calculation and design. This is a substantial tribute to Stewartson's power and foresight, and is his greatest contribution to theoretical fluid mechanics (see, for instance, Stewartson (1974) [137]). For a pertinent comprehensive review of progress in using the asymptotics of the triple-deck theory, see the illuminating paper by Meyer (1983) [138].

Guiraud was fascinated by the breakdown of asymptotic theories at the leading and trailing edges, and in his 1974 study (the first in France) – in particular in his study of separation at the trailing edge of a thin three-dimensional wing – shows that for this complicated case, three different scalings arise, and that the Stewartson layered triple-deck structure is very well adapted.

It should be noted that the triple-deck theory was discovered by Stewartson and Williams in England in 1969, but also independently by Neiland [139] in Soviet Russia. Paul Germain [107] has written:

> Even a genius would not have been able to build the whole of the triple-deck model without the help of matched asymptotic expansion techniques. Triple-deck theory is now a very important building stone in the new "fluid dynamics inspired by asymptotics", and it may be fully included within the heritage of Prandtl.

Part III
Applications of the RAM Approach to Aerodynamics, Thermal Convection, and Atmospheric Motions

Chapter 7
The RAM Approach in Aerodynamics

7.1 Derivation of a Through-Flow Model Problem for Fluid Flow in an Axial Compressor

First, in Sect. 7.1.1, we again consider the simple case examined in Veuillot's thesis [140] devoted to turbomachinery fluid flow, simulated in Sect. 6.3.1; then, in Sect. 7.1.2, the more sophisticated G–Z RAM Approach [141].

It is obvious that the following asymptotic theory of axial flow through a turbine, which is likely be of considerable interest to specialists, is a fascinating application of a complicated engineering problem using the RAM Approach with the basic large parameter being the number of turbine blades per rotor.

7.1.1 The Veuillot Approach

We starting from the system of non-dimensional equations (6.18a)–(6.18c), for u^*, w^*, and $\Gamma^* = rv^*$, and also the relations (6.17a)–(6.17c) relying, u^*, w, and v^*, with the functions ψ^* and Θ.

First, we expand the functions u^*, w^*, Γ^*, and ψ^*, relative to our main small parameter ε:

$$(u^*, w^*, \Gamma^*, \psi^*) = ({u^*}_o, {w^*}_o, \Gamma^*_o, {\psi^*}_o) + \varepsilon\, (u^*_1, w^*_1, \Gamma^*_1, \psi^*_1) + \ldots \tag{7.1}$$

and at zeroth-order we derive the following leading-order approximate equations for the functions $u^*_o = (1/r\Delta)\, \partial\, \psi^*_o/\partial z, w^*_o = -(1/r\Delta)\, \partial\, \psi^*_o/\partial r,\quad \Gamma^*_o = r^2(u^*_o \partial\Theta/\partial r\ + {w^*}_o\, \partial\Theta/\partial z)$: namely,

$$\partial u^*_o/\partial\chi + \ldots.. + (\partial\Theta/\partial r)\, \partial{\Gamma^*}_o/\partial\chi = 0 \tag{7.2a}$$

DOI 10.1007/978-3-642-20746-4_7, © Springer-Verlag Berlin Heidelberg 2012

$$\partial w^*_o/\partial\chi + \cdots + (\partial\Theta/\partial z)\,\partial\Gamma^*_o/\partial\chi = 0, \tag{7.2b}$$

$$(\partial\Theta/\partial z)\,\partial u^*_o/\partial\chi - (\partial\Theta/\partial r)\partial w^*_o/\partial\chi + \cdots = 0. \tag{7.2c}$$

But these three equations are not independent, since the determinant of coefficients for $\partial u^*_o/\partial\chi$, $\partial w^*_o/\partial\chi$, and $\partial\Gamma^*_o/\partial\chi$ is zero! On the other hand, from them we easily derive the following two relations:

$$\partial\Gamma^*_o/\partial\chi = r^2\,[(\partial\Theta/\partial r)\partial u^*_o/\partial\chi + (\partial\Theta/\partial z)\,\partial w^*_o/\partial\chi] \tag{7.3a}$$

$$(\partial\Gamma^*_o/\partial\chi)\,\{1 + r^2[(\partial\Theta/\partial r)^2 + (\partial\Theta/\partial z)^2]\} = 0. \tag{7.3b}$$

Because:

$$1 + r^2[(\partial\Theta/\partial r)^2 + (\partial\Theta/\partial z)^2] \neq 0,\ \partial\Gamma^*_o/\partial\chi \equiv 0,$$

and as a consequence:

$$\partial\Gamma^*_o/\partial\chi = 0,\ \partial u^*_o/\partial\chi = 0, \partial w^*_o/\partial\chi = 0 \tag{7.4}$$

Therefore: when ε tends to zero, and as a consequence of a uniform and constant steady flow far of the row, in the leading-order the approximate limiting through-flow in the row of an axial compressor (turbo-machine) is independent of the short (micro)-scale χ.

Therefore, it is necessary to consider in the starting Eqs. 6.18a–6.18c the next order (terms proportional to ε!) In a such case we obtain, for $u^*{}_1$, $w^*{}_1$, and Γ^*_1, the following three equations:

$$(1/r)\partial\Gamma^*_o/\partial r = (1/r\Delta)[\partial u^*_1/\partial\chi + (\partial\Theta/\partial r)\,\partial\Gamma^*_1/\partial\chi]; \tag{7.5a}$$

$$\partial u^*_o/\partial z - \partial w^*_o/\partial r = (1/\Delta)[\,(\partial\Theta/\partial z)\,\partial u^*{}_1/\partial\chi - (\partial\Theta/\partial r)\,\partial w^*{}_1/\partial\chi]; \tag{7.5b}$$

$$(1/r)\,\partial\Gamma^*_o/\partial z = (1/r\Delta)[\partial w^*_1/\partial\chi + (\partial\Theta/\partial z)\partial\Gamma^*_1/\partial\chi]. \tag{7.5c}$$

From these equations, by elimination of the first-order functions, $u^*{}_1$, $w^*{}_1$, and Γ^*_1, we obtain as a compatibility relation:

$$\partial u^*_o/\partial z - \partial w^*_o/\partial r = (\partial\Theta/\partial z)\,\partial\Gamma^*_o/\partial r - (\partial\Theta/\partial r)\,\partial\Gamma^*_o/\partial z$$

and with

$$u^*_o = (1/r\Delta)\,\partial\psi^*_o/\partial z, w^*_o = -(1/r\Delta)\,\partial\psi^*_o/\partial r$$

we derive an equation for the function $\psi_o^* = \Lambda_o(r, z)$, which characterizes the limiting through-flow in the row. Namely:

$$\partial/\partial r\{(1/r\Delta)\partial\Lambda_o/\partial r\} + \partial/\partial z\{(1/r\Delta)\partial\Lambda_o/\partial z\} = (\partial\Theta/\partial z)\partial\Gamma_o^*/\partial r - (\partial\Theta/\partial r)\,\partial\Gamma_o^*/\partial z. \tag{7.6}$$

This model equation for the function Λ_o is our first main rational and entirely consistent result with the RAM Approach.

Again, due to being far upstream of the row, we have a uniform and constant steady incompressible($\rho_o =$ constant) fluid flow: $(1/2)\rho_o u^2 + p =$ constant. Then the jump,

$$|[p]| \equiv p_{\chi=+1/2} - p_{\chi=-1/2} \tag{7.7a}$$

of the pressure, from blade to blade, gives

$$|[p]| = -\,\varepsilon\,\rho_o\{u^*{}_o|[u_1^*]| + w^*{}_o|[w_1^*]| + (1/r^2)\Gamma_o^*|[\Gamma_1^*]|\} + O(\varepsilon^2) \tag{7.7b}$$

where $|[p]|$ is a quantity of the order ε.

But, according to Eqs. 7.5a–7.5c, obviously $u^*{}_1$, $w^*{}_1$, and Γ_1^* are linear functions of the microscale structure χ.

As a consequence of (7.7b), we have to write:

$$\mathrm{Lim}_{\varepsilon\to 0}(-p/\varepsilon\rho_o) = u_o^*\partial u_1^*/\partial\chi + w_o^*\partial w_1^*/\partial\chi + (1/r^2)\Gamma_o^*\partial\Gamma_1^*/\partial\chi = \Pi_o. \tag{7.8}$$

From (7.5a–7.5c), we now eliminate the terms with the χ – derivatives in the (7.8) relation, and express the function Π_o, in (7.8) simply by:

$$\Pi_o = \Delta[u_o^*\partial\Gamma_o^*/\partial r + w_o^*\partial\Gamma_o^*/\partial z] \tag{7.9a}$$

in the row.

Outside the row, p remains continuous, even in the presence of wakes – which are, in the considered Eulerian fluid flow, only vortex sheets (contact discontinuity surfaces).

Finally, taking into account the periodicity in χ, *outside the row*, we derive in place of (7.9a) the following relation:

$$|[p]| = 0 \Rightarrow \Pi_o = 0 \tag{7.9b}$$

and

$$u_o^*\partial\Gamma_o^*/\partial r + w_o^*\partial\Gamma_o^*/\partial z = 0 \Rightarrow \Gamma_o^* = \Gamma_o^*(\Lambda_o). \tag{7.9c}$$

As a consequence:

$$\text{In the whole outside region upstream ot the row}: \Gamma^*_o = 0 \tag{7.10a}$$

$$\text{In downstream region outside of the row}: \Gamma^*_o = \Gamma^*_o(\Lambda_o) \tag{7.10b}$$

From the above we can formulate the following main results relative to a homogenized through-flow.

The velocity vector $\mathbf{U}^*o = (u^*_o,\ w^*_o,\ v^*_o)$ of the homogenized through-flow is

$$\mathbf{U}^*_o = (1/\Delta)[\nabla(\theta - \Theta) \wedge \nabla\Lambda_o], \tag{7.11}$$

such that the streamlines of the through-flow are obtained by the crossing of *median surfaces*,

$$\theta = \Theta(r,\ z) + \ \text{constant}$$

in the inter-blade row–channel, with the *cylindrical surfaces*, resulting from the rotation around of the z-axis of the turbo-machine of meridian streamline surfaces

$$\Lambda_o = \ \text{constant}$$

For this through-flow we have for the function $\Lambda_o(r,\ z)$, (see Eq. 7.6):

$$\begin{aligned} &\partial/\partial r\{(1/r\Delta)\partial\Lambda_o/\partial r\} + \partial/\partial z\{(1/r\Delta)\partial L_o/\partial z\} \\ &\quad = (\partial\Theta/\partial z)\,\partial\Gamma^*_o/\partial r - (\partial\Theta/\partial r)\,\partial\Gamma^*_o/\partial z, \end{aligned} \tag{7.12a}$$

with, as conditions (if we use (7.11) for the composante $v^*{}_o$):

$$\Gamma^*_o = \ 0, \ \text{upstream of the row}, \tag{7.12b}$$

$$\Gamma^*_o = (r/\Delta)\ \ [\partial\Theta/\partial r)\ \partial\Lambda_o/\partial z - (\partial\Theta/\partial z)\,\partial\Lambda_o/\partial r], \text{in the row}, \tag{7.12c}$$

$$\Gamma^*_o = \Gamma^*_o(\Lambda_o),\ \text{downsream of the row}. \tag{7.12d}$$

This axially symmetric through-flow model, which is dependent only on coordinates r and z, introduces a fictitious force:

$$\boldsymbol{F} = (\Pi_o/\Delta)\nabla(\ \theta - \Theta), \tag{7.12e}$$

which simulates the action of the blades in the row on the turbomachinery flow.

The force $\boldsymbol{F}$, given by (7.12e), is a memory term (a trace) which – via homogenization – replaces (simulates) the (vanishing) effect of the blades in the row.

The first numerical applications of the above through-flow model in a turbomachine blade row was realized by Veuillot [140] at the ONERA and see also his [142] paper.

7.1.2 The G–Z Approach

In a more sophisticated general case, Guiraud and Zeytounian [141] consider in 1971 at the beginning, in the cylindrical coordinates r, θ z, the following starting Eulerian incompressible equation written in the matrix form:

$$\partial \mathrm{T}/\partial \mathrm{t} + \partial \mathrm{R}/\partial \mathrm{r} + \partial \mathrm{Z}/\partial \mathrm{z} + (1/\mathrm{r})\partial \mathrm{S}/\partial \theta + \mathrm{H}/\mathrm{r} = 0, \tag{7.13}$$

and make the change of coordinates from $(\mathrm{t},\ \mathrm{r},\ \theta,\ \mathrm{z})$ to $(\mathrm{t},\ \mathrm{r},\ \mathrm{z},\ \chi)$ as shown, in (7.14a):

$$\theta = \Theta(\mathrm{t},\ \mathrm{r},\ \mathrm{z}) + 2\pi\varepsilon(\mathrm{k} + \chi), \tag{7.14a}$$

with the idea in mind that through-flow will be independent of χ whereas r will appear as a parameter for cascade flow.

Without any approximation the flow has to be periodic in χ, and we enforce this by:

$$U_{\mathrm{k}}(\mathrm{t},\mathrm{r},\mathrm{z},\ \varepsilon;\ \chi+1) = U_{\mathrm{k+1}}(\mathrm{t},\mathrm{r},\mathrm{z},\ \varepsilon;\ \chi), \tag{7.14b}$$

$$U_{\mathrm{k+N}}(\mathrm{t},\mathrm{r},\mathrm{z},\ \varepsilon;\ \chi) = U_{\mathrm{k}}(\mathrm{t},\mathrm{r},\mathrm{z},\ \varepsilon;\ \chi), \tag{7.14c}$$

using for convenience the index k which runs from 1 to N ($\varepsilon = 1/\mathrm{N} << 1$), the number of blades in a row, and accordingly we assume that χ is between zero and one.

We expand, formally, U_{k} as powers of ε, but we obviously need two such expansions, because it is clear that the model through-flow in row is invalid near the locus of the leading/trailing edges of the row.

The first one is a kind of outer expansion (as in (7.1) above):

$$U_{\mathrm{k}} = U_{\mathrm{k,o}} + \varepsilon\, U_{\mathrm{k,1}} + \ldots., \tag{7.15a}$$

and will fail near both ends of the row where (two) inner expansions

$$U_{\mathrm{k}}(\mathrm{t},\mathrm{r},\ \mathrm{z} = \mathrm{h}(\mathrm{r}) + \varepsilon\zeta,\ \varepsilon;\ \chi) = U^{*}_{\mathrm{k,0}} + \varepsilon\, U^{*}_{\mathrm{k,1}} + \ldots, \tag{7.15b}$$

are needed. In (7.15b), $\mathrm{z} = \mathrm{h}(\mathrm{r})$ is the locus of the leading (or trailing) edge of a row.

When the change of coordinates(from θ to χ) is made in the basic matrix form Eq. 7.13 of the inviscid and incompressible Eulerian fluid flow, we obtain (7.16a, b):

$$\partial \Gamma_k / \partial \chi + 2\pi \varepsilon r L_k = 0 \quad (7.16a)$$

with

$$L_k \equiv \partial T_k / \partial t + \partial R_k / \partial r + \partial Z_k / \partial z + H_k / r \quad (7.16b)$$

and we observe that in matrix column Γ_k the following two parameters are present:

$$\lambda = \omega^\circ D / w^\infty, \; \mu = D / w^\infty t^\circ \quad (7.16c)$$

and we have introduced

$$\gamma_k = v_k - r u_k \partial \Theta / \partial r - r w_k \partial \Theta / \partial z - r \partial \Theta / \partial t \quad (7.16d)$$

In (7.16c), t° is a reference time, D is the diameter of the row, ω° is the reference value of the angular velocity (ω) of the row, and w^∞ is the upstream uniform axial velocity.

Two facts should be stressed at the outset. First, if we assume axially symmetric flow, $\partial / \partial \chi = 0$, we obtain $L_k = 0$, and it may be checked that $L_k = 0$ is the matrix form of axially symmetric through-flow. Second, if we use, by "brute force" $\Rightarrow$ $\varepsilon = 0$ in (7.16a), we do not obtain the equations of axially symmetric flow but, rather, the highly degenerate equation $\partial \Gamma_k / \partial \chi = 0$! This is somewhat strange, but is not unexpected.

The way in which χ has been defined substantiates that: when ε is small, variations in the χ direction are magnified by $1/\varepsilon$, in comparison with variations in t, and in the r, or z direction.

Now, substituting the basic outer expansion (7.15a) in Eq. 7.16a we derive a hierarchy of equations. But here we write only first two:

$$\begin{aligned} &\partial \Gamma_{k,0} / \partial \chi = 0, \\ &\partial \Gamma_{k,1} / \partial \chi + 2\pi r L_{k,0} = 0, \end{aligned} \quad (7.17a)$$

which consist of equations to be solved in turn.

We choose, as appropriate to the present problem, the solution of

$$\partial \Gamma_{k,0} / \partial \chi = 0 \quad (7.17b)$$

for which $u_{k,0}$, $w_{k,0}$, $v_{k,0}$, and $p_{k,0}$ are all independent of χ.

At this step we do not know the way in which these functions depend on t, r, and z! Now, if we use the second equation of two equations (7.17a) in order to compute

$u_{k,1}, \ldots$ and so on, we encounter a compatibility condition arising from periodicity, which forces $L_{k,0}$ to be zero!

We have thus obtained a through-flow, axially symmetric theory. The interesting point is that we may go a step further and produce a through-flow theory to order ε inclusively! For this, it is first necessary to define the channel between two consecutive blades:

$$\chi_e \leq \chi \leq \chi_i, \text{ and we define}$$

$$\Delta(r, z) = \chi_i - \chi_e$$

and we then introduce an average, < >, and a jump, [ ], operation, thus:

$$<U> = (1/\Delta) \int U d\chi, \text{ integration from } \chi_e \text{to } \chi_i \tag{7.18a}$$

and

$$[U] = U\chi_i - U\chi_e. \tag{7.18b}$$

Now, if we think of the pressure for U (in (7.18b)), then the bracketed [p] may be viewed as: the pressure difference between the two sides of one and the same blade.

Below, the various equations shows the basic results of the G–Z RAM Approach.

Up to first order in ε, the average of velocity and pressure

$$<\mathbf{V}^{(1)}> = \mathbf{V}_{k,0} + \varepsilon<\mathbf{V}_{k,1}>, <p^{(1)}> = p_{k,0} + \varepsilon<p_{k,1}> \tag{7.19a}$$

satisfies, with an error of order ε^2, axially symmetric through-flow equations:

$$\begin{aligned} &\mathrm{Div}(\Delta<\mathbf{V}^{(1)}>) = O(\varepsilon); \\ &\partial<\mathbf{V}^{(1)}>/\partial t + \{\mathrm{rot}<\mathbf{V}^{(1)}> + 2\Omega \mathbf{e}_z\} \wedge <\mathbf{V}^{(1)}> + \nabla I^{(1)}, \\ &= \boldsymbol{F}^{(1)} + O(\varepsilon^2), \end{aligned} \tag{7.19b}$$

where

$$\begin{aligned} I^{(1)} &= <p^{(1)}> + (1/2)|<\mathbf{V}^{(1)}>|^2 - (1/2)\Omega^2 r^2; \Omega = \lambda\,\omega, \\ \Pi^{(1)} &= (1/2\,\pi)[<p_{k,1}> + \varepsilon<p_{k,2}>], \\ \Sigma &= S + 2\,\pi\,\varepsilon\,[(1/2)(\chi_i + \chi_e)], \; S = \Theta - \theta, \end{aligned} \tag{7.19c}$$

with

$$\partial\Sigma/\partial t + <\mathbf{V}^{(1)}>.\nabla\Sigma = O(\varepsilon^2) \tag{7.19d}$$

$$\boldsymbol{F}^{(1)} \equiv (1/\Delta)\Pi^{(1)}\nabla\Sigma \Rightarrow \boldsymbol{F}^{(1)}.\,\mathrm{rot}\,\boldsymbol{F}^{(1)} = 0. \tag{7.19e}$$

Two points again need to be stressed. First, the breadth of the channel from blade to blade, set as

$$\Delta(\mathrm{r},\ \mathrm{s})$$

enters in the continuity equation in an obvious way. Second, in the momentum equation there is a source term,

$$(1/\Delta)\,\Pi^{(1)}\nabla\Sigma \equiv \boldsymbol{F}^{(1)}$$

which is proportional to the jump in pressure and is orthogonal to $\Sigma =$ constant – a surface which is just in the middle of the channel. This force has long been known in through-flow theory; but in fact, via a very subtle ad hoc consideration (first published, it seems, by Chung-Hua Wu, in NACA TN 2288 (1951); see also Wu [143] paper).

The force $\boldsymbol{F}^{(1)}$ occurs from redistribution (homogenization) of forces acted on the flow by the blades of the row.

The G–Z [141] derivation given above is illuminating with regard to the error involved in the approximation. To order one there is a dependency on χ which may be computed once the through-flow is known.

7.1.3 Transmission Conditions, Local Solution at the Leading/Trailing Edges, and Matching

The above through-flow model in axial turbomachine is invalid near the locus of leading/trailing edges of a row.

According to G–Z theory [144], a local asymptotic analysis is performed by considering the inner expansions (7.15b) and rewriting the starting matrix equation (7.13). We obtain:

$$\partial\Gamma_{\mathrm{k}}^{*}/\partial\chi + 2\pi\mathrm{r}\partial\mathrm{N}_{\mathrm{k}}^{*}/\partial\zeta + 2\pi\varepsilon r M_{\mathrm{k}}^{*} = 0 \tag{7.20a}$$

with

$$\mathrm{N}_{\mathrm{k}}^{*} = \mathrm{Z}_{\mathrm{k}}^{*} - (\mathrm{dh}/\mathrm{dr})R_{\mathrm{k}}^{*} \tag{7.20b}$$

$$M_{\mathrm{k}}^{*} = \partial\mathrm{T}_{\mathrm{k}}^{*}/\partial\mathrm{t} + \partial\mathrm{R}_{\mathrm{k}}{}^{*}/\partial\mathrm{r} + \mathrm{H}_{\mathrm{k}}^{*}/\mathrm{r} \tag{7.20c}$$

and to zeroth order we obtain:

$$\partial\Gamma_{\mathrm{k},0}^{*}/\partial\chi + 2\pi\mathrm{r}\partial\mathrm{N}_{\mathrm{k},0}^{*}/\partial\zeta = 0, \tag{7.21}$$

which is, in fact, the equations of cascade flow – but the configuration is that of semi-infinite cascade flow. In [144] a detailed analysis of (7.21) is performed, adapted to a local frame linked with the curve:

$$\Gamma : \{z = h(r); \theta = \Theta(r, h(r))\}$$

The semi-infinite cascade flow fills the gap between external (outside the row), force-free, axially symmetric through-flow, and internal (in row) through-flow with the source term $\boldsymbol{F}^{(1)}$. Matching provides transmission conditions between these two disconnected through-flows.

The necessity of such conditions appears readily as soon as any numerical treatment of the whole through-flow in a two-row stage is attempted – from upstream to downstream (infinity) of this two-row stage!

To zeroth order these transmission conditions are rather simple – and, indeed, obvious on physical grounds: They mean that mass flow is conserved, as well as the component of momentum parallel to the leading or trailing edge.

Local analysis has also been carried out by Guiraud and Zeytounian [144], to first order, without a simple interpretation of the (rather complicated) result linked with the transmission conditions!

We can consider the singular regions near the entry and exit of the row as planes of discontinuity, if we impose the associated transmission conditions.

7.1.4 Some Complements

We now turn, briefly, to various cases concerning my work devoted with Guiraud during 1969–1978, to turbomachinery fluid flows.

After the axial flow in a turbomachine, with the approximation of ideal incompressible flow, has been analyzed by using an asymptotic method, assuming that the blades are infinitely near one another – [141], and in a companion paper [144] – a local study reveals the nature of the flow in their neighbourhood and leads to a system of transmission conditions, because the partial differential equations of the through-flow (in three different regions: upstream of a row, in a row, and downstream of the row) must be supplemented by them in order to produce a well-posed problem for the whole of the turbomachine (from upstream of the row to downstream of this row) – an application of the concept of multiple scales was considered.

Namely, in [145] an asymptotic theory for the flow in an axial compressor was considered, with the aim of devising a coupling process between the so-called meridian through-flow and the flow around cascades. Again the small parameter ε is the inverse of the (supposed $\gg 1$) number of blades per row and/or number of stages. As a matter of fact, the cascade flow is treated as a small perturbation of the through-flow, and has to be computed, locally, as the two-dimensional unsteady flow around an array of couples of cascades alternately fixed and in motion. The array is constructed by developing on a plane the section of the compressor by a circular cylinder, and continuing, by periodicity, the pair of cascades so obtained, at each location. The coupling between through-flow and cascade flow is part of the analysis. It happens, incidentally, that the equations of through-flow are obtained through an averaging process, completed on a domain of periodicity of the array of cascades flow, while the through-flow appears locally as an unperturbed flow for the linearized problem defining the cascade flow. The 3D nature of the complete flow is built in by the coupling itself, as is visualized by the occurrence of source terms in each of the two sets of equations describing through-flow and cascade flow. This paper [145] is aimed at producing a preliminary answer to the question of: how to devise, as rationally as possible, a way of describing the familiar scheme of cascade flow within the computation of a mean through-flow. The main conclusion is that the concept of cascade flow should be revisited and reassessed as one of unsteady flow around an array of cascades.

In 1978, I published, with Guiraud, a fourth paper [146], entitled "Cascade and through-flow theories as inner and outer expansions". In this, a technique of matched asymptotic expansions is used in order to combine two kinds of approximation. Through-flow theory forms the basis for an outer expansion, while cascade theory forms the basis for an inner one, and matching provides boundary conditions for both flows. It appears that for the downstream through-flow, a technique of multiple scales is necessary – at least in the vibrating case (vibrations induced by harmonic vibrations of the blades) – in order to deal with the unsteady wakes generated by the vibrating blades, and slowly modulated downstream by the steady part of the through-flow.

Although there are very many good papers on the theory of turbomachine flow, we have not found any attempt analogous to that described above in [145] and [146], and it seems difficult to comment on the relation of the work to be presented with others. Concerning the two papers [141] and [144], we observe that in Sirotkin's paper [147] only some results are similar to ours, but the main difference is in the approach, which is less systematic.

Our objective, with Guiraud, concerning the papers [141] and [144–146] was very modest, and we did not solve any problems nor present any results!

What we have proposed in our above-mentioned papers devoted to a rational asymptotic description of turbomachine flows in the framework of a RAM Approach, may be stated as follows.

Considering incompressible non-viscous fluid flow through a one-row machine, assuming that there is a great number of blades, and, in [146], that the corresponding cascade has a chord-to-spacing ratio of order one, we want to show that the first few terms of an asymptotic representation of the 3D flow may be guessed as having the form of an inner and outer multiple-scale expansion.

We confirm our guess – as is usually done with problems not amenable to mathematically rigorous analysis – by an *internal consistency argument*: we show that each term of the expansion, up to the order considered, may be computed by solving well-set problems. We show the rational asymptotic process by which these problems may be extracted from the definition of the original 3D problem, which is a typical RAM Approach.

For *engineering applications* it would have been very useful to find, as partial problems, cascade flow theory as well as through-flow theory across a thick row. Unfortunately, we have been unable to find any asymptotic process leading to such a scheme. As a matter of fact, the obvious way to do so leads to only two significant degeneracies. One is the through-flow of [141] and [144], which leaves no room for cascade flow, and the other is the one considered in [146], which leads to cascade flow but leaves no room for through-flow, including a thick row.

This conclusion inevitably leads to some deception, because there is no way to embed Wu's [143] technique within an asymptotic rational framework.

7.2 The Flow Within a Cavity Which is Changing Its Shape and Volume with Time: Low Mach Number Limiting Case

Concerning the aerodynamics applications it is necessary to distinguish between confined and unconfined flows and, especially, to elucidate the role of the Strouhal number S (unsteadiness).

When we consider the low Mach number case, two distinguished limiting processes emerge. One of them leads, from the full unsteady NS–F equations, to the model of incompressible flow (Navier equations) and, in the case of confined configuration provides a dynamic interpretation of thermostatics. The other one is at the root of acoustics.

The first limiting process (incompressible) corresponds roughly to

$$\mathrm{M} \to 0 \text{ with S fixed}$$

while the second limiting process (of acoustics) corresponds to

$$\mathrm{M} \to 0 \text{ with SM } = \mathrm{O}(1).$$

Curiously enough, the acoustic model enters into the scene even in the situation which is apparently ruled by incompressible aerodynamics.

This occurrence is due to non-uniformities: a spatial one near infinity in the case of unconfined flow, and a temporal one for small time (in particular near time = 0, where the initial data are given) and also for high-frequency oscillations in the case of confined flows.

In [85], the influence of these high-frequency oscillations was taken into account by Zeytounian and Guiraud via a judicious multiple-scales technique, but with an infinity of short acoustic scales! Here, below, we consider, as a physical situation, the low Mach number flow within a cavity which is changing its shape and volume with time. Such a problem presents industrial interest in the case of the compression phase flow in an internal combustion engine (see our short notes in [86]).

We show how a limiting process corresponding to the Mach number going to zero leads to an incompressible unsteady model flow, provided that acoustic waves are averaged out over a great number of periods. This scheme may even describe the case of a gas with a purely temporal variation of density due to substantial changes of volume.

7.2.1 Formulation of the Inviscid Problem

We start from the Euler compressible dimensionless equations:

$$\begin{aligned} &\mathrm{D}\rho/\mathrm{Dt} + \rho\nabla.\mathbf{u} = 0\\ &\mathrm{D}\mathbf{u}/\mathrm{Dt} + (1/\gamma\mathrm{M}^2)\nabla\mathrm{p} = 0\\ &\mathrm{D}S/\mathrm{Dt} = 0\\ &\mathrm{p} = \rho^{\gamma}\exp S \end{aligned} \tag{7.22a}$$

where $\mathrm{D/Dt} = \partial/\partial\mathrm{t} + \mathbf{u}.\nabla$, with standard notation.

The Strouhal number in the first three of these equations is assumed equal to one, such that the reference time is $\mathrm{t}^\circ = \mathrm{L}^\circ/\mathrm{U}^\circ$, with L° a typical length scale of the cavity $\Omega(\mathrm{t})$, and U° a characteristic velocity related to the motion of the wall $\partial\Omega(\mathrm{t}) \equiv \Sigma(\mathrm{t})$.

As boundary condition we write the slip condition:

$$(\mathbf{u}.\mathbf{n})|_{\Sigma(\mathrm{t})} \equiv \mathrm{w}|_{\Sigma(\mathrm{t})} = W(\mathrm{t},\mathbf{P}), \mathbf{P} \in \Sigma(\mathrm{t}) \tag{7.22b}$$

where the (data) velocity $W(\mathrm{t}, \boldsymbol{P})$ characterizes the normal displacement of the wall $\Sigma(\mathrm{t})$, and $\mathbf{n}$ is the unit vector normal to this wall, directed inside Ω (t) – $\boldsymbol{P}$ being the position point vector on the wall $\Sigma(\mathrm{t})$.

But it is also necessary to take into account the conservation of the global mass m° of the cavity (a bounded domain with L° as a diameter). In the dimensionless reduced form we have:

$$\rho = 1/\mathrm{V(t)};\ \ \mathrm{V(t)} = \rho^\circ|\Omega(\mathrm{t})|/m^\circ, \mathrm{V}(\mathrm{t}=0) = 1 \tag{7.22c}$$

where $|\Omega\ (\mathrm{t})|$ is the volume of the cavity – a known function of time t.

As initial conditions we write:

$$\mathrm{t} = 0 : \mathbf{u} = 0,\ \ \mathrm{p} = \rho = 1 \text{ and } \mathrm{S} = 0 \tag{7.22d}$$

More precisely, we consider the case when the motion of the wall, $\Sigma(\mathrm{t})$, is started impulsively from rest, and in a such case,

$$W(\mathrm{t},\boldsymbol{P}) = H(\mathrm{t})\mathrm{W}_{\Sigma}(\boldsymbol{P}), \text{ all along } \Sigma(\mathrm{t}). \tag{7.23a}$$

In this condition, the function H(t) is the Heaviside (or unit) function, such that

$$\mathrm{Lim}_{\mathrm{t}\to 0^+} H(\mathrm{t}) \equiv 1, \text{ but initially} : H(\mathrm{t} = 0^-) \equiv 0. \tag{7.23b}$$

7.2.2 *The Persistence of Acoustic Oscillations*

Asymptotics analysis and the RAM Approach in the formulated problem above is not easy task, mainly due to the persistence of acoustic oscillations in the cavity emerging for $\mathrm{t} = 0^+$.

Therefore, if the Mach number, M, is sufficiently small, the zeroth order approximation leads to the thermostatic isentropic evolution of the gas within the cavity as a whole. Superimposed onto them we have acoustic oscillations which remain undamped as long as viscosity is neglected – when ones takes it into account the Euler equations (7.22a), in place of full unsteady NS–F equations.

When the NS–F equations are considered (in place of Euler equations (7.22a)), then a rather longer time $\mathrm{O}(\mathrm{Re}^{1/2})$ is necessary in order to damp out the oscillations – but a much longer time is necessary in order that the heat exchanges can take place. As a matter of fact, this time is O(Re)!

Indeed, some new features occur (as a consequence of the unsteadiness of the compressible fluid flow) when one deals with internal aerodynamics. The first concerns the leading term in the expansion of pressure which is function of time instead of being a constant. We write:

$$\mathbf{u} = \mathbf{u}_0 + \mathrm{M}\mathbf{u}^* + \mathrm{M}^2(\mathrm{u}^{**} + \mathbf{u}_2) + \ldots, \tag{7.24a}$$

$$\mathrm{p} = \mathrm{P}_0(\mathrm{t}) + \mathrm{M}^2(\mathrm{p}^* + \mathrm{p}_2) + \mathrm{M}^3\mathrm{p}^{**} + \mathrm{M}^4(\mathrm{p}^{***} + \mathrm{p}_4) + \ldots \tag{7.24b}$$

and an expansion similar to the one for p is valid for density ρ.

At the leading order, one finds that $\rho_0(\mathrm{t})$ or $\mathrm{P}_0(\mathrm{t})$ belongs to a family of adiabatic thermostatic evolution of the gas in the cavity (container-bounded domain), and are determined from the overall conservation of mass. That is, they do not depend on position either. Furthermore:

$$\mathrm{P}_0(\mathrm{t}) = [\,\rho_0(\mathrm{t})]^{\gamma} \text{ and } \rho_0(\mathrm{t}) = 1/\mathrm{V}(\mathrm{t}). \tag{7.25}$$

The pair ($\mathbf{u}_0$, p_2) belongs to a so-called "quasi-incompressible" model, but, as a second peculiar feature, we find that it is perturbed by the pair ($\mathbf{u}^*$, p^*), which consists in acoustic oscillations generated during the setting-up of the motion. Therefore, as demonstrated below, we have:

$$\mathbf{u}^* = -\Sigma_{n\geq 1}B_n(t)\sin[\phi_n(t)/M]U_n(t,\mathbf{x}), \tag{7.26a}$$

$$\rho^* = \rho_0(t)[\rho_0/P_0(t)]^{1/2}\Sigma_{n\geq 1}B_n(t)\cos[\phi_n(t)/M]R_n(t,\mathbf{x}), \tag{7.26b}$$

where

$$d\phi_n(t)/dt = [P_0(t)/\rho_0(t)]^{1/2}\omega_n(t). \tag{7.26c}$$

In (7.26c), $\omega_n(t)$ is one of the acoustic frequencies corresponding to the shape of the container at time t, while the pair $\{U_n(t,\mathbf{x}), R_n(t,\mathbf{x})\}$ serves to define the normal mode of oscillation at frequency $\omega_n(t)$ normalized to:

$$\int_{\Omega(t)} [(\mathbf{U}_n)^2 + (R_n)^2]dv = 1$$

where the integral is over the bounded container.

A third peculiar feature is that the first correction due to compressibility corresponds to the pair,

$$(\mathbf{u}_2, p_4 + \omega_4(t)), \text{where } 4\omega_4(t) = \gamma\,\Sigma_{n\geq 1}{B_n}^2|\mathbf{U}_n|^2. \tag{7.26d}$$

The B_n are readily found as function of time t by writing down conservation of acoustic energy. We then obtain, in particular:

$$[\rho_0(t)]^{1/2}Bn(t) = Bn(0). \tag{7.26e}$$

On a time $t = O(Re/M^{1/2})$ this acoustic energy is mainly damped out by viscosity and heat conduction within a Stokes-like boundary layer (see Zeytounian and Guiraud [85]).

Of course, turbulent mixing would be much more efficient, and the long time persistency of acoustic oscillations is mainly a further proof that laminar mixing is very poor!

It must be emphasized that on a laminar basis, even when the transient acoustics has been damped out, $P_0(t)$ and $\rho_0(t)$ remain adiabatically related, but a much longer time would be needed for inducing isothermal evolution.

A final feature should be pointed out. Under resonance conditions ($\mathbf{u}^*$, p^*) gains energy from the motion of the container, and their limiting amplitude is derived from a fairly complicated non-linear process which is understood only in one-dimensional situations (Chester [148], Rott [149]).

7.2.3 *Derivation of an Average Continuity Equation*

First we should use the time t, a slow time, and then we would bring into the solution an infinity of fast times designed to cope with the infinity of periods of free vibrations of the cavity Ω(t). Below we set U for the solution *U* expressed through this variety of time-scales, and in such a case we write

$$\partial U/\partial t = \partial U/\partial t \ + (1/M) D U \tag{7.27}$$

where $\partial U/\partial t$ stands for the time derivative computed when all fast times are maintained constant, while (1/M)*D*U is the time derivative (with *D*, a differential operator) occurring through all the fast times.

We carry such a change into the starting Euler equations (7.22a), and then expand according to:

$$U = U_0 + M\ U_1 + M^2 U_2 + \ldots, \quad \text{with U } \langle U \rangle + U^* \tag{7.28}$$

where ⟨U⟩ is average (over all rapid oscillations and depends only of the slow time t and space position **x**) of U, and U* is the fluctuating (oscillating, which depends of all fast times) part of U. More precisely, the operation

$$U \Rightarrow \langle U \rangle \tag{7.29a}$$

erases all the oscillations associated with the fast times, and obviously

$$D\langle U \rangle = 0. \tag{7.29b}$$

For instance, for the fluctuating parts of $\mathbf{u}^*_0$ and ρ^*_1, we can write, respectively, as a (more complete) solution:

$$\mathbf{u}_0^* = \Sigma_{n\geq 1}[A_n(t)\ Cn - B_n(t)\ Sn]\boldsymbol{U}_n \tag{7.30a}$$

$$\rho_1^* = \rho_0(t)[\rho_0(t)/p_0(t)]^{1/2}\Sigma n \geq 1[A_n(t)\ Sn + B_n(t)\ Cn]R_n \tag{7.30b}$$

with

$$Cn = \cos[(1/M)\varphi n(t)] \text{ and } Sn = \sin[(1/M)\varphi n(t)] \tag{7.31a}$$

$$\langle Cp\ Cq \rangle = \langle Sp\ Sq \rangle = (1/2)\delta pq, \quad \text{and } \delta pq = 0, \text{ if } p \neq q, \delta pq = 1, \text{ if } p \equiv q, \tag{7.31b}$$

$$\langle Cn \rangle = 0, \langle Sn \rangle = 0, \langle Cp\ Sq \rangle = \langle Cq\ Sp \rangle \equiv 0, \tag{7.31c}$$

$$DCn = -(d\varphi_n(t)/dt)Sn,\ DSn = (d\varphi_n(t)/dt)Cn, \tag{7.31d}$$

where again

$$d\phi_n(t)/dt = [p_0(t)/\rho_0(t)]^{1/2}\omega_n; \phi_n(0) = 0. \tag{7.32}$$

In (7.30a, 7.30b) $\boldsymbol{U}_n$ and R_n are the normal modes of vibrations of $\Omega(t)$ with eigen-frequencies ω_n: namely,

$$\omega_n R_n + \nabla.\boldsymbol{U}_n = 0, -\omega_n \boldsymbol{U}_n + \nabla R_n = 0; (\boldsymbol{U}_n.\mathbf{n})_{\Sigma(t)} = 0. \tag{7.33}$$

The relation (7.32) defines the scales of the fast times in relation to the speed of sound in cavity (at the time t) and with the eigenfrequencies of the cavity at the same time.

With (7.27), from the Euler equations (7.22a) we obtain the following equations for the functions $\mathbf{u}$, ρ, p, and S:

$$\begin{aligned}
&D\rho + \mathrm{M}\,(\partial\rho/\partial t + \mathbf{u}.\nabla\rho + \rho\nabla.\mathbf{u}) = 0\\
&(1/\gamma)\nabla p + \mathrm{M}\rho D\mathbf{u} + \mathrm{M}^2\rho[\partial\mathbf{u}/\partial t + (\mathbf{u}.\nabla)\mathbf{u}] = 0\\
&DS + \mathrm{M}\,(\partial S/\partial t + \mathbf{u}.\nabla S) = 0\\
&p = \rho^{\gamma}\exp S;
\end{aligned} \tag{7.34a}$$

with the slip condition

$$(\mathbf{u}.\mathbf{n})_{\Sigma(t)} = W(t, \boldsymbol{P}). \tag{7.34b}$$

From the expansion (7.28), at the zero-order, from the above system (7.34a) we derive:

$$Dp_0 = 0, D\rho_0 = 0, DS_0 = 0$$

which shows that ρ_0 and S_0 are independent of the fast times and, as a consequence of the equation of state, that this is also the case for p_0, which is, in fact, a function of only the slow time t:

$$p_0 = p_0(t), \ \rho_0 = \rho_0(t, \mathbf{x}), S_0 = S_0(t, \mathbf{x}). \tag{7.35a}$$

Now, at the first order, from the equation for S, we derive the equation

$$DS_1 + \partial S_0/\partial t + \mathbf{u}_0.\nabla S_0 = 0$$

and, since $S_0 = S_0(t, \mathbf{x})$ is independent of the fast time and $D\langle S_1\rangle = 0$, we have the following average equation for $S_0(t, \mathbf{x})$:

$$\partial S_0/\partial t + \langle\mathbf{u}_0\rangle.\nabla S_0 = 0. \tag{7.35b}$$

But, close to initial time ($t = 0$), when we consider the Euler equations (7.22a) written with the short time, $\tau = t/M$, in place of the slow time, t, we use the local-in-time asymptotic expansion

$$S = S_{a0} + MS_{a1} + MS_{a2} + \ldots$$

where $S_{ak} = S_{ak}(\tau, \mathbf{x})$, $k = 0, 1, 2, \ldots$, and the initial condition

$$S = 0 \text{ at } \tau = 0$$

We derive

$$\partial S_{a0}/\partial\tau = 0, \partial S_{a1}/\partial\tau = 0 \Rightarrow S_{a0} = 0, S_{a1} = 0$$

and, as a consequence, from (7.35b), by continuity

$$S_0 = 0. \tag{7.35c}$$

On the other hand, with (7.35c) we obtain

$$p_0(t) = [\rho_0(t)]^{\gamma} \tag{7.35d}$$

the function $\rho_0(t)$ being determined by the relation

$$\int_{\Omega(t)} \rho_0(t)\mathrm{dv} = \rho_0(t)\int_{\Omega(t)} \mathrm{dv} \Rightarrow \rho_0(t)|\Omega(t)| = m^\circ$$

where $|\Omega(t)|$ is the volume of the cavity, such that

$$\mathrm{d}|\Omega(t)|/\mathrm{dt} = -\int_{\Sigma(t)} W(t, \boldsymbol{P})\, \mathrm{ds} \tag{7.35e}$$

and m° (= const) is the whole mass of the cavity, and, according to the initial condition for the density we have $|\Omega(0)| \equiv m^\circ$.

If, in particular, we assume that $\rho_0(t) \equiv 1$ (and as a consequence $p_0(t) \equiv 1$ also) then $|\Omega(t)| \equiv m^\circ \equiv$const. Obviously, this is not the case in the various applications!

At the first order, from the first three equations of (7.34a), with the above results, we derive the following two equations:

$$D\, \rho_1 + \mathrm{d}\rho_0/\mathrm{dt} + \rho_0\nabla.\mathbf{u}_0 = 0, \tag{7.36a}$$

$$(1/\gamma)\nabla p_1 + \rho_0 D\mathbf{u}_0 = 0, \tag{7.36b}$$

with, from (7.34b),

$$(\mathbf{u}_0.\mathbf{n})_{\Sigma(t)} = W(t, \boldsymbol{P}). \tag{7.36c}$$

Since, $D\langle \rho_1 \rangle = 0$, from (7.36a) we derive an average (zero-order) continuity equation:

$$(1/\rho_0)d\rho_0/dt + \nabla.\langle \mathbf{u}_0 \rangle = 0. \tag{7.36d}$$

with (from (7.36c))

$$(\langle \mathbf{u}_0 \rangle.\mathbf{n})_{\Sigma(t)} = W(t, \boldsymbol{P}). \tag{7.36e}$$

From (7.36b), since $D\langle \mathbf{u_0} \rangle = 0$, we also have:

$$\nabla\langle p_1 \rangle = 0 \Rightarrow \langle p_1 \rangle = 0; p_1 \equiv p_1^*. \tag{7.36f}$$

7.2.4 *Solution for the Fluctuations* $\mathbf{u}_0^*$ *and* ρ_1^*

For the fluctuations we derive, from Eqs. 7.36a, 7.36b with 7.36c, the following acoustic-type equations with slip condition:

$$D\,\rho_1^* + \rho_0 \nabla.\mathbf{u}_0^* = 0, \tag{7.37a}$$

$$(1/\gamma)\nabla p_1^* + \rho_0 D\mathbf{u}_0^* = 0, \tag{7.37b}$$

with

$$(\mathbf{u}_0^*.\mathbf{n})_{\Sigma(t)} = 0. \tag{7.37c}$$

Concerning the equation for the specific entropy, we have (because $S_0 = 0$):

$$DS_1^* = 0 \Rightarrow S_1^* = 0, \tag{7.37d}$$

and as a consequence, from the equation of state, for the fluctuation of the pressure, we derive

$$p_1^* = \gamma(p_0/r_0)\,\rho_1^*. \tag{7.37e}$$

As a consequence of (7.36f) we also have

$$\langle \rho_1 \rangle = 0 \Rightarrow \rho_1 \equiv \rho_1^* \tag{7.37f}$$

The solution of the two equations for $\mathbf{u}_0^*$ and (ρ_1^*/ρ_0), obtained from (7.37a, 7.37b) when we use (7.37c), is given by (7.30a, 7.30b).

Indeed, if we use the solutions (7.30a, 7.30b) in Eqs. 7.37a and 7.37b, then:

$$D(\rho_1^*/\rho_0) + \nabla.\mathbf{u}_0^* = \Sigma_{n\geq 1}[\mathrm{A_n(t)\ Cn}$$

$$- \mathrm{B_n(t)\ Sn}]\{[\rho_0(t)/\rho_0(t)]^{1/2}(d\varphi_n(t)/dt)R_n + \nabla.\boldsymbol{U}_n\}$$

and

$$(p_0/\rho_0)\nabla(\rho_1^*/\rho_0) + D\mathbf{u}_0^* = \Sigma_{n\geq 1}[\mathrm{A_n(t)\ Sn}$$

$$+ \mathrm{B_n(t)\ Cn}]\{[p_0(t)/\rho_0(t)]^{1/2}\nabla R_n - (d\varphi_n(t)/dt)\boldsymbol{U}_n\}$$

and using (7.31a–7.31d) and (7.32) we determine that the right-hand side of the above equations are quite zero.

We observe also that the eigenfunctions (the normal modes of vibrations of $\Omega(t)$ with eigenfrequencies ω_n), $\boldsymbol{U}_n$ and R_n, are normalized according to:

$$\int_{\Omega(t)} [(\boldsymbol{U}_n)^2 + (R_n)^2]\ dv = 1$$

It is now necessary to determine, from (7.34a, 7.34b), the equations for the second-order approximation, and then derive, first, a system of two equations for the amplitudes, $A_n(t)$ and $B_n(t)$, which present the possibility of considering the long time evolution of the rapid oscillations.

However, it is also necessary to derive an equation for the average value of $\mathbf{u}_0$, which gives, with the average continuity equation (7.36d), a system of two average equations for $\langle \mathbf{u}_0 \rangle$ and $\langle p_2 \rangle$.

7.2.5 The Second-Order Approximation

We return to system of Eqs. 7.34a with 7.34b, and consider the second-order approximation for S_2, p_2, and ρ_2. First, we obtain:

$$DS_2 + \partial\langle S_1\rangle/\partial t + [\langle \mathbf{u}_0\rangle.\nabla)]\langle S_1\rangle = 0$$

but according to (7.37d), $S_1^* = 0$, and also

$$\partial\langle S_1\rangle/\partial t + [\langle \mathbf{u}_0^*\rangle.\nabla]\langle S_1\rangle = 0.$$

With zero initial condition at t = 0, we have (since $\langle S_1 \rangle = 0$):

$$S_1 \equiv 0, \text{ and then : } S_2^* = 0 \tag{7.38a}$$

For the third-order approximation we have

$$DS_3 + \partial\langle S_2 \rangle/\partial t + [\langle \mathbf{u}_0 \rangle.\nabla]\langle S_2 \rangle = 0$$

and again (according to the second relation in (7.38a)) we obtain

$$S_2 = 0 \text{ and } S_3^* = 0 \tag{7.38b}$$

Finally, from the equation of state, when we take into account that

$$S_0 = S_1 = S_2 \equiv 0$$

we derive the following relation between p_2 and ρ_2:

$$p_2 = \gamma(p_0/\rho_0)[\rho_2 + (1/2\rho_0)(\gamma - 1)(\rho_1)^2]. \tag{7.38c}$$

Now, again from the system of Eq. (7.34a), we derive two second-order equations:

$$D\rho_2 + \rho_0\nabla.\mathbf{u}_1 + \partial\rho_1/\partial t + \mathbf{u}_0.\nabla\rho_1 + \rho_1\nabla.\mathbf{u}_0 = 0 \tag{7.39a}$$

$$\nabla(p_2/\gamma\rho_0) + D\mathbf{u}_1 + (\rho_1/\rho_0)D\mathbf{u}_0 + \partial\mathbf{u}_0/\partial t + (\mathbf{u}_0.\nabla)\mathbf{u}_0 = 0 \tag{7.39b}$$

with

$$(\mathbf{u}_1.\mathbf{n})_{\Sigma(t)} = 0 \tag{7.39c}$$

From (7.39b) we now have, first, the possibility of deriving the following average equation for $\langle \mathbf{u}_0 \rangle$:

$$\partial\langle \mathbf{u}_0 \rangle/\partial t + \langle(\mathbf{u}_0.\nabla)\mathbf{u}_0\rangle + \nabla(\langle p_2 \rangle/\gamma\rho_0) + (1/\rho_0)\langle \rho_1 D\mathbf{u}_0 \rangle = 0. \tag{7.39d}$$

This equation is explained below.

On the other hand, using (7.38c), we write Eqs. 7.39a, 7.39b as an inhomogeneous, acoustic-type, system for ρ_2/ρ_0 and $\mathbf{u}_1$: namely,

$$D(\rho_2/\rho_0) + \nabla.\mathbf{u}_1 + G = 0,$$

$$(p_0/\rho_0)\nabla(\rho_2/\rho_0) + D\mathbf{u}_1 + \boldsymbol{F} = 0, \tag{7.40a}$$

$$(\mathbf{u}_1.\mathbf{n})_{\Sigma(t)} = 0,$$

where

$$G = \partial(\rho_1/\rho_0)\partial t + \nabla.[(\rho_1/\rho_0)\mathbf{u}_0] + (1/\rho_0)[d\rho_0/dt](\rho_1/\rho_0); \tag{7.40b}$$

$$\boldsymbol{F} = \partial\mathbf{u}_0/\partial t + (\mathbf{u}_0.\nabla)\mathbf{u}_0 + (\gamma - 2)(p_0/\rho_0)(\rho_1/\rho_0)\nabla(\rho_1/\rho_0). \tag{7.40c}$$

In G and $\boldsymbol{F}$, according to (7.40b, 7.40c), we have three categories of terms:

1. The (average $\langle G\rangle$ and $\langle \boldsymbol{F}\rangle$) terms independent of the scale of the fast times.
2. The terms ($G_L$and $\boldsymbol{F}_L$) which are linearly dependent on the Cn, and Sn.
3. The terms which depend quadratically $\left(G_Q\text{and } \boldsymbol{F}_Q\right)$ on the Cn and Sn, and are proportional to $\cos[(1/M)(\varphi_p(t) \pm \varphi_q(t))]$or $\sin[(1/M)(\varphi_p(t) \pm \varphi_q(t))]$.

As a consequence we write, in system (7.40a) for the inhomogeneous terms G and $\boldsymbol{F}$, the following formal representation:

$$G = \langle G\rangle + [p_0(t)/\rho_0(t)]^{1/2}\Sigma_{n\geq 1}[G_{nC}\mathrm{Cn} + G_{nS}\,\mathrm{Sn}] + G_Q; \tag{7.41a}$$

$$\mathbf{F} = \langle \mathbf{F}\rangle + [p_0(t)/\rho_0(t)]\Sigma_{n\geq 1}[\mathbf{F}_{nS}\mathrm{Sn} + \mathbf{F}_{nC}\,\mathrm{Cn}] + \mathbf{F}_Q, \tag{7.41b}$$

where $\langle G\rangle$ and $\langle \mathbf{F}\rangle$ and also coefficients. $G_{nC}, G_{nS}, \mathbf{F}_{nC}, \mathbf{F}_{nS}$, are determined from (7.40b, 7.40c).

More precisely, in (7.41a, 7.41b), the terms $\langle G\rangle$ and $\langle \mathbf{F}\rangle$ indicate the terms independent of fast times, while in the $\Sigma_{n\geq 1}$ we have the terms with Cn and Sn according to (7.30a, 7.30b). On the other hand, in G_Q and $\mathbf{F}_Q$ we have the terms proportional to

$$\cos[(\varphi_p \pm \varphi_q)/M] \text{ or } \sin[(\varphi_p \pm \varphi_q)/M].$$

Below we assume that the last quadratic terms, G_Q and $\boldsymbol{F}_Q$, are not resonant triads satisfying the relation:

$$|\varphi_p(t) \pm \varphi_q(t)| = \varphi_r(t), \forall p,\ q,\ r \tag{7.41c}$$

Thus, none of the quadratic terms can interfere with any of the terms depending linearly on the Cn and Sn.

As a consequence of the linearity of our system (7.40a), we can, in particular, write the solution for the fluctuations (ρ_2^*/ρ_0) and $\mathbf{u}_1^*$, corresponding only to the terms linearly dependent on the Cn and Sn in (7.41a, 7.41b), in the following form:

$$\rho_2^*/\rho_0 = \Sigma_{n\geq 1}[R_{nC}\mathrm{Cn} + R_{nS}\,\mathrm{Sn}] \tag{7.42a}$$

$$\mathbf{u}_1^* = (p_0/\rho_0)^{1/2}\Sigma_{n\geq 1}[\boldsymbol{U}_{nC}\mathrm{Cn} - \boldsymbol{U}_{nS}\mathrm{Sn}] \tag{7.42b}$$

and, for example, the amplitudes R_{nS} and $\boldsymbol{U}_{nC}$ satisfies the system:

$$\omega_n R_{nS} + \nabla . \boldsymbol{U}_{nC} + G_{nC} = 0$$

$$-\ \omega_n \boldsymbol{U}_{nC} + \nabla R_{nS} + \boldsymbol{F}_{nS} = 0 \tag{7.42c}$$

$$\boldsymbol{U}_{nC}.\mathbf{n} = 0$$
$$\text{on } \Sigma(\text{t}).$$

Obviously, for R_{nC} and $\boldsymbol{U}_{nS}$ we obtain a similar system when in place of R_{nS}, $\boldsymbol{U}_{nC}$, G_{nC} and $\boldsymbol{F}_{nS}$ we write R_{nC}, $\boldsymbol{U}_{nS}$, G_{nS} and $\boldsymbol{F}_{nC}$.

For the existence of a solution of both these inhomogeneous systems it is necessary to use two compatibility relations (which are, in fact, a consequence of the Fredholm alternative), respectively related to $(G_{nC}, \boldsymbol{F}_{nS})$ and $(G_{nS}, \boldsymbol{F}_{nC})$, and for this the system (7.33), for the normal modes $(R_n, \boldsymbol{U}_n)$ of vibrations of the cavity $\Omega(\text{t})$ with eigenfrequencies ω_n, must be taken into account.

Therefore, from (7.33), after an integration by parts, it follows that

$$\begin{aligned} 0 &= \int_{\Omega(\text{t})} \{[\omega_n R_n + \nabla . \boldsymbol{U}_n] R_{nC} - [\ \omega_n \boldsymbol{U}_n - \nabla R_n] \boldsymbol{U}_{nC}\} \text{dv} \\ &= \int_{\Omega(\text{t})} \{[\omega_n R_{nC} + \nabla . \boldsymbol{U}_{nC}] R_n - [\ \omega_n \boldsymbol{U}_{nC} - \nabla R_{nC}] \boldsymbol{U}_n\} \text{dv}, \end{aligned} \tag{7.43a}$$

when we also take into account the boundary (on $\partial\Omega(\text{t}) = \Sigma(\text{t})$), the conditions:

$$\boldsymbol{U}_{nC}.\mathbf{n} = 0, \text{ and } \boldsymbol{U}_{nS}.\mathbf{n} = 0, \text{ on } \Sigma(\text{t}). \tag{7.43b}$$

As a consequence, we derive the following compatibility condition for the resolvability of the above, (7.42c), inhomogeneous system:

$$\int_{\Omega(\text{t})} [G_{nC} R_n - \mathbf{F}_{nS} . \boldsymbol{U}_n] \text{dv} = 0 \tag{7.44}$$

Of course, a compatibility relation similar to (7.44) is verified if we write, in place of G_{nC} and $\boldsymbol{F}_{nS}$, respectively, G_{nS} and $\boldsymbol{F}_{nC}$, after the use of a system similar to (7.42c) for R_{nC}, $\boldsymbol{U}_{nS}$, with G_{nS} and $\boldsymbol{F}_{nC}$.

7.2.6 The Average System of Equations for the Slow Variation

With the average continuity equation (7.36d) and slip condition (7.36e), for $\langle \mathbf{u}_0 \rangle$, we lack sufficient information for the determination of the slow (nearly incompressible) variation! Such information is derived from the average equation (7.39d). Again, therefore, according to solution (7.30a, 7.30b), we first obtain:

$$\langle(\mathbf{u}_0.\nabla)\mathbf{u}_0\rangle = (\langle\mathbf{u}_0\rangle.\nabla)\langle\mathbf{u}_0\rangle + (1/2)\Sigma_{n\geq 1}(A_n{}^2 + B_n{}^2)[\boldsymbol{U}_n.\nabla]\boldsymbol{U}_n \quad (7.45a)$$

and

$$(1/\rho_0)\langle\rho_1 D\mathbf{u}_0\rangle = -(1/2)\Sigma_{n\geq 1}(A_n{}^2 + B_n{}^2)[\boldsymbol{R}_n.\nabla]\boldsymbol{R}_n \quad (7.45b)$$

when we also make use of (7.33). From this equation we also derive the relation:

$$[\boldsymbol{U}_n.\nabla]\boldsymbol{U}_n = (1/2)|\nabla\boldsymbol{U}_n|^2$$

Finally, for $\langle\mathbf{u}_0\rangle$ we derive the following average equation of motion:

$$\partial\langle\mathbf{u}_0\rangle/\partial t + (\langle\mathbf{u}_0\rangle.\nabla)\langle\mathbf{u}_0\rangle + \nabla\Pi = 0 \quad (7.46a)$$

where

$$\Pi = (\langle p_2\rangle/\gamma\rho_0) + (1/4)\Sigma_{n\geq 1}(A_n{}^2 + B_n{}^2)\{|\boldsymbol{U}_n|^2 - |\boldsymbol{R}_n|^2\} \quad (7.46b)$$

is a pseudo-pressure affected by the acoustic perturbations.

But, $\langle\mathbf{u}_0\rangle|_{t=0}$ being irrotational (according to a detailed investigation in [85], Sect. 4 and see also Sect. 7.2.7 below), and according to the average equation (7.46a), for $\langle\mathbf{u}_0\rangle$, it remains irrotational for any time $t > 0$:

$$\langle\mathbf{u}_0\rangle = \nabla\varphi. \quad (7.47)$$

In such a case, the average continuity equation (7.36d) for $\langle\mathbf{u}_0\rangle$, with the slip condition (7.36e) on the wall $\Sigma(t)$, allows us to determine $\langle\mathbf{u}_0\rangle$ due to the following Neumann problem for potential function φ:

$$\Delta\varphi + d\log\rho_0(t)/dt, \quad (7.48a)$$

with

$$(d\varphi/dn)_{\Sigma(\tau)} = W(t, \mathbf{P}). \quad (7.48b)$$

In such a case, for the first term in Π given by (7.46b), we write:

$$\langle p_2\rangle/\gamma\rho_0 = -[\partial\varphi/\partial t + (1/2)|\nabla\varphi|^2] - (1/4)\Sigma_{n\geq 1}(A_n{}^2 + B_n{}^2)\{|\boldsymbol{U}_n|^2 - |\boldsymbol{R}_n|^2\} \quad (7.48c)$$

which take into account, explicitly, the influence of the acoustics on the averaged pressure $\langle p_2\rangle$.

The term

$$-(1/4)\Sigma_{n\geq 1}(A_n{}^2 + B_n{}^2)\{|\boldsymbol{U}_n|^2 - |\boldsymbol{R}_n|^2\},$$

in (7.48c) is a trace of the acoustics, in the model problem (7.48a, 7.48b) with (7.48c) – a sequel (a memory of the acoustic oscillations) of the application of the homogenization technique.

We observe also that as an initial condition for a $\langle \mathbf{u}_0 \rangle$, at $t = 0$, solution of the average equation (7.46a) with (7.46b), according to solution (7.30a) and the starting initial condition (7.22d), we can write:

$$\langle \mathbf{u}_0 \rangle + \Sigma_{n\geq 1} A_n(0) \boldsymbol{U}_n(0, \mathbf{x}) = 0 \text{ for } t = 0 \tag{7.48d}$$

7.2.7 *The Long Time Evolution of the Fast Oscillations*

With the above derivation of the average system of equations for slow variation we have eliminated only part of the secular terms in $\mathbf{u}_1$ and ρ_2. As a consequence it is necessary to consider in detail the system of compatibility conditions (7.44) for G_{nC} and $\mathbf{F}_{nS}$, and similarly for G_{nS} and $\mathbf{F}_{nC}$.

First, we consider G_{nC}, $\mathbf{F}_{nS}$, G_{nS} and $\mathbf{F}_{nC}$, and take into account the relations (7.40b, 7.40c), (7.41a, 7.41b) and the solution (7.30a, 7.30b), for $\mathbf{u}^*_0$ and ρ^*_1, with $\mathbf{u}_0 = \langle \mathbf{u}_0 \rangle + \mathbf{u}_0^*$ and $\rho_1 \equiv \rho_1^*$.

A straightforward but technically long calculation produces the following formulae:

$$G_{nC} = (\rho_0/p_0)\{(dB_n/dt)R_n + B_n[(\partial R_n/\partial t) + (d\log\rho_0/dt)R_n + \nabla.(\langle \mathbf{u}_0 \rangle R_n)]\} \tag{7.49a}$$

$$\mathbf{F}_{nS} = -(\rho_0/p_0)\{(dB_n/dt)\boldsymbol{U}_n + B_n[(\partial \boldsymbol{U}_n/\partial t) + (\boldsymbol{U}_n.\nabla)\langle \mathbf{u}_0 \rangle + (\langle \mathbf{u}_0 \rangle.\nabla)\boldsymbol{U}_n]\} \tag{7.49b}$$

$$G_{nS} = -(\rho_0/p_0)\{(dA_n/dt)R_n + A_n[(\partial R_n/\partial t) + (d\log \rho_0/dt)R_n + \nabla.(\langle \mathbf{u}_0 \rangle R_n)]\} \tag{7.49c}$$

$$\mathbf{F}_{nS} = (\rho_0/p_0)\{(dA_n/dt)\boldsymbol{U}_n + A_n[(\partial \boldsymbol{U}_n/\partial t) + (\boldsymbol{U}_n.\nabla)\langle \mathbf{u}_0 \rangle + (\langle \mathbf{u}_0 \rangle.\nabla)\boldsymbol{U}_n]\} \tag{7.49d}$$

and we observe that in (7.49a) and (7.49c), according to (7.36d):

$$\begin{aligned}\nabla.(\langle \mathbf{u}_0 \rangle R_n) &= R_n(\nabla.\langle \mathbf{u}_0 \rangle) + \langle \mathbf{u}_0 \rangle.\nabla R_n \\ &\equiv \langle \mathbf{u}_0 \rangle.\nabla R_n - (d\log \rho_0/dt)R_n\end{aligned} \tag{7.50}$$

Then, from the compatibility relation (7.44) with (7.49a, 7.49b), and from a similar (to 7.44) compatibility relation, together with (7.49c, 7.49d), we derive the two ordinary differential equations for the amplitudes $A_n(t)$ and $B_n(t)$, taking into account the normalization condition: namely,

$$dA_n/dt + \gamma_n(t)A_n = 0, \tag{7.51a}$$

$$dB_n/dt + \gamma_n(t)B_n = 0, \tag{7.51b}$$

where $\gamma_n(t) = (1/2)\int_{D(t)} \partial/\partial t[|\boldsymbol{U}_n|^2 + |\boldsymbol{R}_n|^2]dv$

$$+ (1/2)\int_{D(t)}\{\langle \mathbf{u}_o\rangle.\nabla[|\boldsymbol{U}_n|^2 + |\boldsymbol{R}_n|^2]\}dv$$

$$+ \int_{D(t)} [(\boldsymbol{U}_n.\nabla)\langle \mathbf{u}_o\rangle].\boldsymbol{U}_n dv$$

which can be rewritten as

$$\gamma_n(t) = (1/2)\text{dlog}\ \rho_o/dt + \int_{D(t)} [(\boldsymbol{U}_n.\nabla)\langle \mathbf{u}_o\rangle].\boldsymbol{U}_n dv. \tag{7.51c}$$

This above relation is derived when we take into account that, respectively:

$$(1/2)\int_{D(t)} \partial/\partial t[|\boldsymbol{U}_n|^2 + |\boldsymbol{R}_n|^2]dv = -(1/2)\int_{\Sigma(t)} [|\boldsymbol{U}_n|^2 + |\boldsymbol{R}_n|^2]W(t,\boldsymbol{P})ds$$

due to normalization and (7.35e), and also that

$$(1/2)\int_{D(t)} \{\langle \mathbf{u}_o\rangle.\nabla[|\boldsymbol{U}_n|^2 + |\boldsymbol{R}_n|^2]\}dv$$
$$= (1/2)\int_{D(t)} \nabla\{[|\boldsymbol{U}_n|^2 + |\boldsymbol{R}_n|^2]\langle \mathbf{u}_0\rangle\}dv$$
$$-(1/2)\int_{D(t)} [|\boldsymbol{U}_n|^2 + |\boldsymbol{R}_n|^2](\nabla.\langle \mathbf{u}_o\rangle)dv.$$

But:

$$(1/2)\int_{D(t)} \nabla[|\boldsymbol{U}_n|^2 + |\boldsymbol{R}_n|^2]\langle \mathbf{u}_o\rangle\}dv$$
$$= (1/2)\int_{\Sigma(t)} [|\boldsymbol{U}_n|^2 + |\boldsymbol{R}_n|^2]W(t,\boldsymbol{P})ds,$$

due to slip condition (7.36e), and

$$-(1/2)\int_{D(t)} [|\boldsymbol{U}_n|^2 + |\boldsymbol{R}_n|^2](\nabla.\langle \mathbf{u}_o\rangle)dv = (1/2)\text{dlog}\ \rho_o/dt$$

according to continuity equation (7.36d) and normalization condition.

At t = 0 we have, as initial conditions from (7.48d),

$$t^{+} = 0 : \Sigma_{n>1} A_n(0) \boldsymbol{U}_n(0, \mathbf{x}) = -\langle \mathbf{u}_o \rangle \tag{7.51d}$$

and

$$B_n(0) = 0, \ n = 1, 2, \ldots \tag{7.51e}$$

We derive the above initial conditions for An(t) and Bn(t) by applying the starting initial conditions (7.22d) for **u** and ρ, and this gives, first, for An the condition (7.48d), because **u** = 0 at t = 0, when we take into account the solution (7.30a) for $\mathbf{u}^*_0$ and also the decomposition (7.28) for $U = \langle U \rangle + U^*$.

The value of $B_n(0) = 0$ is related with the initial condition at t = 0, for ρ(= 1), which is compatible with the leading-order solution:

$$\rho^*_0(t = 0) = 1, \text{ and } \rho^*_1(0, \vec{x}) = 0.$$

Due to Eq. (7.51b) for B_n, obviously:

$$B_n(t) \equiv 0 \text{ for all t.} \tag{7.52a}$$

Concerning $A_n(0)$, its values must be derived from (7.48d/7.51d), and it depends on the value of $< \mathbf{u}_0 >$ at t = 0. On the other hand, obviously, if in condition (7.22b) $W(0, \boldsymbol{P}) = 0$, then $< \mathbf{u}_0 >$ is also zero at t = 0, and

$$A_n(0) = 0$$

which also implies that

$$A_n(t) \equiv 0 \text{ (is zero for all t)} \tag{7.52b}$$

and then the oscillations are absent!

However, *if the motion of the wall of the deformable in time cavity is started impulsively from rest (or accelerated from rest to a finite velocity in a time O(M)), then accordingly we have*:

$$W(0^-, \mathbf{P}) = 0\ldots \text{ but : } W(0^+, \mathbf{P}) \neq 0 \tag{7.53a}$$

and the same holds for the averaged velocity, $< \mathbf{u}_0 >$.

In this case we have $A_n(0^+) \neq 0$, and as consequence:

$$A_n(t) \text{ is also non} - \text{zero, when } t \geq 0^+ \tag{7.53b}$$

7.2.8 Some Concluding Comments

The most important result we obtain is as follows. *If the motion of the wall of the deformable (in time) cavity, where the inviscid gas is confined, is started impulsively from rest, then the acoustic oscillations remain present and have a strong effect on the pressure. Therefore, this pressure would be felt by a gauge, and would not be related to the mean (averaged) motion. The same holds if the motion of the wall is accelerated from rest to a finite velocity in a time O(M).*

We again stress the necessity of building into the structure of the non-viscous solution for $U(\mathbf{u}, \rho, p, S)$, when we consider the Euler equations (7.22a), a multiplicity of times – a family of fast times – in contrast to Müller [150], Meister [151], and Ali [152].

If we deal with a *slightly viscous flow*, when the Mach number $M \ll 1$, we must start from the full unsteady NS–F equations. In such a dissipative (viscous and heat-conducting) case, we bring into the analysis a second small parameter Re^{-1}, the inverse of a (large) Reynolds, $Re \gg 1$, number, and we must then expect that the acoustic oscillations are damped out.

Unfortunately, a precise analytical (when a similarity rule between M and Re^{-1} is assumed) multiple time-scale asymptotic investigation of this damping phenomenon appears to be even more difficult problem, and raises many questions! This damping problem is considered mainly in the framework of the hypothesis (see [13], pp. 148–161):

$$Re >> 1/M \tag{7.54}$$

In Müller [150], the author provides insight into the compressible Navier–Stokes equations at low Mach number when slow flow is affected by acoustic effects in a bounded domain over a long time! As an example of an application, Müller mentions a closed piston-cylinder system in which the isentropic compression due to a slow motion is modified by acoustic waves. Müller uses only a two-time scale analysis, which is obviously insufficient for the elimination of the secular terms in derived approximate systems (as has been mentioned in Sect. 7.3.5).

The results obtained recently by Ali [152] are more interesting than those formally derived by Müller [150], in spite of the fact that in Ali's paper a two-time scale analysis is again used – the Euler equations for a compressible perfect fluid being considered on a bounded time-dependent domain $\Omega_t \in \Re^n$, where Ω_0 denotes the domain at the initial time $t = 0$. The evolution of the bounded time-dependent domain is described by a family of invertible maps:

$$\Phi_t : \Re^n \to \Re^n \tag{7.55a}$$

depending continuously on the time t, such that $\Omega_t = \Phi_t(\Omega_0)$ for all t.

This severe assumption on the domain Ω_t is, nevertheless, general enough to include a moving rigid domain, or a cylinder cut by a fixed surface and a moving surface (piston problem), or a contracting–expanding sphere (star). In the particular case of a moving rigid domain, Ali [152], p. 2023, writes:

$$\Phi_t(\mathbf{x}) = \mathbf{x} + \mathbf{c}(t). \tag{7.55b}$$

The map Φ_t has a geometric meaning and is related neither to the fluid motion nor to the Lagrangian variables. Moreover, Φ_t does not need to be globally unique, since only its restriction to a neighbourhood of the boundary $\partial\Omega_o$ characterizes the motion of the domain's boundary $\partial\Omega_t$.

From the conclusions of Ali [152], pp. 2037–2038, we mention that his analysis is not conclusive, since the theory presented is not capable of providing a full resolution of high-frequency acoustics. Nevertheless, the representation derived, in Sect. 6 of his paper, provides a hint of a partial theoretical comprehension of the acoustic modes generated by the motion of the boundary.

Obviously, the main key point is that one fast time variable is not sufficient to describe the sequence of modes produced by a generic motion of the boundary. Thus we need to extend the "Ali [152] theory" to include a family of fast time variables non-linearly related to the slow time and (eventually) to the space variables.

It is an open question whether the number of independent fast variables for each term of the asymptotic expansion should be increased with the order of the term. This extension, mentioned by Ali, has a theoretical interest in itself, and is a necessary step for the development of efficient numerical schemes for low Mach number flows in a time-dependent bounded domain (as is the case in a combustion problem). It was, in fact, discovered by J.-P. Guiraud and myself 30 years ago, in 1980, and it is formally realized in Sects. 7.2.1–7.2.6 above.

Obviously, the case when the starting equations, in place of (7.22a), are the full unsteady NS-F equations, with

$$\mathrm{Re} = O(1) \text{ and } \mathrm{Pr} = O(1) \text{ fixed, with } M \to 0 \tag{7.56}$$

is a more difficult problem!

Here we mention only a partial result which concerns the derivation of the following averaged reduced system:

$$\begin{aligned} &\nabla[\langle p_2\rangle/\gamma\ \rho_o(t)] + \langle \mathbf{F}\rangle = 0 \\ &\nabla.\langle \mathbf{u}_1\rangle + \langle G\rangle = 0 \\ &\langle H\rangle - (\gamma - 1)T_w(t)\langle G\rangle = 0. \end{aligned} \tag{7.57}$$

This average system merits careful analysis. In the first equation of (7.57), in $< \mathbf{F} >$, the Reynolds number Re is present, while in the third equation, in $< H >$, the Péclet (Pé = PrRe) number is present. From this set of equations a Navier–Fourier-type nearly incompressible average system of equations has been derived (see [13], pp. 149–154).

A difficult problem is also the study of the viscous damping of the acoustic fast oscillations. Obviously, the inviscid theory developed in Sect. 7.2 do not present the possibility of investigating this damping process, and this is also the case when Re = O(1) fixed in the framework of a Navier–Fourier model.

On the other hand, if we deal with a slightly viscous flow (large Reynolds number, Re $\gg$ 1), we must start from NS–F equations, in place of the Euler equations analysed above, and bring into the analysis a second small parameter:

$$\varepsilon^2 = 1/\mathrm{Re} << 1 \tag{7.58a}$$

which is the inverse of the Reynolds number.

We must then expect that the acoustic fast oscillations are damped out. Unfortunately, a precise analysis of this damping phenomenon – which appears, for instance, when a general similarity rule

$$\varepsilon^2 = \mathrm{M}^{\beta}, \quad \beta > 0 \tag{7.58b}$$

is assumed – appears not to be an easy task, and raises many questions.

Therefore, it is first necessary to take into account an acoustic-type inhomogeneous system with a family of very slow times via a new operator ($\mathcal{D}$):

$$\delta\, \mathcal{D}\, \mathrm{U} \text{ in } (7.27) \tag{7.58c}$$

where it is assumed that order $\delta > \mathrm{M}$!

It appears that as a consequence of the inhomogeneity, a boundary–layer analysis is necessary, which is related with a Stokes-layer of thickness

$$\chi^2 = \varepsilon^2\, \mathrm{M}. \tag{7.58d}$$

The analysis of the Stokes-layer equations is rather complicated, but is necessary for the investigation of this damping process.

A matching condition (evaluating the flux outward from the Stokes layer) between the acoustic and Stokes-layer components of the normal velocity gives:

$$\chi = \delta\mathrm{M} \Rightarrow \delta = \chi/\mathrm{M} = [\mathrm{Re\ M}]^{-1/2}. \tag{7.59}$$

However, further investigations are necessary if we want to understand *how viscous damping operates when this relation is not satisfied*, and other points meriting investigation include the *behaviour of the Rayleigh layer*.

For a deeper investigation of dissipative effects in the case of a time-dependent cavity – a problem which has practical interest in the simulation of the *starting process of a space rocket driven by a stream of gases emitted behind it when the fuel is burned inside* – it is necessary to consider the similarity rule (7.58b) for large

Reynolds, Re $>>1(\varepsilon<<1)$, numbers and low Mach, $M \ll 1$, numbers – at least during the starting (at $t^+ = 0$) short time interval.

Various interesting results relating to the above-mentioned "combustion problem" are included in our monograph [13] devoted to low Mach numbers: Chapter 1, pp. 14–15, discusses a simple model for combustion (with various references); Chap. 2, pp. 32–33, presents a brief account of the low Mach number theory applied to combustion (with references); and Sects. 3.3 and 3.4 in Chap. 3 deal with different non-viscous and heat-conducting models in a bounded time-dependent domain.

Concerning the *damping of acoustic oscillations* by viscosity, we observe that the Rayleigh-layer emerges, in the solution of the problem related with the damping phenomenon, because of the conditions on the wall in Stokes-layer equations. The investigations of the evolution of the Rayleigh-layers with time is a difficult problem! If, on the one hand, the Stokes-layer corresponds to acoustic (oscillating) eigenfunctions of the cavity, the Rayleigh-layer corresponds, on the other hand, to conditions on the wall of this cavity. Moreover, the thickness of the Stokes-layer, being given by

$$\chi = [M/Re]^{1/2} \tag{7.60}$$

is independent of the time and the behaviour of the Stokes-layer, for a large time, does not have any influence on the Stokes-layer! Concerning the Rayleigh-layer, however, its thickness grows as the square root of the time, and obviously a deeper analysis of the interaction between these two boundary-layers, when time increase to infinity, is required.

A last remark concerning the adaptation to the initial conditions in a time-dependent bounded container is that our first paper,[1] with Guiraud [85], includes some preliminary results concerning this problem:

> Only with the help of a multiple-scale technique, via an infinity of fast times (designed to cope with the infinity of period of free vibration of the bounded container), do we have the possibility of eliminating the various secular terms in derived model equations.

Unfortunately, a two-time, simple technique, with t and $\tau = t/M$, is not adequate, because such a technique does not provide the possibility of eliminating all seculars terms. The main reason is that the acoustic eigenfrequencies of the bounded container appear in the internal problem, and because the container is a function of the slow time t (the time of the boundary velocity–wall velocity related with the deformation of the container in time), these eigenfrequencies are also functions of the (slow) time τ. More precisely, when the wall, at $t^+ = 0$, is started impulsively from rest ($t^- = 0$), the limiting case

[1] This paper was subject of a communication during the 7th "Colloque d'Acoustique Aerodynamique" in Lyon (France), 4–5 November 1980 and, also, of a very fruitful discussion with D. G. Crighton during this "Lyon's colloque".

$$\mathrm{M} \to 0 \text{ with t fixed} \tag{7.61}$$

is singular near the initial time, and it is necessary to consider a local-in-time limiting case:

$$\mathrm{M} \to 0 \text{ with } \tau \text{ fixed} \tag{7.62}$$

with $\tau = t/M$.

In such a case, close to initial time ($\tau = 0$), we derive the classical equations of acoustics and obtain the corresponding solution (see [13], Sect. 3.3.4) of the Chapter 3. Unfortunately, this solution of the acoustic problem does not tend to a defined limit when τ tends to infinity, which shows that matching is not possible!

Chapter 8
The RAM Approach in the Bénard Convection Problem

8.1 An Introduction

During my time at the University of Lille1 from 1972 to 1996, and while living in retirement at 12 rue Saint-Fiacre, Paris, from 1997 to 2009, I published various papers devoted to the well-known Bénard convection problem. As an introduction to this chapter 8, I present a short account of some of these results, which – at least from my point of view – seem valuable.

A *first result*, obtained in 1983, was published in a short note [153], where a rigorous RAM formulation of the Rayleigh–Bénard (RB) thermal convection problem is discussed. This result opened the way for a consistent derivation of the second-order approximate model equations for the Bénard problem of thermal instability with non-Boussincsq effects.

In 1989, by means of a careful dimensionless analysis of the exact Bénard problem of thermal instability for a weakly expansible liquid heated from below, as a *second new result* [154] I show that:

> If you intend to take into account, in approximate model equations for the Bénard problem, the viscous dissipation term in the equation for the temperature, then it is necessary to replace the classical shallow convection (RB) equations by a new set of equations, called the deep convection – DC Zeytounian – equations, which contain a 'depth' parameter.

A *third result*, which appears as a quantitative criterion for the valuation of the rôle of the buoyancy in the Bénard problem, is the following alternative [155] obtained in 1997:

> *Either the buoyancy is taken into account, and in this case the free-surface deformation effect is negligible and we rediscover the classical leading-order Rayleigh–Bénard (RB) shallow convection, rigid-free, problem, or the free-surface deformation effect is taken into account, and in this case at the leading-order the buoyancy does not play a significant rôle in the Bénard–Marangoni (BM) thermocapillary instability problem.*

R. Kh. Zeytounian, *Navier–Stokes–Fourier Equations*,

DOI 10.1007/978-3-642-20746-4_8,

This alternative is related to the value of the reference Froude number $\mathrm{Fr}_d = (\nu_d/d)/(gd)^{1/2}$ based on the thickness d of the liquid layer, magnitude of the gravity g, and constant kinematic viscosity ν_d, and for:

$$\text{RB model problem}: \ \mathrm{Fr}_d \ll 1 \Rightarrow d \gg (\nu_d{}^2/\mathrm{g})^{1/3}, \tag{8.1}$$

while for the

$$\text{BM model problem}: \ \mathrm{Fr}_d \approx 1 \Rightarrow d \approx (\nu_d{}^2/\mathrm{g})^{1/3} \approx 1\ \text{mm}. \tag{8.2}$$

On the other hand, a small effect of the viscous dissipation, in the RB model problem, produces a complementary criterion for the thickness d: namely,

$$d \approx C(\mathrm{T}_d)(\Delta\mathrm{T}/\mathrm{g}) \tag{8.3}$$

such that

$$(\nu_d{}^2/\mathrm{g})^{1/3} \ll d \approx C(\mathrm{T}_d)(\Delta\mathrm{T}/\mathrm{g}) \equiv d_{Sh}, \tag{8.4}$$

where $C(\mathrm{T}_d)$ is the specific heat at constant temperature T_d, on $\mathrm{z} = d$, and $\Delta\mathrm{T}$ is the difference between the temperature T_w, on $\mathrm{z} = 0$, and T_d.

A *fourth result* is my survey of 1998 [156], devoted to various facets of the BM thermocapillary instability problem.

A *fifth result* is linked with my lecture notes for the Summer Course (coordinated by M. G. Velarde and myself) held at CISM, Udine, Italy, in July 2000: "Theoretical aspects of interfacial phenomena and the Marangoni effect – modelling and stability", in [70], pp. 123–190.

Finally, in 2009 I published my book *Convection in Fluids: A Rational Analysis and Asymptotic Modelling* [27].

8.2 Some Unexpected Results for the Bénard Problem of an Expansible Liquid Layer Heated from Below

Consider a horizontal layer of expansible liquid and assume that an adverse temperature gradient (β_s) is maintained by heating the underside – a lower horizontal rigid, $\mathrm{z} = 0$, heated plane at temperature $\mathrm{T_w}$.

The occurrence of the phenomena seems to be associated with cooling of the liquid at its deformable (temperature-dependent) free surface, which in a conduction motionless state is at level $\mathrm{z} = \mathrm{d}$. This free surface is exposed to the air, above, at constant temperature $\mathrm{T_A}$ and constant pressure $\mathrm{p_A}$.

A very slight excess of temperature in the heated liquid layer on z = 0, above that of the surrounding air, $T_W > T_A$, institutes the “tessellated” changing structure (according to Thompson (1881–1882)).

More precisely, the conduction adverse temperature gradient in liquid,

$$T_s(z) = T_W - \beta_s\, z \Rightarrow \beta_s = -\, dT_s(z)/dz, \tag{8.5}$$

is directly determined by the difference $(T_W - T_A)$, via a Newton's cooling law of heat transfer with a conduction unit constant thermal surface conductance q_s:

$$\beta_s = (T_W - T_A)/[(k/q_s) + d] \tag{8.6}$$

where k is thermal conductivity of the heated liquid, and d is the thickness of the liquid layer, both constant in a conduction motionless state.

Concerning, more precisely, the Newton's cooling law of heat transfer, written for the basic particular case of a motionless conduction temperature $T_s(z)$, we have:

$$k(T_d)\, dT_s(z)/dz + q_s(T_d)[T_s(z) - T_A] = 0, \text{ at } z = d. \tag{8.7}$$

the relation (8.6), for β_s, being a direct consequence of (8.7).

If we introduce a conduction Biot number:

$$\mathrm{Bi}_s(T_d) = dq_s(T_d)/k(T_d), \tag{8.8}$$

then for β_s we obtain the following relation in place of (8.6):

$$\beta_s = \{\mathrm{Bi}_s(T_d)/[1 + \mathrm{Bi}_s(T_d)]\}[(T_W - T_A)/d]. \tag{8.9}$$

In the simplest Bénard problem of a liquid heated from below, with β_s, we have four main driving effects:

1. The buoyancy directly related to the thermal shallow convection.
2. The temperature-dependent surface tension which is responsible for the thermocapillary convection.
3. The viscous dissipation, in the equation for the temperature of the liquid layer, which leads to a consideration of deep thermal convection.
4. The effect related to the influence of the deformable free surface.

These four main effects affect the Bénard convection phenomenon, and it is necessary, from the start of the mathematical formulation of the full Bénard problem for an expansible, viscous, and heat-conducting liquid, to take all of them into account.

The significant interconnections between the three main facets of Bénard convection are shown in a sketch (with pecked lines) in Fig. 8.1 of our FMIA 90 [27].

RB thermal shallow convection – without viscous dissipation; and temperature-dependent surface tension.
Deep – *à la* Zeytounian – thermal convection with viscous dissipation.
Thermocapillary – Marangoni – convection with temperature-dependent surface tension.

8.2.1 The Questionable Davis (1987) Upper, Free Surface, Temperature Condition, and the Problem of Two Biot Numbers

In the framework of a mathematical formulation of the full Bénard problem, the derivation of a consistent rational boundary condition for the temperature of the liquid layer, T(t, x, y, z), on a deformable free surface, simulated by the dimensionless Cartesian equation:

$$z = 1 + \eta h(t', x', y'), \tag{8.10}$$

assuming that the gravity vector $\mathbf{g} = -g\mathbf{k}$ acts in the negative z direction and where ('a' being an amplitude)

$$\eta = a/d, \tag{8.11}$$

is a decisive step – but a highly controversial issue!

From 1987, after the publication of the paper by Davis [157], his derived (p. 407), upper condition was used systematically in nearly all works relative to thin films, where the so-called Marangoni effect is taking into account (see, for instance, our *Convection in Fluids* [27], Chaps. 7 and 8).

Only in 1996, in Parmentier et al. paper [158], was the problem of two Biot numbers discussed; and in various parts of our [27] this problem is considered and a corrected condition – in place of Davis's [157] erroneous condition – is derived.

It now seems appropriate to consider, first, the physical nature of the Bénard problem!

The lower heated plate temperature,

$$T = T_w \equiv T_s(z = 0),$$

being given data, the adverse conduction temperature gradient β_s appears (according to (8.9)) as a known function of the (positive) temperature difference $(T_w - T_A)$, where $T_A < T_w$ is also a known constant temperature of the passive (motionless) air, far above the free surface, when the conduction constant Biot number, $Bi_s(T_d)$, is assumed known.

But for this it is necessary to also consider the conduction unit constant thermal surface conductance $q_s(T_d)$ as starting data for the Bénard problem. If so, then $T_s(z = d) = T_d(\equiv T_w - \beta_s d)$ is assumed to be determined. On the other hand, it should also be realized that β_s is always different from zero, in the framework of the Bénard convection problem heated from below!

As a consequence, the above, defined by (8.8), constant conduction Biot number, $Bi_s(T_d)$, is also always different from zero. It characterizes the Bénard conduction stage, and makes it possible to determine the purely static basic temperature gradient $\beta_s \equiv \Delta T/d$ with $\Delta T = T_w - T_d$.

It is also crucial that with this above purely static basic temperature gradient $\beta_s \equiv \Delta T/d$, in all published papers relative to thermocapillary (Marangoni) convection, there are defined various dimensionless parameters ε, Gr or Ra, Ma, and Bo, and dimensionless temperature θ. Therefore (the subscript "$_d$" in various temperature-dependent coefficients and dimensionless parameters or numbers is relative to constant temperature T_d):

$$\varepsilon = \alpha(T_d)\Delta T \text{ (expansibity parameter)} \tag{8.12a}$$

where $\alpha(T_d) \equiv \alpha_d$ is the constant (at $T = T_d$) coefficient of thermal expansion of the liquid,

$$Gr = \varepsilon/(Fr_d)^2 \text{(Grashof number)} \tag{8.12b}$$

or

$$Ra = PrGr \text{(Rayleigh number)} \tag{8.12c}$$

where

$$Pr = \nu_d/\kappa_d \text{(Prandtl number)} \tag{8.12d}$$

with, $\kappa_d = k_d/\rho_d C_d$

$$Ma = \gamma_d d\Delta T/\nu_d{}^2 \rho_d \text{ (Marangoni number)} \tag{8.12e}$$

$$Bo = gd/\Delta T C_d \text{ (a similar to Boussinesq number)} \tag{8.12f}$$

and, as dimensionless temperature, we introduce:

$$\theta = (T - T_d)/\Delta T \tag{8.13}$$

For the convective in-motion stage, in principle Newton's cooling law (which is a third-type boundary condition (2.42c) on a solid heated wall) can again be used, as this is the case in almost all papers devotes to BM problems. Therefore, we write (in dimensional quantities):

$$- k(\mathrm{T})\partial \mathrm{T}/\partial \mathrm{n} = q_{\mathrm{conv}}(\mathrm{T} - \mathrm{T_A}) + \mathrm{Q_0}, \text{at } \mathrm{z} = d + a\, \mathrm{h}(\mathrm{t}, \mathrm{x}, \mathrm{y}), \qquad (8.14)$$

with $\partial \mathrm{T}/\partial \mathrm{n} = \nabla \mathrm{T}.\mathbf{n}$, where $\mathrm{Q_0}$ is an imposed heat flux to the environment and is to be defined.

This condition (8.14) is, in fact, also the starting condition in Davis [157]. In (8.14) our introduced q_{conv} is, indeed, an unknown convection heat transfer coefficient, strongly different from the constant conduction heat transfer coefficient, $q_s(\mathrm{T}_d)$, which appears in (8.7) and (8.8).

As observed in Joseph's monograph [159], the heat transfer coefficient q_{conv}, in the convection stage, depends in general on the free surface properties of the expansible liquid, the unknown motion of the ambient air near the air surface, and also the spatio-temporal structure of the temperature field. q_{conv} is therefore a very complicated function. Obviously, from a practical point of view, the above upper condition (8.14) for the temperature T does not seems not have, in its general form, any interesting perspectives in various practical applications, simply because q_{conv} is an unknown. Nevertheless, from a theoretical point of view it seems preferable to derive, from (8.14), a correct upper condition for the dimensionless temperature function θ, defined by (8.13). This condition (8.14), in dimensionless form, for θ, is written, without any approximation, in the following form:

$$\partial\theta/\partial \mathrm{n}' + \mathrm{Bi_{conv}}\{[(\mathrm{T}_d - \mathrm{T_A})/(\mathrm{T_w} - \mathrm{T_d})] + \theta\} + \mathrm{Q_0}/k_d\beta_s = 0,$$

or more precisely (n′ is dimensionless)

$$\partial\theta/\partial \mathrm{n}' + [\mathrm{Bi_{conv}}/\mathrm{Bi_s}(\mathrm{T}_d)]\{1 + \mathrm{Bi_s}(\mathrm{T}_d)\theta\} + \mathrm{Q_0}/k_d\beta_s = 0, \text{ at } \mathrm{z}' = 1 + \eta\, \mathrm{h}'(\mathrm{t}', \mathrm{x}', \mathrm{y}'), \qquad (8.15a)$$

with a Biot convective number:

$$\mathrm{Bi_{conv}} = dq_{\mathrm{conv}}/k_d. \qquad (8.15b)$$

Condition (8.15a), where not only $\mathrm{Bi_{conv}}$ is present but also $\mathrm{Bi_s}(\mathrm{T}_d)$, is a direct and exact consequence of (8.14), when we take into account Newton's cooling law in the conduction stage (8.7), which leads to:

$$d\beta_s = \mathrm{Bi_s}(\mathrm{T}_d)(\mathrm{T}_d - \mathrm{T_A}) \Rightarrow (\mathrm{T}_d - \mathrm{T_A})/(\mathrm{T_w} - \mathrm{T_d}) \equiv 1/\mathrm{Bi_s}(\mathrm{T}_d).$$

Only after the confusion of Biot convection, $\mathrm{Bi_{conv}}$, with Biot conduction, $\mathrm{Bi_s}(\mathrm{T}_d)$, do we obtain the (approximate?) Davis [157] reduced condition (with $\mathrm{Q_0} = 0$):

$$\partial\theta/\partial \mathrm{n}' + 1 + \mathrm{Bi_s}(\mathrm{T}_d)\theta = 0, \text{ at } \mathrm{z}' = 1 + \eta\, \mathrm{h}'(\mathrm{t}', \mathrm{x}', \mathrm{y}'), \qquad (8.15c)$$

where (in (8.15c)), in place of $\mathrm{Bi_s}(\mathrm{T}_d)$, in front of θ, Davis writes a "B" – the meaning of which is unclear. Perhaps, with his ambiguous temperature free surface condition, he thinks the problem too difficult to solve!

The fact is, however, that over many years (and today) many interesting papers have been published with Davis's (8.15c) "questionable" upper condition for temperature! In particular, this seems important for the papers devoted to linear theory (see, for example, Takashima's two papers [160][1]). I therefore pose a simple question: What is the real value of these papers? Several ideas and aspects of this "two Biot" problem are discussed in various chapters of my [27].

8.2.2 *The Mystery of the Disappearance of Influence of the Free Surface in the RB Leading-Order Shallow Thermal Convection Model*

First, taking into account the definition of the Grahof number, Gr (8.12b), as a ratio of $\varepsilon/\mathrm{Fr}_d^2$, when $\varepsilon<<1$ – our main (expansibility) small parameter – we see that only if $\mathrm{Fr}_d^2<<1$ do we have the possibility, at the leading order, to take into account the buoyancy effect directly related to the thermal shallow convection via Gr.

Namely, in the *RB* case it is necessary to consider the following limiting process:

$$\mathrm{Gr} = \varepsilon/\mathrm{Fr}_d^2 \text{fixed, when } \varepsilon \to \mathrm{o} \text{ and } \mathrm{Fr}_d^2 \to 0, \tag{8.16}$$

simultaneously.

---

[1] In [160] Takashima derives, from Davis's upper condition (8.15c), the following linear upper condition:

$$\partial\theta_L/\partial z' + \mathrm{Bi_s}(\theta_L - h') = 0, \text{ at } z' = 1, \tag{8.*}$$

where θ_L is the perturbation of θ $(= \theta_0 + \eta\theta_L + \ldots)$ relative to a steady-state solution, $\theta_O = 1 - z'$, when we assume $\eta << 1$, and in a such case, $\partial\theta/\partial n' \approx \partial\theta_L/\partial z' + O(\eta)$. If now we consider our corrected upper condition (8.15a),we derive, in place of Takashima's linear upper condition (*), for θ_L, the following linearized upper condition:

$$\partial\theta_L/\partial z' + [\mathrm{Bi_{conv}}]_0(\theta_L - h') = 0, \text{ at } z' = 1, \tag{8.**}$$

but only if we assume that $Q_0 = k_d\beta_s\{1 - -([\mathrm{Bi_{conv}}]_0/\mathrm{Bi_s})\}$, and $[\mathrm{Bi_{conv}}]_0 =$ constant. Another possibility, when we assume that $Q_0 = 0$, is to consider a $\mathrm{Bi_{conv}}$ dependent of $H = 1 + \eta\, h'$, and write $\mathrm{Bi_{conv}}(H) = \mathrm{Bi_{conv}}(1) + \eta\Lambda(H = 1)\, h'$, with $\Lambda(H = 1) \equiv d\mathrm{Bi_{conv}}(H)/dH$. If now $\mathrm{Bi_{conv}}(1) \equiv \mathrm{Bi_s}$, then at the order $O(\eta)$ we derive the following upper (at $z' = 1$) linearized condition:

$$\partial\theta_L/\partial z' + \mathrm{Bi_s}(\theta_L - h') + \Lambda(H = 1)\, h' = \mathrm{o}, \text{ at } z' = 1 \tag{8.***}$$

but now we have $\mathrm{Bi_s}$ in front of $(\theta_L - h')$, and $\mathrm{Bi_s} \neq 0$, which was assumed equal to a constant value of $\mathrm{Bi_{conv}}$ at $z' = 1$. In (***) we also have a third term proportional to h'.

The mathematical dimensionless formulation of the Bénard problem, heated from below, is the subject of Chap. 4 of [27], where can be found a dimensionless system of three dominant dimensionless equations for $\mathbf{u}'$, π, and θ, where

$$\pi = (1/\ \mathrm{Fr}_d{}^2)\{[(\mathrm{p}\ - \mathrm{p_A})/gd\rho_d] +\ \mathrm{z}' - 1\} \tag{8.17}$$

which is a companion dimensionless perturbation pressure to θ.

In particular, if we take into account that for the density ρ, at the leading order, we can write the following approximate equation:

$$\rho =\ \rho_d[1 -\ \ \varepsilon\theta]. \tag{8.18a}$$

Then, for $\mathbf{u}'$, we obtain as a leading-order equation:

$$\mathbf{du}'/\mathrm{dt}' + \nabla'\pi - (\varepsilon/\mathrm{Fr}_d{}^2)\theta\mathbf{k} = \Delta'\mathbf{u}' + (1/3)\nabla'(\nabla'.\mathbf{u}'), \tag{8.18b}$$

with

$$\nabla'.\mathbf{u}' = \varepsilon(\mathrm{d}\theta/\mathrm{dt}'). \tag{8.18c}$$

On the other hand, it is necessary, at the upper free surface, to take into account a jump condition for the difference of pressure (p–$\mathrm{p_A}$). With (8.17), in dimensionless form, we obtain for π as an upper, free surface the following dimensionless condition (at $\mathrm{z}' = 1 + \eta\ \mathrm{h}'$):

$$\begin{aligned}\pi = (\eta/\mathrm{Fr}_d{}^2)\ \mathrm{h}'(\mathrm{t}', \mathrm{x}', \mathrm{y}') + \left[\partial \mathrm{u}'_\mathrm{i}/\partial \mathrm{x}'_\mathrm{j} + \partial \mathrm{u}'_\mathrm{j}/\partial \mathrm{x}'_\mathrm{i}\right]\mathbf{n}'_\mathrm{i}\mathbf{n}'_\mathrm{j} \\ +[\mathrm{We} - \mathrm{Ma}\ \theta](\nabla'_\parallel \cdot \mathbf{n}') - (2/3)\ \varepsilon(\mathrm{d}\theta/\mathrm{dt}')\end{aligned} \tag{8.18d}$$

where – (1/2) $(\nabla'_\parallel \,\mathbf{.}\, \mathbf{n}')$ is the mean curvature of the free surface.

Because $\mathrm{Fr}_d{}^2 \to 0$, in (8.18b) and also in (8.18d), if we want, for $\varepsilon << 1$, to take into account the buoyancy effect, we see that, necessarily, we have the following similarity rule between η and $\mathrm{Fr}_d{}^2$

$$\eta/\mathrm{Fr}_d{}^2{=}\eta{*} =\ \ \mathrm{O}(1),\ \text{when}\ \eta\ \text{and}\ \mathrm{Fr}_d{}^2,\ \text{both} \to 0 \tag{8.19}$$

As a conclusion, in leading-order RB thermal shallow convection, driven by the buoyancy force, the deformation of the upper, free surface is negligible.

In fact, at leading order we have the possibility, as free surface in the RB model problem, of considering the plane $\mathrm{z}' = 1$.

But a pertinent (open) question remains. What is to be done for the determination of the deformation $\mathrm{h}'(\mathrm{t}', \mathrm{x}', \mathrm{y}')$ of the free surface? It seems that such an equation for $\mathrm{h}'(\mathrm{t}', \mathrm{x}', \mathrm{y}')$ was first discovered thanks to our RAM Approach (see [27], pp. 106, 135).

Therefore, with the above upper, free surface (exact) condition (8.18d), written at level $z' = 1$ (because of the limiting process (8.19)), it is necessary to also consider that when $\eta \rightarrow o$ the second term (proportional to $\mathbf{n}'_i\ \mathbf{n}'_j$) in (8.18d) gives only a term:

$$2(\partial u'_3/\partial x'_3), \tag{8.20a}$$

and from a third term in (8.18d), in place of $(\nabla'_{\|}\ .\ \mathbf{n}')$, we derive:

$$-\ \eta[\partial^2 h'/\partial x_1'^2 + \partial^2 h'/\partial x_2'^2]. \tag{8.20b}$$

Finally, from the condition (8.18d) and relation (8.20b), when we assume that (which is usually the case)

$$\mathrm{We}\ = \sigma_d d/\ {v_d}^2 \rho_d\ (\text{Weber number}) >> 1, \tag{8.21a}$$

such that

$$\eta\ \mathrm{We} = \mathrm{We}^* = O(1) \tag{8.21b}$$

we derive the following equation for the determination of the free surface deformation $h'(t', x', y')$ in the leading-order $\varepsilon \rightarrow 0$:

$$\begin{aligned}\partial^2 h'/\partial x_1'^2 + \partial^2 h'/\partial x_2'^2 - (\eta^*/\mathrm{We}^*)\ h' \\ = -(1/\mathrm{We}^*)\pi(t',\ x', y',\ z' = 1),\end{aligned} \tag{8.22}$$

It should be noted that in the definition of Marangoni, Ma, and Weber. We, numbers respectively by (8.12e) and (8.21a), the coefficients (function of T_d) γ_d and σ_d are related with a linear law for the temperature-dependent surface tension:

$$\sigma(T) = \sigma_d - \gamma_d(T\ - T_d)\ \text{with}\ \gamma_d = -\ [d\sigma(T)/dT]_d. \tag{8.23}$$

8.2.3 *Influence of the Viscous Dissipation*

Only an attentive examination of the dominant dimensionless energy equation, for dimensionless temperature θ, presents the possibility of understanding the influence of the viscous dissipation phenomenon in thermal convection.

Indeed, Turcotte et al. [161] observe that the viscous dissipation term, in the dimensionless energy equation, is linked with a "dissipation number":

$$\mathrm{Di}^* = (1/2[({v_d}^2/d^2)/\Delta T C_d]$$

which can be rewritten in the following form (Zeytounian [155]):

$$\text{Di}^* = (1/2)\ \text{Bo}\ (\text{Fr}_d)^2 = \ \varepsilon\text{Bo}/2\text{Gr}, \tag{8.24}$$

where Bo is defined in (8.12f).

On the one hand, we see that when we assume that the Grashof number Gr is 0(1), in thermal convection (buoyancy effect), according to (8.24), where always $\varepsilon << 1$, the constraint:

$$\text{Bo} \ \approx 1 \tag{8.25a}$$

in (8.24), leads to

$$\text{Di}^* \to 0, \tag{8.25b}$$

and we derive the classical RB, thermal shallow convection model problem, where in the energy equation for the dimensionless temperature θ, the viscous dissipation is absent (neglected at the leading-order).

In this case we derive, for the thickness d_{Sh}, in (8.4) of the liquid layer in the RB model problem, precisely the relation:

$$\text{C}(\text{T}_d)(\Delta\text{T}/\text{g}) \approx d,$$

mentioned in (8.4).

On the other hand, again with $\varepsilon << 1$, Gr $=$ O(1), obviously, only the case

$$\text{Bo} >> 1, \tag{8.26a}$$

with

$$\varepsilon\text{Bo} = \text{O}(1) \tag{8.26b}$$

allows us to take into account the viscous dissipation at leading-order in the dominant dimensionless energy equation for dimensionless temperature θ, where the term proportional to Di* is present.

In such a case we derive the deep thermal convection model problem (Zeytounian [154]).

Section 8.3 of [27] presents these deep thermal convection model, leading-order, equations, together with a new ("depth") parameter:

$$\text{Di} \ = \ \alpha(\text{T}_d)gd/C(\text{T}_d). \tag{8.26c}$$

8.3 The Marangoni Effect

Marangoni (1865) provided a wealth of detailed information on the effects of variations of the potential energy of liquid surfaces and, in particular, flow arising from variations in temperature and surfactant composition. Among the phenomena involving Marangoni flows we observe that associated with the name of Bénard (1900), which refers to the formation of a cellular structure in a thin liquid layer heated from below.

Chapter 7 of our recent monograph (2009) [27], presents (in the framework of 67 pages) detailed formulation, discussions, and various reflections related to Bénard–Marangoni thermocapillary convection.

In fact, today, a detailed understanding of flows in thin viscous liquid films with a deformable free surface with a temperature dependent tension is important for a wide range of modern engineering processes. For this it is necessary to build mathematical models that can predict the various performance of these processes, in order to have confidence in the predictions of the derived models. For this, our RAM Approach is obviously a well-adapted and invaluable tool!

A recent special issue of *J. Engng. Maths.* [162] includes various interesting results related to the dynamics of thin liquid film. However, many of these papers include results based on ad hoc non-rational procedures, and do not have any relation to our RAM Approach rigorously based on the postulate formulated in Chap. 6.

It is necessary to have in mind that in the Bénard expériments (thin layer of an expansible liquid heated from below) the influence of the free surface is present, and obviously generates the thermocapillary–Marangoni effect.

On the other hand, in the simple Rayleigh thermal convection problem the free surface is replaced by the plane $z = d$. Because Lord Rayleigh, in his well-known paper of 1916 [163], considered a liquid layer with a constant thickness d, the Marangoni–Biot effects are obviously absent in the exact formulation (*à la* Rayleigh) of the thermal convection problem. The effect of the constant surface tension is also absent, and the Weber number does not appear in Rayleigh's formulation of the thermal convection problem. As a consequence, we see that the Rayleigh analytical problem (considered in 1916) has no relation with the physical experimental problem considered by Bénard in his various experiments in 1900. In the Rayleigh analytical problem the main driving force, which gives a bifurcation from a conduction motionless regime to a convective motion regime, is the buoyancy!

Nevertheless, Rayleigh's theoretical problem, leading to the well-known "RB instability problem", is a typical problem in hydrodynamic instability and represents a transition to turbulence (chaos) in a fluid system. It is now well known (mainly thanks to Pearson [164]) that:

> Bénard convective cells are primarily induced by the temperature-dependent surface tension gradients resulting from the temperature variations along the free surface – the so-called Marangoni thermocapillary effect.

For the Bénard–Marangoni (BM) problem, in leading order, the equations are those which govern an incompressible viscous liquid (a Navier–Fourier-type equations) – both the buoyancy force and viscous dissipation effect being neglected – but with an energy equation for θ.

The main difference with the RB thermal convection model problem is the influence of the free surface deformations, which are taken into account, and as a consequence the upper, free surface conditions, in the BM model problem, are very complicated, and it is just in these upper conditions that the Marangoni, Weber, and two Biot numbers appear! (Chaps. 6 and 7 of [27] discuss a detailed derivation of this BM model problem).

In fact, the inappropriateness of Lord Rayleigh's (1916) analytical model to Bénard's experiments was not adequately explained until Pearson, in 1958 [164] showed (in a simplified ad hoc linear theory) that:

> Rather than being a buoyancy-driven flow, Bénard cells are a direct consequence of a temperature-dependent surface tension.

Curiously, although Bénard (as a physicist) initially assumed that surface tension at the upper free surface of a thin layer film was an important factor in his discovered cell formation, this idea was abandoned by Bénard for some time as the result of the work of Rayleigh in 1916 [163], where he in fact analyzed the buoyancy-driven natural thermal convection of a layer of fluid heated from below.

Rayleigh found that if hexagonal cells are formed, the ratio of the spacing to cell depth almost exactly equalled that measured by Bénard – an agreement which we now know to have been fortuitous!

Indeed, it was the experimental work of Block [165] which put to end to the confusion surrounding the interpretation of Bénard's experiments, and which demonstrated conclusively that Bénard's results were not a consequence of buoyancy but were induced by (temperature-dependent) surface tension. Finally, Block concluded that:

> For thin film of thickness less than 1 mm, variation in surface tension due to temperature variations (Marangoni effect) were the cause of Bénard cell formation and not buoyancy as postulated by Rayleigh in his 1916 paper.

Only 15 years later, via an "alternative" [155], I presented an asymptotic rational well-argued formulation for the Block (1956) conclusion:

> Either the buoyancy is taken into account, and in such a case the free surface deformation effect is negligible and we have the possibility to take into account in the Rayleigh-Bénard (RB), leading-order, shallow thermal convection model problem the Marangoni thermocapillary effect only partially, or the free surface deformation effect is taken into account, and in such a case the buoyancy does not play a significant leading-order role in the Bénard–Marangoni full thermocapillary model problem.

We observe also that for a given temperature difference, the ratio between buoyancy (Archimedean effect) and the surface tension gradient (Marangoni effect) – Gr/Ma – varies with d^2, and as a result (according to physicists), "the Marangoni effect dominates for small thickness of the liquid layer, and buoyancy effects dominate for very thick liquid layers!" Therefore (Bd is the "dynamic Bond number"):

$$Gr/Ma = Bd = [\rho_d \, \alpha_d/\gamma_d]gd^2. \tag{8.27}$$

Twenty-five years ago, when I first read the above sentence in Guyon, Hulin, and Petit's *Hydrodynamique Physique* [166] – which was mainly inspired by de Gennes' various courses at the École Supérieure de Physique et Chimie Industrielle in Paris – I understood that these two effects should (certainly) be related to two particular values of a single dimensionless reference parameter. A little later, I discovered that the Grahof number Gr ($= \alpha_d \Delta \mathrm{T} g d^3/{\nu_d}^2$) is, in fact, a ratio of two dimensionless parameters:

$$\text{expansibility parameter } \varepsilon (= \alpha_d \Delta \mathrm{T})$$

$$\text{to squared Froude number } {\mathrm{Fr}_d}^2 (= (\nu_d/d)^2/gd)$$

For me, these two facts have been an illuminating indication that ${\mathrm{Fr}_d}^2$ in the ratio $\mathrm{Gr} = \varepsilon/\,{\mathrm{Fr}_d}^2$, where the thickness d of the liquid layer is present, must play a decisive role, because usually $\varepsilon << 1$! This observation allowed me to formulate the above-cited "alternative", published in 1997 [155] (and see also our survey of 1998 [156]).

8.3.1 The Long-Wave Approach

The full BM thermocapillary problem formulated in Sect. 7.2 of [27] – even in the framework of a numerical simulation – is a very difficult, awkward, and tedious problem, mainly because of the complicated form of upper, free surface boundary conditions.

It is clear that simplifications in a rational approach are necessary, obviating the need for computationally expensive (in time and money) fully numerical simulations – while at the same time preserving essential elements of the physics of the formulated BM thermocapillary convection model problem.

Among various approaches linked with this BM model problem, the formation of long waves – with respect to a very thin film layer – at the surface of a falling film is a challenging problem (which is the case for a free-falling film down a uniformly heated vertical plane).

In a very thin film, obviously, a typical length λ of the (long) waves is large in comparison with the thickness $d << \lambda$ of the thin film, so that the slope of the free surface is always small. In such a case we have the advantage of introducing a long-wave dimensionless parameter:

$$\delta = d/\lambda << 1 \tag{8.28}$$

The essential advantage of the long-wave approximation is a drastic simplification of the full dimensionless BM model problem, mainly because the upper, free surface conditions are significantly simplified.

Section 7.3 of our [27] includes some "BM long-wave" reduced convection model problems. These simplified model problems are derived via the introduction of new coordinates

$$X = \delta x, Y = \delta y, \ Z \equiv z, \tag{8.29a}$$

and new time

$$T = \delta \mathrm{Re}_d t, \tag{8.29b}$$

where

$$\mathrm{Re}_d = \mathrm{U_c} d/\nu_{\mathrm{A}}, \tag{8.29c}$$

$\mathrm{U_c}$ being determined by a similarity relation (see below) and $\nu_{\mathrm{A}} = \nu(\mathrm{T_A})$.

On the other hand, in place of leading-order (when $\varepsilon \to o$) dimensionless velocity components u′, v′, w′, the following new components are also introduced:

$$\mathrm{U} = \mathrm{u}'/\mathrm{Re}_d, \ \mathrm{V} = \mathrm{v}'/\mathrm{Re}_d, \ \mathrm{W} = \mathrm{w}'/\delta\mathrm{Re}_d \tag{8.30a}$$

In place of π a

$$\Pi = \pi/(\mathrm{Re}_d)^2 \tag{8.30b}$$

and in place of θ (see (8.13)), a (where in place of T_d we write $\mathrm{T_A}$):

$$\Theta = (\mathrm{T} - \mathrm{T_A})/(\mathrm{T_w} - \mathrm{T_A}) \tag{8.30c}$$

We observe also that the various coefficient functions of temperature T are fixed at $\mathrm{T} = \mathrm{T_A}$, and ε and Ma are defined by $(\mathrm{T_w} - \mathrm{T_A})$ which replace $\Delta\mathrm{T}$.

Finally, via a laborious and tedious transformation, and via the long-waves limiting process:

$$\delta \to 0 \text{ and } \mathrm{Re}_d \to \infty, \text{ such that } \delta\,\mathrm{Re}_d = Re^*, \tag{8.31}$$

we derive, first ('*0*' subscript) the following *reduced* system of equations for the BM long-wave problem:

$$\mathbf{D}.\mathbf{V}_0 + \partial \mathrm{W}_o/\partial \mathrm{Z} = 0, \tag{8.32a}$$

$$\mathrm{D}\mathbf{V}_0/\mathrm{D}T + \mathbf{D}\Pi_0 - (1/Re^*)\partial^2\mathbf{V}_0/\partial \mathrm{Z}^2 = 0, \tag{8.32b}$$

$$\partial\Pi_o/\partial Z = 0, \tag{8.32c}$$

$$\Pr D\Theta_0/DT - (1/Re^*)\partial^2\ \Theta_0/\partial Z^2 = 0, \tag{8.32d}$$

where

$$D/DT = \partial/\partial T + \ U\partial/\partial X + V\partial/\partial Y + W\partial/\partial Z,$$

$$D = (\partial/\partial X,\ \partial/\partial Y),\ \mathbf{D}^2 = \partial^2/\partial X^2 + \partial^2/\partial Y^2,\ V = (U,\ V).$$

Equation (8.32c), $\partial\Pi_o/\partial Z = 0$, is typically a "boundary layer equation", and the system (8.32a)–(8.32d) is certainly singular near the initial time $T = 0$, where it is necessary to write initial data for $\mathbf{V}_O$, W_O and Θ_o – the system (8.32a)–(8.32d) being, in fact, only an outer system relative to time valid far to $T = 0$.

As boundary dimensionless conditions, we have, first:

$$\mathbf{V}_0 = 0, W_0 = 0, \text{and } \Theta_0 = 1 \text{ at } Z = 0. \tag{8.33a}$$

If, now, $Z = H(T, X, Y)$ is the equation of the deformable surface, the kinematic upper condition at $Z = H : W_0 = \partial H/\partial T + \mathbf{V}_0.\mathbf{D}H$, with $W_O = 0$ at $Z = 0$, leads to the following averaged evolution equation for $H(T, X, Y)$:

$$\partial H/\partial T + \mathbf{D}.\left(\int_{Z=0}^{Z=H} \mathbf{V}_O\ dZ\right) = 0, \tag{8.33b}$$

which plays a central role in lubrication theory.

Above, in (8.29c), we introduced $Re_d(= dU_c/\nu_A)$ via the characteristic velocity U_c. If, by analogy, we define a new squared Froude number, $Fr^2 = U_c/gd$, and corresponding, modified, Weber (We) and Marangoni (Ma) number via U_c, and then if we assume that:

$$Re^*/Fr^2 = G^* \approx 1,\ \delta^2\text{We}\ = W^* \approx 1 \text{ and Ma} \approx 1, \tag{8.33c}$$

at leading-order we obtain, for the above Eqs. (8.32a)–(8.32d), the following *free surface*, at $Z = H$, upper, reduced conditions:

$$\Pi_0 = (G^*/Re^*)(H - 1) - W^*\mathbf{D}^2H \equiv\ \Pi_o^*(H), \tag{8.33d}$$

$$\partial\mathbf{V}_o/\partial Z = -Re^*\text{Ma}[\mathbf{D}\Theta_o + (\mathbf{D}H)\partial\Theta_o/\partial Z], \tag{8.33e}$$

$$\partial\Theta_o/\partial Z + \text{Bi}_{\text{conv}}\Theta_o = 0. \tag{8.33f}$$

This derived simplified equations, (8.32a–8.32d) with the conditions, (8.33d–8.33f), seems to be a very convenient reduced form, and it is not a bad start as a *reduced BM thermocapillary model problem*, subject to numerical simulation!

8.3.2 Towards a Lubrication Equation

We observe also that as a consequence of Eq. 8.32b for $\mathbf{V}_0$, with (8.32a), and (8.32c) for Π_0, subject to the upper condition (8.33d), we derive for $\mathbf{V}_0$ the following problem (8.34), with the problem (8.35) for Θ_0:

$$\begin{aligned}
&\partial\mathbf{V}_0/\partial T-(1/Re^*)\partial^2\mathbf{V}_0/\partial Z^2-\left(\int_{z=0}^{Z=H}(\mathbf{D}.\mathbf{V}_0)\mathrm{d}Z\right)\partial\mathbf{V}_0/\partial Z+(\mathbf{V}_0.\mathbf{D})\mathbf{V}_0\\
&\qquad\qquad=-(G^*/Re^*)\mathbf{D}H+W^*\mathbf{D}(\mathbf{D}^2H),\\
&Z=o:\mathbf{V}_o=o,\\
&Z=H:\partial\mathbf{V}_0/\partial Z=-Re^*Ma[\mathbf{D}\Theta_0+(\mathbf{D}H)\partial\Theta_0/\partial Z],
\end{aligned}\tag{8.34}$$

where the function Θ_0 is the solution of the model problem:

$$\begin{aligned}
&\mathrm{Pr}\ \mathrm{D}\Theta_0/\mathrm{DT}-(1/Re^*)\partial^2\Theta_0/\partial Z^2=0,\\
&\Theta_0=1 \text{ at } Z=0,\\
&\partial\Theta_0/\partial Z+\mathrm{Bi}_{\mathrm{conv}}\Theta_0=0, \text{ at } Z=H.
\end{aligned}\tag{8.35}$$

Our monograph [27], pp. 213–218, includes a more simplified case, when $\mathrm{Pr}\to 0$ – which decouples problem (8.35) from problem (8.34).

Then, if in the derived decoupled problems (when $\mathrm{Pr}\to 0$) we assume

$$Re^*\to 0 \text{ or}: \delta\to o, \text{ with } \mathrm{Re}_d \text{ fixed},\tag{8.36a}$$

and as a consequence

$$\lambda>>(\mathrm{U_c}/\nu_{\mathrm{A}})d^2,\tag{8.36b}$$

assuming also that

$$W^*Re^*=W^{**}\approx 1 \text{ and } Re^*\mathrm{Ma}=Ma^*\approx 1,\tag{8.36c}$$

we have the possibility of deriving the following (rather awkward) lubrication equation, assuming that $\mathrm{Bi}_{\mathrm{conv}}$ is a function of H denoted by $B(H)$: namely:

$$\begin{aligned}
&\partial H/\partial T+(1/3)\mathbf{D}.\{H^3[W^{**}\mathbf{D}(\mathbf{D}^2H)-G^*\mathbf{D}H]\\
&\quad+Ma^*\underline{\mathrm{B}(H)}H^2\mathbf{D}H/[1+HB(\mathrm{H})]^2\\
&\quad+Ma^*H^3(\mathrm{d}B(H)/\mathrm{d}H)\mathbf{D}H/[1+HB(\mathrm{H})]^2\}=0.
\end{aligned}\tag{8.37}$$

In the case of a vanishing convective Biot number, $B(H)\to 0$, we see that if in (8.37) the second term proportional to B(H) disappears, this is not necessarily the case with the term $dB(H)/dH$, which need not be zero, and the influence of a large Marangoni effect remains operative.

Unfortunately, usually in classical lubrication equations, if we consider a vanishing Biot number, then the Marangoni effect also disappears! This non-physical (from my point of view) consequence is practically always encountered in all derived lubrication equations.

At this point it is opportune to observe that in a short paper by VanHook and Swift [167] it is clearly mentioned that the Pearson result has two Biot numbers (one for the conduction state and one for the perturbation), while the distinction between the two Biot numbers has not been made in some experimental papers. A theoretical analysis, however, should maintain the distinction!

In the unsteady one-dimensional case (T, X), taking into account that

$$H = 1 + \eta h(\mathrm{T},\ \mathrm{X})\text{with } \eta << 1, \tag{8.38}$$

the linearization of (8.37) produces, at the order η, a linear equation for the thickness h (T, X), when $B(H) \to 0$:

$$\partial h/\partial T + (1/3)[W^{**}\partial^4 h/\partial X^4 + Ma^*(dB(H)/dH)_{\mathrm{H}=1}\partial^2 h/\partial X^2 - G^*\partial^2 h/\partial X^2] = 0, \tag{8.39}$$

From (8.39) we obtain for the cut-off wave number k_c (when $k > k_c$ there is a linear instability) the relation:

$$k_c = \{[Ma^*/W^{**}](dB(H)/dH)_{\mathrm{H}=1} - (\mathrm{G}^*/\mathrm{W}^{**})\}^{1/2}. \tag{8.40a}$$

Finally, we observe that in (8.39) the terms proportional to – G^* and W^{**} are a stabilizing effect in evolution of the free surface, in the Bénard convection problem of heating from below – and in particular, that the thicker the film, the stronger the gravitational stabilization.

Conversely, the term proportional to Ma^*, linked with the thermocapillary (large Marangoni) effect, has a stabilizing effect on the free surface if

$$\mathrm{dB(H)/dH} > 0! \tag{8.40b}$$

In particular (see the 1997 survey [168]), thermocapillary destabilization is explained by examining the fate of an initial corrugated free surface in the linear temperature field by a thermal condition.

Where the free surface is depressed, it lies in a region of higher temperature than its neighbours. Therefore, if surface tension is a decreasing function of temperature, free surface stresses drive liquid on the free surface away from the depression, because the liquid is viscous, causing the depression to deepen further. Hydrostatic and capillary forces cannot prevent this deepening!

8.4 From Deep to Shallow Thermal Convection Model Problems

Thanks to a detailed analysis, we now see that the RAM Approach presents the possibility, first, to determine the various dimensionless parameters driving the main four physical effects which govern the mathematical formulation of the Bénard convection problem of a liquid layer heated from below and limited by a deformable free surface from overlying ambient passive air:

ε	Characterizes the expansibility of the viscous liquid.
Fr_d	The Froude number, which characterizes the thickness, *d*, of the liquid layer.
Bo	The Bousssinesq number, which characterizes the importance of the viscous dissipation..
Ma	The Marangoni number, which characterizes the thermocapillarity.
We	The Weber number, which characterizes the effect of a constant surface tension.
$\mathrm{Bi}_{\mathrm{conv}}$	The convective Biot number, which characterizes the transfer via deformable free surface in the convection regime.

Then, in a second step, a dimensionless Bénard problem is formulated where the above parameters take their respective place in equations and in lower fixed boundary and in upper deformable free surface.

Below we consider the case of thermal convection when the buoyancy is the main driven force. For this case, the buoyancy is linked with the Grashof number, Gr = O(1).

In our RAM Approach the main working hypothesis is relative to a weakly expansible liquid, $\varepsilon << 1$, with

$$\mathrm{Gr} = \varepsilon/\mathrm{Fr}_d^{\,2} = \mathrm{O}(1), \tag{8.41a}$$

We then see that it is necessary for Gr = O(1), the condition:

$$\mathrm{Fr}_d^{\,2} << 1 \Rightarrow d >> (\nu_d^{\,2}/g)^{1/3}. \tag{8.41b}$$

On the other hand, the possibility of taking into account the effect of the viscous dissipation is linked with the condition:

$$\mathrm{Di} = \varepsilon \mathrm{Bo} = \mathrm{O}(1). \tag{8.42a}$$

This is possible only if the condition

$$\mathrm{Bo} >> 1 \Rightarrow d >> C(\mathrm{T}_d)\Delta\mathrm{T}/g, \tag{8.42b}$$

is satisfied.

As a consequence of the relations of (8.41a)–(8.42b), we can write the following "deep thermal convection with viscous dissipation" equations for the leading-order functions:

$$\lim_{\text{deep}}(\mathbf{u}, \theta, \pi) = (\mathbf{u}_D, \theta_D, \pi_D), \tag{8.43a}$$

Namely

$$\nabla.\mathbf{u}_D = o; \tag{8.43b}$$

$$\begin{aligned} &\partial \mathbf{u}_D/\partial t + (\mathbf{u}_D.\nabla)\mathbf{u}_D + \nabla\pi_D - Gr\theta_D\mathbf{k} = \nabla^2\mathbf{u}_D; \\ &[1 - Di\ (p_d + 1 - z)]\{\partial\theta_D/\partial t + (\mathbf{u}_D.\nabla)\theta_D\} = (1/Pr)\nabla^2\theta_D \end{aligned} \tag{8.43c}$$

$$+ (1/2)(Di/Gr)[\partial(u_D)_i/\partial x_j + \partial(u_D)_j/\partial x_i]^2, \tag{8.43d}$$

where

$$Di = \alpha(T_d)gd/C(T_d). \tag{8.44a}$$

As boundary conditions for these deep thermal convection equations we can write:

$$\text{at } z = 0 : \mathbf{u}_D = 0 \text{ and } \theta_D = 1, \text{ at } z = 1 : w_D = 0, \tag{8.45a}$$

$$\text{at } z = 1 : \partial^2 w_D/\partial z^2 = Ma[\partial^2\theta_D/\partial x^2 + \partial^2\theta_D/\partial y^2], \tag{8.45b}$$

$$\text{at } z = 1 : \partial\theta_D/\partial z + [Bi_{conv}/Bi_S(T_d)]\{1 + Bi_S(T_d)\theta_D\} = 0, \tag{8.45c}$$

where $z \equiv x_3, w_D \equiv (u_D)_3 = \mathbf{u}_D.\mathbf{k}$.

We observe that in conditions (8.45c), written at $z = 1$, for θ_D, it seems possible (because (8.45c) is satisfied on $z = 1$) to identify Bi_{conv} with $Bi_S(T_d)$ – but, in fact, this is only a conjecture. In such a case we recover at $z = 1$ the Davis (1987) condition!

Now, if we consider the following (RB!) limiting process:

$$\lim_{Di\downarrow 0}(\mathbf{u}_D, \theta_D, \pi_D) = (\mathbf{u}_{RB}, \theta_{RB}, \pi_{RB}) \tag{8.46a}$$

we find from the deep convection equations (8.43b–8.43d) the usual RB equations for the shallow thermal convection:

$$\nabla.\mathbf{u}_{RB} = 0 \tag{8.47a}$$

$$\partial\mathbf{u}_{RB}/\partial t + (\mathbf{u}_{RB}.\nabla)\mathbf{u}_{RB} + \nabla\pi_{RB} - Gr\theta_{RB}\mathbf{k} = \nabla^2\mathbf{u}_{RB}, \tag{8.47b}$$

$$\partial\theta_{RB}/\partial t + (\mathbf{u}_{RB}.\nabla)\theta_{RB} = (1/Pr)\nabla^2\theta_{RB}, \tag{8.47c}$$

For the RB equations (8.47a–8.47c), the boundary conditions (8.45a–8.45c) are also true – but written for, $\mathbf{u}_{RB}$, w_{RB} and θ_{RB}.

Chapter 9
The RAM Approach in Atmospheric Motions

In Sect. 9.1 of this chapter we consider the 2D steady lee waves problem, in the framework of a non-viscous but compressible and adiabatic fluid flow. The starting equations are Euler two-dimensional steady equations. From these we derive a single, rather awkward, but very convenient equation for the stream function in the case of low Mach numbers fluid flow theory, for the application of the RAM Approach. From this single equation for the stream function we derive a family of model equations for the lee waves problem by considering various limiting cases. This example very well illustrates the possibility of a theoretical investigation before the use of the RAM Approach.

In Sect. 9.2, as an application of the RAM Approach to very difficult, atmosphere–meteo-fluid motions, we consider the derivation of a simplified (but rather realistic) and consistent "meteo–fluid–dynamic model": namely, the "low Kibel number asymptotic model" derived from the dissipative hydrostatic equations.

In particular, our RAM Approach presents the possibility of solving the difficult but decisive singular problem relative to initial conditions encountered by meteorologists during their weather forecasting, and also take into account the influence of the Ekman boundary layer near the Earth's surface.

9.1 Some Models for the Lee-Waves Problem

With the dimensions, the steady two-dimensional Euler equations are written in the following form, for the velocity components u and w, pressure p and density ρ – all assumed depending on the coordinates x and z:

$$\rho[u\ \partial u/\partial x + w\ \partial u/\partial z] + \partial p/\partial x = 0, \qquad (9.1a)$$

$$\rho[u\ \partial w/\partial x + w\ \partial w/\partial z] + \partial p/\partial z + g\ \rho = 0, \qquad (9.1b)$$

R. Kh. Zeytounian, *Navier–Stokes–Fourier Equations*,
DOI 10.1007/978-3-642-20746-4_9, © Springer-Verlag Berlin Heidelberg 2012

$$\partial(\rho u)/\partial x + \partial(\rho w)/\partial z = 0, \tag{9.1c}$$

$$[u\, \partial/\partial x + w\, \partial/\partial z](p/\rho^{\gamma}) = 0. \tag{9.1d}$$

We consider a mountain, and write as equation for the gravity ($\mathbf{g} = -g\mathbf{k}$) plane:

$$z = h^{\circ}\eta(x/l^{\circ}), \; -1/2 \leq x/l^{\circ} \leq +1/2. \tag{9.2a}$$

The slip condition gives:

$$w = u d\eta/dx \text{ on } z = h^{\circ}\eta(x/l^{\circ}). \tag{9.2b}$$

Again, our non-viscous and non-heat-conducting fluid is a thermally perfect gas, and the equation of state is with dimensions:

$$T = p/R\rho, \tag{9.3}$$

with $R = Cp - Cv$ and $\gamma = C_p/C_v$.

First, from the equation of continuity (9.1c), we introduce a stream function $\psi(x, z)$ such that:

$$\rho u = -\partial \psi/\partial z \text{ and } \rho w = +\partial \psi/\partial x. \tag{9.4}$$

9.1.1 First Integrals

Using the relations (9.4), from the equation of the adiabaticity (9.1d), we derive the following first integral

$$p = \rho^{\gamma}\, \Pi(\psi). \tag{9.5a}$$

where the function $\Pi(\psi)$ is arbitrary and conservative along each streamline $\psi = \text{constant}$.

Then, from equations (9.1a, 9.1b) we obtain the well-known Bernoulli first integral, when we exclude the pressure p:

$$|\mathbf{u}|^2/2 + [\gamma/(\gamma-1)]\, \rho^{\gamma-1}\, \Pi(\psi) + gz = I(\psi), \tag{9.5b}$$

where the function $I(\psi)$ is a second arbitrary function, also conservative along each streamline. From these two equations (9.1a, 9.1b) we can also derive a relation for the vorticity ω:

$$\omega \equiv \partial u/\partial z - \partial w/\partial x = -\rho[dI/d\psi - [1/(\gamma-1)](p/\rho) d\mathrm{Log}\,\Pi/d\psi]. \tag{9.5c}$$

The two arbitrary functions $I(\psi)$ and $\Pi(\psi)$ are determined from the behaviour conditions in the upstream unperturbed region (subscript '$_\infty$'), when $x \to -\infty$ and where z is z_∞ the altitude of the unperturbed streamline at upstream infinity. Therefore, at $x \to -\infty$, we assume:

$$u = U_\infty(z_\infty), \; w = 0, \; p = p_\infty(z_\infty), \; \rho = \rho_\infty(z_\infty), \tag{9.6a}$$

$$T = T_\infty(z_\infty), \text{and } S_\infty(z_\infty) = C_v \mathrm{Log}(p_\infty/(\rho_\infty)^\gamma). \tag{9.6b}$$

With (9.6a, 9.6b) we obtain, from (9.5c), the following relation:

$$\begin{aligned} dI/d\psi - [1/(\gamma-1)](p/\rho)d\mathrm{Log}\Pi/d\psi \\ \equiv -(1/\rho_\infty)\{dU_\infty/dz_\infty - [\gamma R/(\gamma-1)U_\infty]S_\infty\}, \end{aligned} \tag{9.6c}$$

where

$$S_\infty \equiv N^2(z_\infty)(T - T_\infty), \tag{9.7a}$$

and

$$N^2(z_\infty) \equiv (1/T_\infty)\{dT_\infty/dz_\infty + g[(\gamma-1)/\gamma R]\} > 0. \tag{9.7b}$$

9.1.2 *An Equation for the Vertical Deviation* $\delta(x, z)$

Since the functions $I(\psi)$ and $\Pi(\psi)$ are both conservative along each streamline, then from the Bernoulli integral (9.5b) we determine the temperature T in the following form:

$$T = T_\infty - [(\gamma-1)/\gamma R]\{(1/2)[|\mathbf{u}|^2 - U^2{}_\infty(z_\infty)] + g(z - z_\infty)\}, \tag{9.8a}$$

and for the density we obtain the relation

$$\rho = \rho_\infty\{1 + [(T - T_\infty)/T_\infty]\}^{1/\gamma-1}. \tag{9.8b}$$

Finally, with the above results we obtain an equation for $\psi(x, z)$ in an "awkward" form (derived by Zeytounian in [12], pp. 315–330), although here we do not write this equation.

For our purpose (the *lee waves problem*) we introduce, in place of ψ, the vertical deviation of a streamline, $\delta(x, z)$, in the perturbed flow over a mountain, relative to its unperturbed altitude at upstream infinity. We therefore write, for the altitude of a perturbed streamline:

$$z = z_\infty(\psi) + \delta(x, z) \tag{9.9a}$$

and in a such case

$$\partial\psi/\partial x = -\rho_\infty U_\infty \partial\delta/\partial x, \ \partial\psi/\partial z = \rho_\infty U_\infty [1 - \partial\delta/\partial z]. \tag{9.9b,c}$$

In place of the slip condition (9.2b) we obtain:

$$\delta(x, \ h^\circ\eta(x/l^\circ)) = h^\circ\eta(x/l^\circ). \tag{9.10}$$

We observe that at upstream infinity we have the relation

$$\psi = -\int_0^{z_\infty} \rho_\infty U_\infty dz \ \equiv \ \psi(z_\infty) \Leftrightarrow z_\infty{}^{-1}(\psi) \tag{9.11}$$

where $z_\infty{}^{-1}(\psi)$ is the inverse function of $\psi(z_\infty)$.

As a consequence of relations (9.9a, 9.9b,c), for the function $\delta(x, z)$ we derive the following partial second-order differential equation:

$$\begin{aligned}
&\partial^2\delta/\partial x^2 + \partial^2\delta/\partial z^2 + (\rho/\rho_\infty)^2 (g/U^2{}_\infty) N^2(z_\infty)\,\delta \\
&\quad = -(1/2)(\rho/\rho_\infty)^2 d/dz_\infty [\mathrm{Log}[U^2{}_\infty \exp(-S_\infty/C_p)]] \\
&\quad + (1/2) d/dz_\infty [\mathrm{Log}[\rho_\infty{}^2 U^2{}_\infty \exp(-S_\infty/C_p)]]\{(\partial\delta/\partial x)^2 \\
&\quad + (\partial\delta/\partial z)^2 - 2\partial\delta/\partial z + 1\} \\
&\quad + (\partial \mathrm{Log}\ \rho/\partial x)\partial\delta/\partial x + (\partial \mathrm{Log}\ \rho/\partial z)[\partial\delta/\partial z - 1].
\end{aligned} \tag{9.12a}$$

The above equation for the deviation in altitude,

$$\delta(x, z) = z - z_\infty(\psi) \tag{9.12b}$$

is rather awkward, but very convenient for further analysis – in particular when we assume that the upstream, constant, Mach number, $M^0{}_\infty \ll 1$.

For equation (9.12a), with the slip condition (9.10), we also have the following three conditions:

$$\delta(x = -\infty, z_\infty) = 0, \tag{9.13a}$$

$$\delta(x, \ z = H_\infty) = 0, \tag{9.13b}$$

$$|\ \delta(x = +\infty, \ z)| < \infty, \tag{9.13c}$$

where H_∞ is the altitude of the upper level (for instance, the tropopause assumed as a flat horizontal plane), where the streamlines are undeflected.

The last condition (9.13c) is the only possible physical one, because of the lee-waves regime downstream of the mountain. However, in equation (9.12a) we also have, as an unknown function, the density ρ, and consequently we must return to the two relations (9.8a, 9.8b), which we transform to an relation for ρ in which δ is present:

$$(\rho/\rho_\infty)^{\gamma-1} = 1 + (U^2{}_\infty/2C_pT_\infty)[\rho_\infty/\rho]^2[(\partial\delta/\partial x)^2 + (\partial\delta/\partial z)^2 - 2\partial\delta/\partial z + 1] + (U^2{}_\infty/2C_pT_\infty)[(2g/U^2{}_\infty)\,\delta - 1], \tag{9.14a}$$

and we have the following upstream infinity condition:

$$\rho \to \rho_\infty(z_\infty) \text{ when } x \to -\infty. \tag{9.14b}$$

The above problem – (9.12a) with (9.10), (9.13a, 9.13b, 9.13c), (9.14a), and (9.14b) – for two functions δ and ρ, is strongly non-linear. Below we consider a simplified case when

$$U_\infty \equiv U^0{}_\infty = \text{ const} \tag{9.15a}$$

and

$$-\,dT_\infty/dz_\infty \equiv \Gamma^o{}_\infty = \text{ const}, \tag{9.15b}$$

such that

$$T_\infty(z_\infty) = T_\infty(0)[1 - (\Gamma^0{}_\infty/T_\infty(0))z_\infty]. \tag{9.15c}$$

This linear (9.15c) distribution for $T_\infty(z_\infty)$ is very well justified for the usual meteorological situation in the troposphere, where the lee-waves regime is considered – H_∞ being the height of the whole troposphere.

The parameter

$$\mu^0{}_\infty = \Gamma^0{}_\infty H_\infty/T_\infty(0) \tag{9.15d}$$

is a reference parameter for the temperature profile at upstream infinity, and in dimensionless form we have

$$T_\infty(z_\infty)/T_\infty(0) \equiv \Theta(\zeta_\infty) = 1 - \mu^0{}_\infty\,\zeta_\infty \tag{9.15e}$$

where

$$\zeta_\infty \equiv z_\infty/H_\infty$$

is the dimensionless altitude far upstream.

9.1.3 Dimensionless Formulation

First we introduce the non-dimensional density perturbation:

$$\varpi = (\rho - \rho_\infty)/\rho_\infty, \tag{9.16a}$$

and the non-dimensional vertical displacement of the streamline,

$$\Delta = \delta/\mathrm{h}^\circ. \tag{9.16b}$$

In place of the relation (9.12b) – since far ahead of the mountain there is assumed to be a uniform flow with velocity components ($\mathrm{U}^0{}_\infty = \mathrm{const}, 0$) – we write:

$$\zeta_\infty = \mathrm{Bo}[\zeta - (1/\nu^\circ)\,\Delta],\ \zeta = \frac{\mathrm{z}}{H_o} \tag{9.16c}$$

where z is reduced with the vertical length scale H_0, characterizing the lee-wave process, and z_∞ is reduced with $\mathrm{H}_\infty \equiv \mathrm{RT}_\infty(0)/g$. In this case we have two ratios:

$$\nu^\circ = \mathrm{H}_0/\mathrm{h}^\circ,\ \mathrm{Bo} = \mathrm{H}_0/\mathrm{H}_\infty, \tag{9.17}$$

where ν° is the "linearization" parameter, and Bo is the Boussinesq number.

In place of the slip condition (9.10) we have the following dimensionless slip condition:

$$\Delta(\xi,\ \zeta = (1/\nu^\circ)\eta(\xi)) = \eta(\xi), \tag{9.18a}$$

where $\xi = \mathrm{x}/l^\circ$, and the condition (9.13b) gives

$$\Delta(\xi,\ \zeta = 1/\mathrm{Bo}) = 0. \tag{9.18b}$$

As a "long-wave" approximation parameter we have

$$\varepsilon = \mathrm{H}_0/l^\circ. \tag{9.19}$$

Finally, for ϖ, defined by (9.16a), we obtain from (9.14a) the following dimensionless relation:

$$\begin{aligned}(1+\varpi)^{\gamma-1} = 1-(1/\Theta)\{&(1/2)(\gamma-1)(\mathrm{M}^0{}_\infty/\nu_0)^2[1/(1+\varpi)^2][\varepsilon^2(\partial\Delta/\partial\xi)^2\\ &+(\partial\Delta/\partial\zeta)^2-2\nu_0\partial\Delta/\partial\zeta+\nu_0{}^2]\\ &+[(\gamma-1)/\gamma](\mathrm{Bo}/\nu_0)\Delta-(1/2)(\gamma-1)(\mathrm{M}^o{}_\infty)^2\},\end{aligned} \tag{9.20a}$$

where

$$\Theta = \Theta(\zeta,\Delta) = 1 - \mathrm{Bo}\mu^0{}_\infty[\zeta - (1/\nu_0)\Delta]. \tag{9.20b}$$

Then, for the function $\Delta(\xi,\zeta)$ we obtain, from equation (9.12a), the following dimensionless main equation:

$$\begin{aligned}&\Theta(\zeta,\Delta)\{\varepsilon^2\partial^2\Delta/\partial\xi^2+\partial^2\Delta/\partial\zeta^2-[1/(1+\varpi)][\varepsilon^2(\partial\Delta/\partial\xi)\partial\varpi/\partial\xi\\&+(\partial\Delta/\partial\zeta)\partial\varpi/\partial\zeta-\nu_0\partial\varpi/\partial\zeta]\}+(\mathrm{Bo}^2/\gamma \mathrm{M}^{02}_{\infty})S_0(1+\varpi)^2\Delta\\&=(\nu_0/2)\mathrm{Bo}S_0\{\varpi(2+\varpi)-[\varepsilon^2(\partial\Delta/\partial\xi)^2+(\partial\Delta/\partial\zeta)^2\\&-2\nu_0\partial\Delta/\partial\zeta]\},\end{aligned}\tag{9.21}$$

In equation (9.21), S_0 is the "hydrostatic stability" parameter:

$$S_0=[(\gamma-1)/\gamma]-\mu^0{}_{\infty},\tag{9.22}$$

which characterizes the stratification of the unperturbed flow at upstream infinity.

For normal meteorological values in the troposphere, for dry air, we have the following atmospheric values:

$$\begin{aligned}&\gamma=1.4,\\&\mathrm{T}_{\infty}(0)=288^{\circ}\mathrm{C},\\&\mu^0{}_{\infty}=(\mathrm{R}/g)\Gamma^0{}_{\infty}\approx 0.19037,\\&[(\gamma-1)/\gamma]-\mu^0{}_{\infty}\approx 0.09534.\end{aligned}$$

Conversely,

$$[\gamma \mathrm{R}\ \mathrm{T}_{\infty}(0)]^{1/2}\approx 340,17\mathrm{m/sec},$$

and if

$$34\mathrm{m/sec}\geq \mathrm{U}^0{}_{\infty}\geq 10\,\mathrm{m/sec},$$

then we obtain

$$0.03\leq \mathrm{M}^0{}_{\infty}\leq 0.1.$$

We observe that for the unknown function $\Delta(\xi,\zeta)$, related with the lee-waves problem – over and downstream of the mountain – which is the solution of equation (9.21), we have the following conditions:

$$\Delta(\xi,(1/\nu_0)\eta(\xi))=\eta(\xi),\ \text{when}\ \xi\in[-(1/2),+(1/2)],\tag{9.23a}$$

$$\Delta(\xi=-\infty,\zeta_{\infty})=0,\tag{9.23b}$$

$$\Delta(\xi,\zeta=1/\mathrm{Bo})=0,\tag{9.23c}$$

$$|\Delta(\xi = +\infty, \zeta)| < \infty, \tag{9.23d}$$

In main equation (9.21) the more important parameter is

$$K_0^2 = S_0(\mathrm{Bo}^2/\gamma \mathrm{M}^{02}_{\infty}) = g[(\Gamma_A - \Gamma^0{}_{\infty})/\mathrm{T}_{\infty}(0)](\mathrm{H}_0/\mathrm{U}^0{}_{\infty})^2 \tag{9.24}$$

– the *Dorodnitsyn–Scorer* parameter, where

$$\Gamma_A = (g/R)[(\gamma - 1)/\gamma], \tag{9.25a}$$

is the dry adiabatic temperature gradient, which plays a fundamental role when

$$\mathrm{M}^0{}_{\infty} = \mathrm{U}^0{}_{\infty}/[\gamma R \mathrm{T}_{\infty}(0)]^{1/2} \ll 1, \text{ but } \mathrm{Ko}^2 = \mathrm{O}(1). \tag{9.25b}$$

An important characteristic of the modelling, via the RAM Approach, of the above lee-waves problem, governed by the main equation (9.21), with the relations (9.20a, 9.20b) and conditions (9.23a–9.23d), is linked with the requirement that *without fail the parameter* Ko^2 *must be bounded*!

9.1.4 Four Distinguished Limiting Cases

We also observe that the strong constraint mentioned above, $\mathrm{Ko}^2 = \mathrm{O}(1)$, does not present the possibility of considering the case:

$$\mathrm{M}^{02}_{\infty} \downarrow 0, \text{ solely},$$

because in a such case,

$$\mathrm{Ko}^2 = (S_0 \mathrm{Bo}^2/\gamma)/\mathrm{M}^{02}_{\infty} \uparrow \infty!$$

The asymptotic analysis for a such case is very complicated and deserves another approach.

9.1.4.1 The Deep Convection Case

The first case, which is linked to the *deep convection*, is valid in the whole troposphere when

$$\mathrm{Bo} = 0(1), \tag{9.26a}$$

but with the following similarity rule between low Mach number and "hydrostatic stability" parameter:

$$\mathrm{M}^0{}_{\infty} \ll 1 \text{ and } S_0 \gg 1, \tag{9.26b}$$

assumed large, such that

$$S_0 = S^* \mathrm{M}^{02}_{\infty} \Leftrightarrow \Gamma^0{}_{\infty} \approx \Gamma_{\mathrm{A}} - (g/\mathrm{R}) S^* \mathrm{M}^{02}_{\infty}, \tag{9.26c}$$

with

$$S^* = 0(1). \tag{9.26d}$$

When the "linearization" parameter ν_0 is fixed (not considered as a small parameter, for the case of an important elevated mountain) the limiting case

$$\mathrm{M}^{02}_{\infty} \downarrow 0, \text{ with } (8.26\mathrm{a}-\mathrm{c}), \text{ and fixed } \gamma \text{ and } \nu_0, \tag{9.27a}$$

leads, in place of the equation (9.21) and relations (9.20a, 9.20b), at the leading order for Δ_{deep} and ϖ_{deep}, in expansions

$$\Delta = \Delta_{\mathrm{deep}} + \mathrm{M}^{02}_{\infty} \Delta^* + \ldots, \; \varpi = \varpi_{\mathrm{deep}} + \mathrm{M}^{02}_{\infty} \varpi^* + \ldots\;, \tag{9.27b}$$

the following reduced system of two equations:

$$\begin{aligned}(1+\varpi_{\mathrm{deep}})^{\gamma-1} &= 1/\{1-\ \mathrm{Bo}((\gamma-1)/\gamma)[\zeta-(1/\nu_0)\Delta_{\mathrm{deep}}]\} \\ &- [\mathrm{Bo}((\gamma-1)/\gamma)\ \zeta]/\{1-\mathrm{Bo}((\gamma-1)/\gamma)[\zeta-(1/\nu_{\mathrm{o}})\Delta_{\mathrm{deep}}];\end{aligned} \tag{9.27c}$$

$$\begin{aligned}&\{1-[\mathrm{Bo}(\gamma-1)/\gamma][\ \zeta-(1/\nu_0)\Delta_{\mathrm{deep}}]\}\{[\ \varepsilon^2\partial^2\Delta_{\mathrm{deep}}/\partial\xi^2+\partial^2\Delta_{\mathrm{deep}}/\partial\zeta^2 \\ &-[1/(1+\varpi_{\mathrm{deep}})][\ \varepsilon^2(\partial\Delta_{\mathrm{deep}}/\partial\xi)\partial\varpi_{\mathrm{deep}}/\partial\xi+(\partial\Delta_{\mathrm{deep}}/\partial\zeta)\partial\varpi_{\mathrm{deep}}/\partial\zeta \\ &-\ \nu_{\mathrm{o}}\partial\varpi_{\mathrm{deep}}/\partial\zeta]\}\ \mid\ (\mathrm{Bo}^2/\gamma)S^*(1+\varpi_{\mathrm{deep}})^2\Delta_{\mathrm{deep}} = 0.\end{aligned} \tag{9.27d}$$

As boundary conditions for Δ_{deep}, we have the full conditions (9.23a–9.23d).

Combining (9.27c) and (9.27d), we can derive a single equation for the single function, Δ_{deep}. But this single equation is how much complex and strongly non-linear, that we do not write a such equation here.

If we assume a complementary (linearization) constraint:

$$\begin{aligned}\nu_0 \gg\ 1 \text{ such that } :\ (1/\nu_0) &=\ \nu^*\mathrm{M}^{\mathrm{o}2}_{\infty}, \\ \text{and } \nu^* &=\ \mathrm{O}(1) \Rightarrow\ \mathrm{h}^{\circ} \approx \mathrm{U}^{02}_{\infty}/\gamma g,\end{aligned} \tag{9.27e}$$

then we derive a linear problem, in place of the non-linear deep convection problem (equations (9.27c, 9.27d) with the full conditions (9.23a–9.23d)), for the limit function:

$$\sigma_{\mathrm{deep}} = [1-((\gamma-1)/\ \gamma)\mathrm{Bo}\ \zeta]^{1/2(\gamma-1)}\Delta'_{\mathrm{deep}}. \tag{9.27f}$$

where $\Delta'_{\text{deep}} = \lim_{\nu 0 \uparrow \propto} \Delta_{\text{deep}}$.

Therefore, we obtain the following linear model equation:

$$\varepsilon^2 \partial^2 \sigma_{\text{deep}}/\partial\xi^2 + \partial^2 \sigma_{\text{deep}}/\partial\zeta^2 + D(\text{Bo}\,\zeta) = 0, \tag{9.28a}$$

where the coefficient $D_0(\text{Bo}\zeta)$ is given by:

$$\begin{aligned} D_0(\text{Bo}\zeta) = &\{(\text{Bo}/\gamma)S^*/[1 - \text{Bo}((\gamma-1)/\gamma)\zeta]\} \\ &-(\text{Bo}/2\gamma)^2(2\gamma-1)/[1 - \text{Bo}((\gamma-1)/\gamma)\zeta]^2. \end{aligned} \tag{9.28b}$$

As (linearized) conditions for the above linear equation (9.28a) we have:

$$\begin{aligned} &\sigma_{\text{deep}}(\xi, 0) = \text{h}(\xi) \text{ with } \xi \in [-(1/2), +(1/2)], \\ &\sigma_{\text{deep}}(\xi, 1/\text{Bo}) = \sigma_{\text{deep}}(-\propto, \zeta_\propto) = 0, \\ &|\sigma_{\text{deep}}(\xi \to +\propto, \zeta)| < \propto. \end{aligned} \tag{9.28c}$$

This above linear case, (9.28a)–(9.28c), is very similar to the one considered by Dorodnitsyn in 1950 [169]. The reader can find in Zeytounian [12], pp. 324–328, some results concerning the solution of this equation (9.28a), with (9.28b), under the associated conditions (9.28c).

9.1.4.2 The Boussinesq Case

The second case is the Boussinesq case (considered in Chap. 4) when Bo and the Mach number, M, are both small parameters:

$$\text{Bo} \ll 1 \text{ and } \text{M}^{\text{o}}_\infty \ll 1, \text{ but } S_{\text{o}} = 0(1), \tag{9.29a}$$

such that

$$\text{Bo}/\text{M}^{\text{o}}_\infty = B^* = 0(1), \Leftrightarrow \text{H}_{\text{o}} \approx (\text{U}_{\text{o}}/\text{g})[\text{RT}_\infty(0)/\gamma],^{1/2} = \text{H}_{\text{B}} \tag{9.29b}$$

where H_{B} is the characteristic vertical displacement of the Boussinesq lee-waves which is only of the order of 1 km!

9.1.4.3 The Isochoric Case

The third case is the isochoric case when γ is large and the Mach number, M, is low:

$$\text{M}^{\text{o}}_\infty \ll 1 \text{ such that } S_{\text{o}} = 0(1) \text{ and } \text{Bo} = 0(1), \tag{9.30a}$$

but

$$\gamma M^{o2}_{\infty} = M^* = O(1), \Leftrightarrow U^o{}_{\infty} \approx [RT_{\infty}(o)]^{1/2}. \tag{9.30b}$$

9.1.4.4 The Very Thin Layer Case

The fourth case is relative to a very thin atmospheric layer, when we assume:

$$S_o Bo \approx 1, \text{ with } Bo \ll 1, \text{ and } \gamma M^{o2}_{\infty} \ll 1, \tag{9.31a}$$

such that

$$Bo/\gamma M^{o2}_{\infty} \approx 1 \Leftrightarrow H_0 \approx U^{o2}_{\infty}/g \text{ and } \Gamma_A - \Gamma^0{}_{\infty} = U^{o2}_{\infty}/gT_{\infty}(0). \tag{9.31b}$$

For each of the above cases – from the formulated full non-linear problem – equation (9.21) for the function $\Delta(\xi, \zeta)$, with the relations (9.20a, 9.20b), (9.22), for ϖ, Θ, S_0, and boundary conditions (9.23a–9.23d) – we derive a consistent low Mach number model problem with $Ko^2 = O(1)$!

The fourth case is also considered in Zeytounian [12], pp. 328–330. The second (Boussinesq) case is considered here in Chap. 4, and also in [12], chapter 8. The third (isochoric) case was considered in our thesis [2]. In [2], and also in our 1969 paper [170], the reader can find various results of computations of 2D steady lee-waves over and downstream of several mountains. In our book [37], the 3D steady problem is also considered, and in pp. 168–170 two typical figures and some comments are presented. Section 5.4.2, of [27], pp. 164–166, concerning the isochoric 2D steady case, contains various configurations of streamlines.

9.2 The Low Kibel Number Asymptotic Model

In the framework of an asymptotic modelling theory for the atmospheric motions, it is first necessary to formulate a physically realistic mathematical problem written in a dimensionless form. For this, in a coordinates frame rotating with the Earth we consider the full Navier–Stokes–Fourier (NS–F) equations for a perfect gas which is viscous, compressible, and heat-conducting dry atmospheric air.

We take into account the Coriolis force $(2\boldsymbol{\Omega}\wedge\mathbf{u})$, the gravitational acceleration $\mathbf{g} = -g\mathbf{k}$, with $\mathbf{k}$ the unit vector directed to the zenith (force of gravity modified by the centrifugal force), and the effect of the radiative heat transfer.

The above velocity vector $\mathbf{u} = (\mathbf{v}, w)$ is the (relative) velocity vector as observed in the Earth's frame rotating with the angular velocity $\boldsymbol{\Omega} = \Omega°\mathbf{e}$, with

$$\mathbf{e} = \mathbf{k}\sin\phi + \mathbf{j}\cos\phi \tag{9.32}$$

The algebraic latitude of the origin-point $\boldsymbol{P}^\circ$ of the observation on the Earth's surface is ϕ, and around $\boldsymbol{P}^\circ$ (in prediction domain D with a diameter L°) the atmospheric flow is analyzed.

The unknown thermodynamic functions are the density of air ρ, the atmospheric pressure p, and the absolute temperature T, such that the equation of state for thermally perfect gas is taken into account (the dry atmospheric air being assumed a trivariate baroclinic fluid):

$$p = R\,\rho T, \tag{9.33}$$

where R is the perfect gas constant.

A main feature of the atmospheric motions is the existence of the standard atmosphere, which is motionless and dependent only on the vertical coordinate z_S, the "standard altitude" directed in the opposite direction from the force of gravity **g**. The standard thermodynamic functions, p_S, ρ_S, and T_S, for the standard atmosphere, satisfy the following equations:

$$\begin{aligned} dp_S/dz_S + g\rho_S &= 0, \\ p_S &= R\rho_S T_S, \\ k_S dT_S/dz_S + R(T_S(z_S)) &= 0, \end{aligned} \tag{9.34}$$

where $R(T_S)$ is the radiative heat transfer in the standard atmosphere (with 's' as subscript and assumed a function of $T_S(z_S)$).

Below, we work mainly with dimensionless quantities, and in particular the thermodynamic functions, p, ρ, T and $R(T_S(z_S))$, are reduced relative to p_S, ρ_S, T_S, and $R(T_S)$ at the ground $z_S = 0$. The dimensionless horizontal velocity $\mathbf{v} = (u, v)$ and vertical velocity w are non-dimensionalized with U° and εU°, respectively, and we assume that, in the hyodrostatic parameter,

$$\varepsilon = H^\circ/L^\circ, \tag{9.35}$$

the vertical length scale H° is of the order of the height of the standard atmosphere $H_S(= RT_S(0)/g \ll L^\circ)$, assumed homogeneous, such that the Boussinesq number (ratio of two vertical length scale),

$$B_0 = H^\circ/H_S \approx 1. \tag{9.36}$$

In (9.35), L° is a horizontal length scale of a domain on the Earth's surface ground, such that the influence of the Coriolis force is taken into account.

For dimensional time we write $t^\circ t$, and consider that the Strouhal number:

$$S = L^\circ/U^\circ t^\circ \equiv 1 \Rightarrow t^\circ = L^\circ/U^\circ. \tag{9.37a}$$

The (turbulent) kinematic viscosity coefficient is $\nu^\circ\ \nu = (\mu^\circ/\rho_S(0))[\mu/\rho]$, and

$$\mathrm{Re} = \mathrm{U}^\circ\mathrm{L}^\circ/\ \nu^\circ, \tag{9.37b}$$

is the Reynolds number, while the (turbulent) heat conduction is represented by a Prandtl number,

$$\mathrm{Pr} = \mathrm{C_p}\ \mu^\circ/\mathrm{k}^\circ, \tag{9.37c}$$

and we set $\mathrm{k}^\circ\mathrm{k}$ for the (turbulent) heat-conduction coefficient.

We will also use, as non-dimensional parameters, a Mach number,

$$\mathrm{M} = \mathrm{U}^\circ/(\gamma \mathrm{RT_S}(0))^{1/2} \tag{9.37d}$$

where $\gamma = \mathrm{C_p}/\mathrm{C_v}$ is the ratio of specific heat capacities at constant pressure ($\mathrm{C_p}$) and at constant volume($\mathrm{C_v}$). Related with the Coriolis force we also have a Rossby number,

$$\mathrm{Ro} = \mathrm{U}^\circ/f^\circ\mathrm{L}^\circ \equiv \frac{1}{\mathrm{f}^\circ\mathrm{t}^\circ} \equiv \mathrm{Ki}, \tag{9.37e}$$

if we take into account (9.37a), where Ki is the Kibel number (introduced in the Soviet Union; see, for instance, Monin [94][1] and Kibel [171]) linked with the Rossby number, since the Strouhal number, $\mathrm{S} \equiv 1$. In (9.37e),

$$f^\circ = 2\Omega^\circ \sin\phi^\circ, \tag{9.37f}$$

is the Coriolis parameter, where ϕ° is a reference latitude. According to Obukhov, the typical horizontal length scale for the synoptic processes is on the order of ($c^\circ = \sqrt{\gamma \mathrm{RT_S}(0)}$ being the sound speed)

$$L_{\mathrm{Ob}} = c^\circ/f^\circ \approx 3000\ \mathrm{km}. \tag{9.37g}$$

On the other hand, the ratio

$$\mathrm{L}^\circ/L_{\mathrm{Ob}} = \mathrm{M}/\mathrm{Ki}, \tag{9.37h}$$

is related to the horizontal compressibility of the atmosphere.

Finally, as a radiative heat transfer parameter we have

$$\sigma = \mathrm{R}(\mathrm{T_S}(0))/(\mathrm{gk}^\circ/\mathrm{R}), \tag{9.37i}$$

[1] Monin's book (English translation, 1972) includes a concise introduction to physical and mathematical thinking in meteorology. In Chapter 2, pp. 14–78, Monin presents an exposition of the "Hydrodynamic Theory of Short-Range Weather Prediction".

and our Ekman number is related to

$$\mathrm{Re}_{\perp} = \varepsilon^2 \mathrm{Re} \tag{9.37j}$$

by the relation

$$\mathrm{Ek}_{\perp} = \mathrm{Ki}/\mathrm{Re}_{\perp}. \tag{9.37k}$$

Now, for our RAM Approach below, it is very convenient to introduce the following transformations from the spherical to the Cartesian system of coordinates (see the relations (3.16c)):

$$x = a_0 \cos\phi^\circ\, \lambda;\ y = a_0(\phi - \phi^\circ);\ z = r - a_0 \tag{9.38a}$$

and for $\phi^\circ \approx 45^\circ$ we have for the radius of the Earth, $a_o \approx 6300$ km.

The origin of the above right-handed curvilinear coordinates system (x, y, z) (9.38a) lies on the Earth's surface (for a flat ground, where $r = a_0$) at latitude ϕ° and longitude $\lambda = 0$. Obviously, the sphericity parameter

$$\delta = L^\circ / a_0, \tag{9.38b}$$

plays also an important role – see the dimensionless equations (9.39a–9.39d) below – and for $\delta = O(1)$ the equations for the atmospheric motions are more complicated, as is obvious from these equations. However, in the case when δ is assumed $\approx 1 (L^\circ \approx a_0 \approx 6300$ km), we obviously have

$$\varepsilon \ll 1. \tag{9.38c}$$

We assume, therefore, that the atmospheric motion occurs in a mid-latitude region, distant from the equator, around some central latitude ϕ°, and $\sin\phi^\circ$, $\cos\phi^\circ$, and $\tan\phi^\circ$ are all of order unity.

After a careful dimensional analysis (see, for instance, [19], Chap. 2) a set of dimensionless dominant (relative to $\varepsilon \ll 1$) atmospheric equations are derived, and below our functions and time–space variables are dimensionless.

9.2.1 The Dissipative (Viscous and Non-adiabatic) NS–F atmospheric Equations

In the general case, when the Reynolds number Re is different to infinity, we derive the following dissipative non-hydrostatic starting dominant dimensionless NS–F atmospheric equations:

$$\begin{aligned}\rho_D\{d\mathbf{v}_D/dt &+ [(1/Ro)(\sin\phi/\sin\phi^\circ) \\ &+ \delta\tan\phi u_D/(1+\varepsilon\,\delta z)](\mathbf{k}\wedge\mathbf{v}_D)\} \\ &+ [1/\gamma M^2(1+\varepsilon\,\delta\, z)]\mathbf{D}p_D \\ &= (1/\varepsilon^2 Re)\partial/\partial z(\,\mu\partial\mathbf{v}_D/\partial z) + \,O(\varepsilon);\end{aligned} \tag{9.39a}$$

$$\begin{aligned}\rho_D\{\varepsilon^2 dw_D/dt &- (\varepsilon/Ro)(\cos\phi/\cos\phi^\circ)u_D\} \\ &+ (1/\gamma M^2)[\partial p_D/\partial z + \,Bo\rho_D] \\ &= (4/3Re)\partial/\partial z(\mu\partial w_D/\partial z) + \,O(\,\varepsilon^2) + \,O(\varepsilon\,\delta);\end{aligned} \tag{9.39b}$$

$$\begin{aligned}d\rho_D/dt &+ \rho_D\{\partial w_D/\partial z \\ &+ [1/(1+\varepsilon\,\delta z)][\mathbf{D}\cdot\mathbf{v}_D - \delta\tan\phi v_D \\ &+ 2\,\varepsilon\,\delta w_D]\} = \,0;\end{aligned} \tag{9.39c}$$

$$\begin{aligned}\rho_D dT_D/dt - [(\gamma-1)/\gamma]dp_D/dt &= (1/\varepsilon^2 RePr)\{\partial/\partial z(k\partial T_D/\partial z) \\ &+ \,Pr\,\,\mu(\gamma-1)M^2[1/(1+\varepsilon\,\delta z)]|\partial/\partial z[\mathbf{v}_D/(1+\varepsilon\,\delta z)]|^2] \\ &+ \sigma dR/dz\} + \,O(\,\varepsilon^2),\end{aligned} \tag{9.39d}$$

with $p_D = \rho_D T_D$.

In the above dimensionless equations (9.39a–9.39d), for

$$U_D = (\mathbf{v}_D, w_D, \rho_D, p_D, T_D) \text{ with } \mathbf{v}_D = (u_D, v_D)$$

we have, as dimensionless material time-derivative operator (with the same notations for dimensionless time and space coordinates):

$$d/dt = \partial/\partial t \,+ [1/(1+\varepsilon\,\delta z)]\mathbf{v}_D.\mathbf{D} + w_D\partial/\partial z \tag{9.40a}$$

where, as horizontal gradient operator,

$$\mathbf{D}{=}(\cos\phi^\circ/\cos\phi)(\partial/\partial x)\mathbf{i} + (\partial/\partial y)\mathbf{j}. \tag{9.40b}$$

such that

$$\mathbf{D.k} = 0. \tag{9.40c}$$

Obviously, the dominant equations (9.39a–9.39d), with (9.40a–9.40c), are very convenient when we assume that

$$\delta = O(1) \text{ but } \,\varepsilon \ll \,1 \text{ and } Re \,\gg 1, \tag{9.41}$$

which is the case for the rather large synoptic dissipative atmospheric motions. For equations (9.39a–9.39d) it is also necessary to write boundary and initial conditions.

Here, as boundary conditions, for the velocity ($\mathbf{v}_D$, w_D) and temperature T_D, we write, on the flat ground,

$$\text{on } z = 0 : \mathbf{v}_D = 0, w_D = 0, \text{ and } k\partial T_D/\partial z + \sigma R = 0, \tag{9.42a}$$

according to equation (9.39d) for T_D.

Concerning the initial conditions, we observe that the initial (proper) data for the four above evolution equations (9.39a–9.39d) need not fit the dimensionless hydrostatic balance,

$$\partial p_D/\partial z + B_0 \rho_D = 0, \tag{9.42b}$$

because, in general, the dimensionless vertical velocity w need not be $O(\varepsilon)$ with respect to the horizontal one, as is the case in the hydrostatic approximate equations. Hence, in order to consider the most general case, we must assume (with dimensionless quantities) that at the initial time $t \leq 0$, εw is of order $O(1)$, and accordingly we obtain as initial conditions for the four evolution dominant NS–F equations (9.39a–9.39d):

$$\text{at } t \leq 0 : \mathbf{v}_D = \mathbf{V}_D{}^0, \quad \varepsilon w_D = W_D{}^0, T_D = T_D{}^0, \rho_D = R_D{}^0, \tag{9.42c}$$

where data

$$\mathbf{V}_D{}^0, W_D{}^0, T_D{}^0, R_D{}^0, P_D{}^0 = R_D{}^0 T_D{}^0 \text{ are given (dissipative) data.}$$

Obviously, these data are dependent (in reality) on horizontal and vertical coordinates in a prediction domain D with a diameter L°. But it is not obvious that these dimensionless coordinates, in data, are precisely the coordinates x, y, and z.?

This is an important problem (at least from my point of view) when we consider the unsteady adjustment problem to hydrostatic balance in the set of a dissipative hydrostatic, DH, approximate equations – a problem which is actually still open.

Here we leave unspecified the behaviour conditions at high altitude when $z \uparrow \infty$, and far off in the horizontal directions. These behaviour conditions are strongly linked to the numerical simulations, and can be changed in order to ensure the stability of the numerical scheme used (see, for instance, the recent paper by Cullen [172], pp. 202–287).

Usually, it is necessary that:

Total energy density must decay
sufficiently rapidly at infinity. (9.42d)

The initial-boundary value dominant, dissipative non-hydrostatic problem – (9.39a–9.39d), with (9.40a–9.40c) and (9.42a–9.42d) – seems well-posed, at least from a fluid dynamics point of view (see, for instance the review paper by Gresho [173], p. 52), and deserves further investigation – for example, in the spirit of the recent paper by White [172], pp. 1–100.

The above formulation – which is rather general for atmospheric motions – clearly shows the difficulty with these motions in the framework of a RAM Approach. For this it is no wonder that weather prediction is a very complicated task, mainly because it is very difficult to extract from the above general formulation a model sufficiently and adequately simplified, and especially realistic, for a simulation/forecast via a high-speed computer: "What the weather will be like ... tomorrow evening or for the next few days?"

9.2.2 Hydrostatic Limiting Processes

The above set of dimensionless dominant dissipative non-hydrostatic equations (9.39a–9.39d) are very complicated, and here we consider, for their simplification, the main hydrostatic limiting case:

$$\varepsilon \to \mathrm{o} \text{ and } \mathrm{Re} \to \infty, \text{ with } \varepsilon^2\mathrm{Re} \equiv \mathrm{Re}_\perp = 0(1), \tag{9.43a}$$

assuming that $\delta = \mathrm{O}(1)$, as in (9.41), which reinforces the hydrostatic constraint, $\varepsilon \ll 1$. We assume that in this limiting process (9.43a) the parameters, Ro (or Ki), M, Bo, Pr, and υ are fixed and O(1).

The corresponding limiting equations below – (9.45) and (9.46a–9.46d) – are evidently very valuable for weather forecasting of synoptic processes (*à la* Obukhov[2]), when (see (9.37h)):

$$\mathrm{L}^\circ \approx L_{\mathrm{ob}} \Rightarrow \text{such that } \mathrm{M}/\mathrm{Ki} = 0(1), \tag{9.43b}$$

and we observe that the case of the low Mach number ($\mathrm{M} < < 1$) for these synoptic process is also not a bad idea! But with $\mathrm{M} < < 1$ we must also assume that $\mathrm{Ki} < < 1$; and low Kibel number approximation, according to (9.43b), is, in fact, a necessary condition!

On the one hand, in the framework of singular perturbation problems, we see that the hydrostatic limiting process (9.43a) is also strongly dependent on the considered time–space region in weather prediction domain D, with the point $\boldsymbol{P}^\circ(\mathrm{x}^\circ, \mathrm{y}^\circ)$ as

[2] During several decades up to the early 1990s, A. M. Obukhov was the Director of the Institute of Physics of the Atmosphere of the Academy of Sciences in Moscow – actually named the "Obukhov Institute." He is a discoverer (with Kolmogorov) of the well-known (1962) "Kolmogorov–Obukhov law 2/3" in turbulence theory.

origin, and at least is certainly singular near the initial time where the dissipative data in (9.42c) are given. On the other hand, when Ki < < 1, in the framework of quasi-geostrophic modelling, in the (Ekman) layer near the ground we have also a singular behaviour.

Therefore, an asymptotic analysis shows that we can at least consider the following two hydrostatic limiting processes linked with (9.43a) in the framework of the non-hydrostatic dominant equations (9.39a–9.39d):

$$\mathrm{Lim}^{\mathrm{DH}} = [(9.39\mathrm{a} - 9.39\ \mathrm{d}) + (9.43\mathrm{a}),\ \text{with t, x, y, z fixed}]; \tag{9.44a}$$

and

$$\mathrm{Lim}^{\mathrm{DAdj}} = [(9.39\mathrm{a} - 9.39\mathrm{d}) + (9.43\mathrm{a}),\ \text{with}\ \theta = \mathrm{t}/\varepsilon^2, \mathrm{x}, \mathrm{y}, \zeta = \mathrm{z}/\varepsilon\ \text{fixed}]. \tag{9.44b}$$

9.2.2.1 Dissipative Hydrostatic (DH) Large-Scale Equations

From the limiting process (9.44a) we derive the dissipative hydrostatic (DH) large-scale, non-tangent equations for the functions

$$\begin{aligned}[\mathbf{v}_{\mathrm{DH}} &= (\mathrm{u}_{\mathrm{DH}}, \mathrm{v}_{\mathrm{DH}}), \mathrm{w}_{\mathrm{DH}}, \mathrm{p}_{\mathrm{DH}}, \rho_{\mathrm{DH}}, \mathrm{T}_{\mathrm{DH}}] \\ &= \mathrm{Lim}^{\mathrm{DH}}[\mathbf{v}_{\mathrm{D}} = (\mathrm{u}_{\mathrm{D}}, \mathrm{v}_{\mathrm{D}}), \mathrm{w}_{\mathrm{D}}, \mathrm{p}_{\mathrm{D}}, \rho_{\mathrm{D}}, \mathrm{T}_{\mathrm{D}}],\end{aligned} \tag{9.45}$$

dependent on the time-space coordinates, t, x, y, and z. Therefore, with (9.43a) only, in place of (9.39a–9.39d) we obtain the following DH model approximate leading-order equations:

$$\begin{aligned}\rho_{\mathrm{DH}}\{\mathrm{d}\mathbf{v}_{\mathrm{DH}}/\mathrm{dt} &+ [(1/\mathrm{Ki})(\sin\phi/\sin\phi^\circ) + \delta\tan\phi\mathrm{u}_{\mathrm{DH}}](\mathbf{k}\wedge\mathbf{v}_{\mathrm{DH}})\} \\ &+ (1/\gamma\mathrm{M}^2)\mathbf{D}\mathrm{p}_{\mathrm{DH}} = (1/\mathrm{Re}_\perp)\partial/\partial\mathrm{z}(\mu\partial\mathbf{v}_{\mathrm{DH}}/\partial\mathrm{z});\end{aligned} \tag{9.46a}$$

$$\partial\mathrm{p}_{\mathrm{DH}}/\partial\mathrm{z} + \mathrm{Bo}\rho_{\mathrm{DH}} = 0; \tag{9.46b}$$

$$\mathrm{d}\rho_{\mathrm{DH}}/\mathrm{dt} + \rho_{\mathrm{DH}}\{\partial\mathrm{w}_{\mathrm{DH}}/\partial\mathrm{z} + \mathbf{D}.\mathbf{v}_{\mathrm{DH}} - \delta\tan\phi\mathrm{v}_{\mathrm{DH}}\} = 0; \tag{9.46c}$$

$$\begin{aligned}\rho_{\mathrm{DH}}\mathrm{d}\mathrm{T}_{\mathrm{DH}}/\mathrm{dt} &- [(\gamma-1)/\gamma]\mathrm{d}\mathrm{p}_{\mathrm{DH}}/\mathrm{dt} \\ &= (1/\mathrm{Re}_\perp)(1/\mathrm{Pr})\{\partial/\partial\mathrm{z}(\mathrm{k}\partial\mathrm{T}_{\mathrm{DH}}/\partial\mathrm{z}) \\ &+ \mathrm{Pr}\ \mu(\gamma-1)\mathrm{M}^2|\partial\mathbf{v}_{\mathrm{DH}}/\partial\mathrm{z}|^2 + \sigma\mathrm{dR}/\mathrm{dz}\},\end{aligned} \tag{9.46d}$$

with

$$\mathrm{p}_{\mathrm{DH}} = \rho_{\mathrm{DH}}\mathrm{T}_{\mathrm{DH}},\ \text{and}\ \mathrm{d}/\mathrm{dt} = \partial/\partial\mathrm{t} + \mathbf{v}_{\mathrm{DH}}.\mathbf{D} + \mathrm{w}_{\mathrm{DH}}\partial/\partial\mathrm{z}. \tag{9.47}$$

In (9.46a–9.46d) the parameters δ, M, $Re_\perp$, Pr, Ki, Bo, σ, and $\sin\phi$, and $\cos\phi$ are fixed. These DH non-tangent model equations constitute a very significant approximate system of equations for large-scale atmospheric motions ($\delta = O(1)$) in a thin layer, such as the troposphere, around the Earth's sphere.

On the flat ground, from (9.42a), we write:

$$\mathbf{v}_{DH} = 0, w_{DH} = 0, \text{and } k\partial T_{DH}/\partial z + \sigma R = 0, \text{ on } z = 0 \tag{9.48a}$$

and here again we leave unspecified the behaviour conditions at high altitude when $z \uparrow \infty$, and far off, in the horizontal x and y directions.

Concerning the initial conditions for the evolution equations (9.46a, 9.46c, 9.46d), we must give, in fact, only the initial values for $\mathbf{v}_{DH}$ and T_{DH} (or p_{DH}):

$$\text{at } t \leq 0: \mathbf{v}_{DH} = \mathbf{v}^0{}_{DH}, T_{DH} = T^0{}_{DH}. \tag{9.48b}$$

Here, the dissipative hydrostatic data $\mathbf{v}_{DH}{}^0$, $T_{DH}{}^0$ have (strictly speaking) nothing to do with the corresponding given proper dissipative non-hydrostatic initial data $\mathbf{V}_D{}^0$, $T_D{}^0$ in (9.42c), for the atmospheric dominant non-hydrostatic dissipative equations (9.39a) and (9.39d).

We observe that from hydrostatic balance (9.46b), with

$$p_{DH} = \rho_{DH} T_{DH},$$

we also have

$$\partial \log p^0{}_{DH}/\partial z = -\left(Bo/T^0{}_{DH}\right) \text{ and } \rho^0{}_{DH} = p^0{}_{DH}/T^0{}_{DH}, \tag{9.48c}$$

and indeed, two of the initial conditions (related with the given proper dissipative non-hydrostatic initial data $W_D{}^0$ and $R_D{}^0$) in (9.42c) have been lost during the dissipative hydrostatic limiting process (9.44a) with (9.45), taking into account (9.43a).

As a consequence, a primary question arises: "How are dissipative hydrostatic data $\mathbf{v}_{DH}{}^0$ and $T_{DH}{}^0$ related to given proper dissipative non-hydrostatic initial data $\mathbf{V}_D{}^0$, $W_D{}^0$, $T_D{}^0$ and $R_D{}^0$?" The answer to this decisive question, for weather forecasting, must be derived from the dissipative non-hydrostatic (where the parameter ε is present) equations (9.39a, 9.39b), via (see (9.44b)) an unsteady adjustment of the hydrostatic balance (9.46b) in the DH system of equations (9.46a–9.46d).

The significant non-hydrostatic equations (where acoustics is present) which govern this adjustment problem are derived below (see equations (9.50) and (9.51a–9.51d)) just through the limiting process (9.44b).

For a "physical" introduction to adjustment (adaptation) of meteorological fields, see the discussion in §6 of Monin [94] – a small but important book which presents a unique account of the early evolution (up to 1968) of the dynamic and physical bases for modelling and appropriate simulation of atmospheric motions spanning the large spectrum of time-scales.

9.2.2.2 Adjustment to Hydrostatic Balance in the Framework of the DH Equations

For the DH equations (9.46a–9.46d) with the no-slip and temperature boundary conditions (9.48a), singular near time $t = 0$, it is necessary to consider a dissipative unsteady adjustment problem of hydrostatic balance (9.46b), which is significant, simultaneously, close to initial time and near the ground. This problem is derived from the dissipative non-hydrostatic dominant equations (9.39a–9.39d), via the hydrostatic limiting process (9.44b), with

$$[\mathbf{v}_{DAdj}, w_{DAdj}, p_{DAdj}, \rho_{DAdj}, T_{DAdj}] = \mathrm{Lim}^{DAdj}[\mathbf{v}_D, \varepsilon w_D, p_D, \rho_D, T_D], \tag{9.49}$$

dependent on the time–space coordinates, θ, x, y, and ζ.

In this case we derive the following dissipative (non-hydrostatic) adjustment equations, which are significant close to initial time $\theta = 0$ and near the ground $\zeta = 0$:

$$\rho_{DAdj}[\partial \mathbf{v}_{DAdj}/\partial\theta + w_{DAdj}\partial \mathbf{v}_{DAdj}/\partial\zeta] = (1/Re_\perp)\partial^2\mathbf{v}_{DAdj}/\partial\zeta^2; \tag{9.50}$$

$$\partial\rho_{DAdj}/\partial\theta + \partial(\rho_{DAdj} w_{DAdj})/\partial\zeta = 0; \tag{9.51a}$$

$$\rho_{DAdj}[\partial w_{DAdj}/\partial\theta + w_{DAdj}\partial w_{DAdj}/\partial\zeta] + (1/\gamma M^2)\partial p_{DAdj}/\partial\zeta = (3/4Re_\perp)\partial^2 w_{DAdj}/\partial\zeta^2; \tag{9.51b}$$

$$\rho_{DAdj}[\partial T_{DAdj}/\partial\theta + w_{DAdj}\partial T_{DAdj}/\partial\zeta] - [(\gamma-1)/\gamma][\partial p_{DAdj}/\partial\theta + w_{DAdj}\partial p_{DAdj}/\partial\zeta] = (1/PrRe\perp)\partial^2 T_{DAdj}/\partial\zeta^2 + (\gamma-1)(M^2/Re_\perp)[|\partial\mathbf{v}_{DAdj}/\partial\zeta|^2 + (4/3)|\partial w_{DAdj}/\partial\zeta|^2], \tag{9.51c}$$

$$p_{DAdj} = \rho_{DAd} T_{DAdj}. \tag{9.51d}$$

The above DAdj equations (9.50) and (9.51a–9.51d) are derived when we assume that the two dissipative coefficients μ and k do not have a vertical structure dependent on ζ! For these above DAdj (unsteady, *à la* Rayleigh) compressible, viscous, and heat-conducting equations it is necessary to associate initial and boundary conditions.

According to starting conditions (9.42c) at $t = 0$, for (9.39a–9.39d) as initial conditions we write:

$$\text{at } \theta \le 0: \mathbf{v}_{DAdj} = \mathbf{V}_D{}^0, w_{DAdj} = W_D{}^0, T_{DAdj} = T_D{}^0, \rho_{DAdj} = R_D{}^0 \tag{9.52a}$$

At flat, thermally non-homogeneous ground, according to (9.42a), as boundary conditions we have:

$$\text{on } \zeta = 0 : \mathbf{v}_{\mathrm{DAdj}} = 0, \mathrm{w}_{\mathrm{DAdj}} = 0, \text{ and } \partial \mathrm{T}_{\mathrm{DAdj}}/\partial\zeta = 0 \tag{9.52b}$$

The question concerning the structure in altitude of the initial data in (9.52a) for DAdj equations (9.50) and (9.51a–9.51d), requires further and more detailed investigation, because it is clear that in the dissipative case the unsteady adjustment to hydrostatic balance raises many unsolved problems.

On the other hand, from our above RAM Approach we see that a numerical simulation of atmospheric motions, via DH model equations (9.46a–9.46d), must be coupled with the above DAdj equations (9.50) and (9.51a–9.51d), which presents the possibility of taking into account the consistent initial conditions for (9.46a) and (9.46d), linked with the unknown data in (9.48b).

During a numerical simulation of an atmospheric motion via an approximate model derived asymptotically for more complete starting equations, it seems inevitable that some partial derivatives relative to time in derived approximate simplified models equation are lost (they disappear). In such a case, only via an unsteady adjustment problem is it consistent, via a *matching*, to take into account the influence of real data, written for the starting problem, on formulation of a well-posed initial-boundary value problem for this derived simplified model!

9.2.3 The Dissipative Hydrostatic Equations in p*-System*

We now consider the dissipative hydrostatic (DH) system of equations (9.46a–9.46d), with the boundary conditions (9.48a) on $\mathrm{z} = 0$ and initial conditions (9.48b, 9.48c) at $\mathrm{t} \leq 0$. For simplicity, however, we do not write the subscript 'DH' in these equations,

It is well known that a convenient hypoproduct of the hydrostatic approximation is the possibility of using variables other than z as the vertical coordinate (for example, pressure, potential temperature, and so on).

In particular, if pressure p is used as the vertical coordinate (according to (9.46b), with $\mathrm{Bo} \equiv 1$) such that

$$\partial/\partial \mathrm{z} = -\rho\partial/\partial p \text{ and } \mathrm{z} = H(\mathrm{t}, \mathrm{x}, \mathrm{y}, p), \tag{9.53}$$

where here (x, y) now denotes the horizontal coordinates on constant pressure (isobaric) surfaces, and $H(\mathrm{t}, \mathrm{x}, \mathrm{y}, p)$ is the local height of an isobaric surface above the flat ground surface, then, in place of DH equations (9.46a–9.46d), without any approximation nor ambiguity, we derive, for the horizontal velocity vector

$\mathbf{v}$(u, v), temperature T, $\omega = dp/dt$, and local height H, as a function of t, x, y, and p, the following DH equations written in p-system:

$$\begin{aligned} d\mathbf{v}/dt + [(1/\mathrm{Ki})(\sin\phi/\sin\phi^{\circ}) + \delta\tan\phi\ u](\mathbf{k}\wedge\mathbf{v}) \\ + (1/\gamma M^2)\mathbf{D}H = (1/\mathrm{Re}\perp)\partial(\mu\rho\partial\mathbf{v}/\partial p)/\partial p; \end{aligned} \tag{9.54a}$$

$$\begin{aligned} dT/dt - [(\gamma-1)/\gamma](T/p)\omega = (1/\mathrm{Re}\perp)(1/\mathrm{Pr})\{\partial(\rho k\partial T/\partial p)/\partial p \\ + \mathrm{Pr}(\mu/\gamma)(\gamma-1)M^2\rho|\partial\mathbf{v}/\partial p|^2 - \sigma dR/dp\}; \end{aligned} \tag{9.54b}$$

$$\partial\omega/\partial p + \mathbf{D}\cdot\mathbf{v} - \delta\tan\phi\,v = 0; \tag{9.54c}$$

$$\partial H/\partial p + T/p = 0; \tag{9.54d}$$

$$\omega = (p/T)[\partial H/\partial t + \mathbf{v}.\mathbf{D}H - w]. \tag{9.54e}$$

In these equations the operator $\mathbf{D}$ is the horizontal gradient on the isobaric surface, $p =$ const, with the components $[((\cos\phi^{\circ}/\cos\phi)\partial/\partial x; \partial/\partial y))]$, and the material derivative operator is:

$$d/dt = \partial/\partial t + \mathbf{v}.\mathbf{D} + \omega\partial/\partial p. \tag{9.55}$$

In this case we assume that μ, k, and R are known functions of p. The Eliasen p-system has the advantage that the region of numerical integration has a limited vertical extent, instead of $0 \le z < +\infty$, for the original DH system (9.46a–9.46d).

One disadvantage is that the lower boundary conditions (9.48a) on flat ground becomes a condition at the unknown isobaric surface $H = 0$! Therefore, it is necessary to impose, in place of boundary conditions (9.48a), the following boundary conditions:

$$\mathbf{v} = 0, \omega = (p/T)\partial H/\partial t, \text{ on } H = 0, \tag{9.56a}$$

$$kp\partial \mathrm{Log}T/\partial p = \sigma R, \text{ on } H = 0. \tag{9.56b}$$

As a boundary condition "at infinity in altitude" we can assume that at the upper end of the atmosphere, when

$$p = 0, \text{the total energy density decays sufficiently rapidly.} \tag{9.56c}$$

The DH set of equations (9.54a–9.54e) in p-system, with the boundary conditions (9.56a–9.56c) and the initial conditions:

$$\text{at } t \le 0: \mathbf{v} = \mathbf{v}^0{}_{DH},\ T = T^0{}_{DH} \tag{9.56d}$$

can be used as a theoretical basis for the various investigations of features of atmospheric dissipative hydrostatic motions depending on parameters Ki, $\mathrm{Re}\perp$, δ, Pr and M^2, and (from my point of view) it seems very interesting to take into account, as a starting model, the DH model problem ((9.54a–9.54e), (9.56a–9.56d)), written in *p*-system, for a discussion of the various approximations, new mathematical developments, and their application to computer simulations.

For instance, in Norbury and Roulstone's *Large-Scale Atmospheric–Ocean Dynamics* [172], the first paper, by Cullen, and the second, by White, discuss various interesting developments in the theoretical investigation of atmospheric motions. Among other things, we observe that the smallness of the Mach number, M, for the usual atmospheric motions, poses many unresolved difficult problems – the main reason being that from the so-called non-divergent (quasi-solenoidal) approximation, in leading order, we obtain a very degenerate limit system of approximate equations (see, for instance, Monin [94], §8, and in [12], the Chap. 12).

9.2.3.1 The $\boldsymbol{\beta}$, $\boldsymbol{\lambda}°$, and $\boldsymbol{\kappa}°$ Effects

In equation (9.54a) we assume that the sphericity parameter $\delta \ll 1$, such that

$$(\sin\phi/\sin\phi°) = 1 + (\delta/\tan\phi°)\ y\ +\ \mathrm{O}(\delta^2)$$
$$\tan\phi = \tan\phi°[1 +\ \mathrm{O}(\delta)]$$

since in dimensionless form, from relation for y in (9.38a), we have:

$$\phi = \phi° +\ \delta y.$$

As a consequence we can write

$$(1/\mathrm{Ki})(\sin\phi/\sin\phi°) \approx (1/\mathrm{Ki}) + \beta y, \tag{9.57a}$$

with an error of $\mathrm{O}(\delta^2)$, where (β-effect)

$$\beta =\ \delta/\mathrm{Ki}\ \tan\phi°. \tag{9.57b}$$

In reality, (9.57b) is a similarity relation between the two small parameters Ki and the small sphericity parameter δ, when $\beta = \mathrm{O}(1)$. On the other hand, when $\tan\phi° \approx 1$, for $\phi° \approx 45°$, if we consider a "low Kibel number" limiting process.

A well-adapted form of the DH equations (9.39a–9.39d), with β-effect, for the low Kibel number asymptotic theory, is obtained when we assume that the following similarity relation between the Kibel and Mach numbers, both assumed small, is realized:

$$\lambda^\circ = (1/\gamma)[\mathrm{Ki/M}]^2 \tag{9.57c}$$

and we observe that $\lambda^\circ > 1$ is in fact related with the "Mach number, low compressibility, effect", in the framework of low Kibel number flows. The relation (9.57c) is related with (9.37h) and gives an estimation for the horizontal length scale L°.

On the other hand, the similarity relation

$$\mathrm{Ek}_\perp = \kappa^\circ\ \mathrm{Ki}^2, \tag{9.57d}$$

with $\kappa^\circ = O(1)$ – as a measure of the "viscous effect" – is motivated by the fact that it presents the possibility of consistently deriving the so-called "Ackerblom's problem" for the Ekman boundary-layer in the vicinity of the ground, $p = 1$, in the framework of low Kibel number flows.

9.2.3.2 Kibel Equations

In the "Kibel equations" (9.58a–9.58d) below, the Kibel number is present but is assumed small, such that β, λ°, and κ° are O(1) – the Mach number M also being assumed a small parameter but do not appear in (9.58a) and (9.58b) since (9.57c) is taken into account. In these Kibel equations the term proportional to $\delta << 1$ are neglected since $\beta = 0(1)$.

The dissipative coefficients are assumed constant (dimensionless μ and $k \equiv 1$). As consequence the starting Kibel equations, for **v**, T, H and ω, are written in the following form:

$$\begin{aligned}\mathrm{Ki}\{\partial\mathbf{v}/\partial t + [(\mathbf{v}.\mathbf{D})\mathbf{v} + \omega\,\partial\mathbf{v}/\partial p]\} + [1 + \beta\mathrm{Ki}\ y](\mathbf{k}\wedge\mathbf{v})\\ + \lambda^\circ(1/\mathrm{Ki})\mathbf{D}H = \kappa^\circ\ \mathrm{Ki}^2\partial(\rho\,\partial\mathbf{v}/\partial p)/\partial p;\end{aligned} \tag{9.58a}$$

$$\begin{aligned}\mathrm{Ki}\{\partial\mathrm{T}/\partial t + (\mathbf{v}.\mathbf{D})\ \mathrm{T}\ + \omega[\partial\mathrm{T}/\partial p - [(\gamma-1)/\gamma](\mathrm{T}/p)]\}\\ = (1/\mathrm{Pr})\ \kappa^\circ\ \mathrm{Ki}^2\{\partial(\rho\,\partial\mathrm{T}/\partial p)/\partial p\\ +\mathrm{Pr}(1/\lambda^\circ)(\gamma-1)\mathrm{Ki}^2\rho|\partial\mathbf{v}/\partial p|^2 - \sigma\mathrm{dR/d}p\};\end{aligned} \tag{9.58b}$$

$$\partial\,\omega/\partial p + \mathbf{D}.\mathbf{v} = 0 \tag{9.58c}$$

$$\partial H/\partial p + \mathrm{T}/p = 0 \tag{9.58d}$$

When $\mathrm{Ki} \to 0$, the complete derivation of a consistent limit reduced model is linked concurrently with a "principal" expansion and two "local" expansions. A detailed derivation was performed by Guiraud and Zeytounian in [173] (see the sketch below).

$\uparrow p$ $|$ $I-$ Quasi $-$ geostrophic region (t, p)
$|\tau = \mathrm{t}/\mathrm{Ki}$ $II-$ Local adjustment region $(\tau = t/\mathrm{Ki}, p)$
$|$ $II \Rightarrow^{\text{initial condition}} \rightarrow I$ $III-$ Ekman BL region $(\mathrm{t}, \zeta = (\mathrm{p}-1)/\mathrm{Ki})$
$|\mathrm{O(Ki)}|_{\text{condition on the ground}} \uparrow$ $IV-$ Corner region (τ, ζ)
$|---|----\Uparrow----$
$|$ IV $|\mathrm{O(Ki)}$ III $\zeta = (p-1/\mathrm{Ki}$
$\mathrm{t} = 0$ $|\_\_\_|\_\_\_\_\_\_\_\_\_\_\_\_\_\_ \rightarrow_{-\mathrm{t}}$
$p = 1$

For the above Kibel equations (9.58a–9.58d) we have, as conditions:

$$\mathbf{v} = 0, \omega = (p/\mathrm{T})\partial H/\partial \mathrm{t}, \text{on } H = 0. \tag{9.59a}$$

$$-p\partial \mathrm{Log}[p\partial \mathrm{H}/\partial p]/\partial p = \sigma \mathrm{R}, \text{on } H = 0. \tag{9.59b}$$

$$\text{at } \mathrm{t} \leq 0 : \mathbf{v} = \mathbf{v}^{\mathrm{o}}, \ \mathrm{T} = \mathrm{T}^{\mathrm{o}}. \tag{9.59d}$$

The above sketch presents a view of the *three regions* I, II, III, where the main (QG), adjustment (Adj), and Ekman BL (Ek) expansions are considered. The corner region IV plays a crucial role in the derivation of a second-order "*ageosprophic*" asymptotic model.

Region II, via the unsteady adjustment process and matching, allows us to obtain the consistent initial condition (at t = 0) that must be applied to the QG single unsteady model equation.

Region III, via the solution of the Ackerblom's problem in the steady Ekman problem, produces (by matching) the boundary condition (at the ground $p = 1$) that must be supplied to the QG single non-viscous model equation.

9.2.4 The Leading-Order QG Model Problem

In considering the QG model problem it is first necessary to derive, from the above Kibel equations (9.58a–9.58d), the main ("outer") QG leading-order model equation valid in main region I, via the following limiting process:

$$\begin{aligned} &\mathrm{Ki} \to 0 \text{ with t, x, y, } p, \text{ fixed,} \\ &\beta, \lambda^{\circ} \text{ and } \kappa^{\circ} \text{ being } 0(1), \end{aligned} \tag{9.60a}$$

with the following (9.61a–9.61d) asymptotic expansion for the functions $\mathbf{v}, H$, ω and T, in the main region I:

$$\mathbf{v} = \mathbf{v}_{QG} + \text{Ki}\ \mathbf{v}_{AG} + \dots, \tag{9.61a}$$

$$H = H_S(p) + \text{Ki}\ H_{QG} + \text{Ki}^2 H_{AG} + \dots, \tag{9.61b}$$

$$\omega = \omega_{QG} + \text{Ki}\ \omega_{AG} + \dots, \tag{9.61c}$$

$$\text{T} = \text{T}_S(p) + \text{Ki}\ \text{T}_{QG} + \text{Ki}^2 \text{T}_{AG} + \dots, \tag{9.61d}$$

where we have assumed that $H_S(p)$ and $\text{T}_S(p)$ are functions only of p! Although this does not follow (concerning the dependence of time t) directly from the Kibel equations (9.58a–9.58d), it will be found to be consistent with the constancy (relative to time t) of

$$\text{d}/\text{d}p[p\text{dlog}\text{T}_S/\text{d}p] = \sigma\text{dR}/\text{d}p, \tag{9.62a}$$

which is a consequence of (9.58b). Of course, we have

$$\text{T}_s(p) = -\text{pd}H\text{s}/\text{d}p, \tag{9.62b}$$

$$\rho_S(p) = \text{p}/\text{T}_s(p), \tag{9.62c}$$

but we do not yet know how $\text{T}_S(p)$ depends on p.

Then, from the boundary condition (9.59b), written for $p = 1$, assuming that $p = 1$ is the solution of $\text{H}(p) = 0$, it is found that (9.62b) allows us to compute $H\text{s}(p)$, and (9.62a) allows us to compute $\text{R}(\text{T}_s(p))$.

From our point of view we assume that $\text{T}_s(p)$ is a given function and

$$[(\gamma - 1)/\gamma]\text{T}_S - p\ \text{dT}_S/\text{d}p \equiv \text{K}_S(p) \neq 0, \tag{9.62d}$$

which represents data for the derived, in region I, QG model equation (see (9.65) with (9.66) below).

In such a case, from (9.58b) we find at leading-order:

$$\omega_{QG} \equiv 0. \tag{9.63a}$$

Now, from (9.58a) we find at leading-order the well-known geostrophic balance:

$$(\mathbf{k} \wedge \mathbf{v}_{QG}) + \lambda^0 \mathbf{D}\text{H}_{QG} = 0 \Leftrightarrow \mathbf{v}_{QG} = \lambda^0(\mathbf{k} \wedge \mathbf{D}\text{H}_{QG}), \tag{9.63b}$$

and the companion divergenceless equation for $\mathbf{v}_{QG}$, from (9.58b)

$$\mathbf{D}.\mathbf{v}_{QG} = 0. \tag{9.63c}$$

With (9.63a) from (9.58b) we derive for T_{QG} the following equation:

$$\partial T_{QG}/\partial t + \mathbf{v}_{QG}.\mathbf{D}T_{QG} - [K_S(p)/p]\ \omega_{AG} = 0. \tag{9.63d}$$

Going to higher order, we derive from the Kibel equations (9.58a, 9.58c, 9.58d) the following three equations:

$$\mathbf{v}_{AG} = \mathbf{k} \wedge [\partial \mathbf{v}_{QG}/\partial t + (\mathbf{v}_{QG}.\mathbf{D})\mathbf{v}_{QG} + \lambda o \mathbf{D} H_{AG}] - \beta\ y\ \mathbf{v}_{QG}, \tag{9.64a}$$

$$\partial\, \omega_{AG}/\partial p + \mathbf{D}.\mathbf{v}_{AG} = 0; \tag{9.64b}$$

$$T_{QG} = -\ p\partial H_{QG}/\partial p, \tag{9.64c}$$

From the continuity equation (9.64b), with (9.64a), and the expression for ω_{AG} – obtained through the elimination of T_{QG} and $\mathbf{v}_{QG}$, from (9.63d), using (9.63b) – as a function of H_{QG}, we obtain the following single quasi-geostrophic potential vorticity model equation:

$$\begin{aligned} \partial L H_{QG}/\partial t &+ \lambda^0[(\partial H_{QG}/\partial x)\partial L H_{QG}/\partial y - (\partial H_{QG}/\partial y)\partial L H_{QG}/\partial x] \\ &+ \beta \partial H_{QG}/\partial x = 0, \end{aligned} \tag{9.65}$$

where

$$L H_{QG} = \lambda^0 \mathbf{D}^2 H_{QG} + \partial/\partial p\{[p^2/(K_S(p)]\partial H_{QG}/\partial p\}. \tag{9.66}$$

We observe that the QG model equation (9.65) contains one derivation with respect to time t, and as a consequence only one initial condition must be supplied for H_{QG}, via an unsteady adjustment problem, which we shall derive in the next section 9.2.4.1. The boundary condition that must be supplied on the ground $p = 1$, for QG equation (9.65), will be derived below in section 9.2.4.2.

Finally, concerning the boundary conditions that must be applied at the upper end of the atmosphere, $p = 0$, and far off in the horizontal plane, we can again assume that the total energy density

$$[p^2/(K_S(p))][\partial H_{QG}/\partial p]^2 + \lambda_0|\mathbf{D}^2 H_{QG}|^2 \tag{9.67}$$

decays sufficiently rapidly at infinity.

9.2.4.1 Adjustment to Geostrophic Balance (9.63b)

It is not difficult to verify, by trial, that the inner in time unsteady adjustment equations to geostrophic balance (9.63b) are derived by setting

$$\tau = t/\mathrm{Ki}, \tag{9.68a}$$

and applying an initial limiting process

$$\mathrm{Ki} \downarrow 0, \text{ with } \tau, \text{ x, y, and } p \text{ fixed.} \tag{9.68b}$$

Concerning, more precisely, the introduction of a short adjustment time (like τ) in the case of the various unsteady adjustment problems and matching of meteorological fields, I would here like to exemplify my difficulty, as a young mathematician, in realizing the subtlety of arguments related to the adjustment to geostrophic balance.

I very well remember – when a 1957 graduate (PhD) student in Kibel's dynamic meteorology department at the (now) Obukhov Institute of Atmospheric Physics, in Moscow – reading Kibel's newly published *An Introduction to the Hydrodynamical Methods of Short Period Weather Forecasting* [171], Chap. 4, how I did not in any way seem to understand in what manner appear, simultaneously, this adjustment short time τ (= t/Ki), in the case of the adjustment to geostrophic balance in the framework of the low Kibel number asymptotics, and of an evolution prediction time t – both times being denoted, in Kibel's book, by the same symbol, t.

Only after September 1967, while working in the Aerodynamics Department at ONERA, did I have the possibility – due to Van Dyke's *Perturbation Methods in Fluid Mechanics* [14] – of really understanding the profound significance of these two (inner and outer) times in the asymptotic, outer–inner expansions with matching ($\tau \to \infty \Leftrightarrow t \to 0$) machinery.

Let us set f^τ for any quantity f considered as a function of $\tau = t/\mathrm{Ki}$ instead of t. First, we rewrite the Kibel equations (9.58a–9.58d), with, in place of

$$\partial/\partial t \Rightarrow (1/\mathrm{Ki})\partial/\partial\tau,$$

for the new functions, $\mathbf{v}^\tau$, $\omega^\tau, H^\tau, T^\tau$, and then expanding

$$(\mathbf{v}^\tau, \omega^\tau, H^\tau, T^\tau) = (\mathbf{v}^\tau{}_0, \omega^\tau{}_0, H^\tau{}_0, T^\tau{}_0) + \mathrm{Ki}\,(\mathbf{v}^\tau{}_1, \omega^\tau{}_1, H^\tau{}_1, T^\tau{}_1)$$

$$+ \mathrm{Ki}^2(\mathbf{v}^\tau{}_2, \omega^\tau{}_2, H^\tau{}_2, T^\tau{}_2) + \ldots, \tag{9.69a}$$

one first finds

$$(H^\tau{}_0, T^\tau{}_0) = (H_S(p), T_S(p)),$$

with

$$T_S(p) = -p\partial H_S/\partial p. \tag{9.69b}$$

In order to find equations for $(\mathbf{v}^{\tau}{}_0,\ \omega^{\tau}{}_0, H^{\tau}{}_1, \mathrm{T}^{\tau}{}_1)$ we have to go to higher order:

$$\partial \mathbf{v}^{\tau}{}_0/\partial\tau + \mathbf{k}\wedge \mathbf{v}^{\tau}{}_0 + \lambda^{\mathrm{o}}\mathbf{D}H^{\tau}{}_1 = 0, \tag{9.70a}$$

$$\partial \omega^{\tau}{}_0/\partial p + \mathbf{D}.\mathbf{v}^{\tau}{}_0 = 0; \tag{9.70b}$$

$$\mathrm{T}^{\tau}{}_1 = -p\partial H^{\tau}{}_1/\partial p, \tag{9.70c}$$

$$\partial \mathrm{T}^{\tau}{}_1/\partial\tau - [\mathrm{K_S}(p)/p]\ \omega^{\tau}{}_0 = 0, \tag{9.70d}$$

where $\mathrm{K_S}(p)$ is given by (9.62d) as a function of p alone.

We observe that $\mathrm{K_S}(p)$ has a suggestive interpretation-namely, if we introduce a dimensionless specific entropy S, then with

$$S = S^{\tau}{}_0(\mathrm{p}) + \ \mathrm{O(Ki)}:\ \mathrm{d}S^{\tau}{}_0(p)/\mathrm{d}p = -(1/p\mathrm{T_S}(\mathrm{p}))\mathrm{K_S}(p). \tag{9.71}$$

The above system of equations (9.70a–9.70d) is the system governing the unsteady process of adjustment to geostrophic balance (9.63b).

We observe, first, that the derivation of the unsteady adjustment local equations significant close to initial time, in the case of the low Kibel number asymptotics, is not a consequence of the linearization of the Kibel equations (9.58a–9.58d) (as is written in [171], p. 83).

These inner, local-in-time, unsteady adjustment (linear) equations to geostrophic balance are rationally derived, in the general case, from full unsteady Kibel equations (9.58a–9.58d), as a significant limit when $\mathrm{Ki}\to 0$, with short time τ fixed in place of time t fixed, the system of equations (9.70a–9.70d), being, in fact, a significant degeneracy of Kibel equations (9.58a–9.58d) near the initial time.

From the last two equations of (9.70a and 9.70d) we derive the following relation:

$$\omega^{\tau}{}_0 = -\left[p^2/\mathrm{K_S}(p)\right]\partial(\partial H^{\tau}{}_1/\partial p)/\partial\tau, \tag{9.72a}$$

and going back to the first two of equations of the system (9.70a and 9.70b) we find a couple of equations for $\mathbf{v}^{\tau}{}_0$ and $H^{\tau}{}_1$: namely,

$$\partial \mathbf{v}^{\tau}{}_0/\partial\tau + \mathbf{k}\wedge \mathbf{v}^{\tau}{}_0 + \lambda^{\mathrm{o}}\mathbf{D}H^{\tau}{}_1 = 0; \tag{9.72b}$$

$$\mathbf{D}.\mathbf{v}^{\tau}{}_0 - \partial/\partial \mathrm{p}\{[p^2/\mathrm{K_S}(\mathrm{p})]\partial^2 H^{\tau}{}_1/\partial\tau\partial p\} = 0. \tag{9.72c}$$

But for the two evolution (in time τ) equations (9.72b) and (9.72c), for $\mathbf{v}^{\tau}{}_0$ and $H^{\tau}{}_1$, it is necessary to give an initial condition for

$$\mathbf{v}^{\tau}{}_0 \text{ and for } H^{\tau}{}_1 \text{ at } \tau = 0,$$

according to (9.59d). Concerning $\mathbf{v}^{\tau}{}_{0}$ we may use the initial data, $\mathbf{v}^{0}$, assuming:

$$\tau = 0 : \mathbf{v}^{\tau}{}_{0} = \mathbf{v}^{0}. \tag{9.73a}$$

Concerning the initial data for $H^{\mathrm{t}}{}_{1}$, it is necessary to assume that the data T^{0} (in (9.59d)) may be set in the form

$$\mathrm{T}^{0} = \mathrm{T_S}(p) + \ \mathrm{Ki}\ \mathrm{T}^{0}{}_{1} + \ldots . \Rightarrow \quad \tau = 0 : H^{\tau}{}_{1} = H^{0}{}_{1}, \tag{9.73b}$$

when we use the relation between T^{0} and H^{0} : $\mathrm{T}^{0} = -p\partial H^{0}/\partial p$. Whenever the data $\mathrm{T}^{0} = -p\partial H^{0}/\partial p$, in (9.59d), cannot be put into the above form, we must expect that another adjustment process holds!

There is an important observation, which was known to Kibel in 1955 (see [171]), and which concerns the way in which

$$\lim_{\tau\uparrow\propto} H^{\tau}{}_{1} \equiv (H^{\tau}{}_{1})^{\propto}$$

is related to the initial values (9.73a, 9.73b). For this, we start from equations (9.72b), (9.72c), and first deduce the equation:

$$\partial/\partial\tau\{\mathbf{k}.(\mathbf{D} \wedge \mathbf{v}^{\tau}{}_{0}) + \frac{\partial}{\partial p}\{[p^{2}/\mathrm{K_S}(p)]\partial H^{\tau}{}_{1}/\partial p\}\} = 0. \tag{9.74a}$$

If we now integrate this last equation between $\tau = 0$ and $\tau = \propto$, and if we use the geostrophic balance,

$$(\mathbf{v}^{\tau}{}_{0})^{\propto} = \ \lambda^{\mathrm{o}}[\mathbf{k} \wedge \mathbf{D}(H^{\tau}{}_{1})^{\propto}],$$

for limiting values of $\mathbf{v}^{\tau}{}_{0}$ and $H^{\tau}{}_{1}$, when $\tau \uparrow\propto$, we obtain for $(H^{\tau}{}_{1})^{\propto}$ the following relation:

$$\begin{aligned} &\lambda^{\mathrm{o}}\mathbf{D}(H^{\tau}{}_{1})^{\propto} + \frac{\partial}{\partial p}\{[\mathrm{p}^{2}/\mathrm{K_S}(p)](\partial H^{\tau}{}_{1})^{\propto}/\partial p\} \\ &\qquad = \mathbf{k}.(\mathbf{D} \wedge \mathbf{v}^{0}) + \frac{\partial}{\partial p}\{[p^{2}/\mathrm{K_S}(p)]\partial H^{0}{}_{1}/\partial p\}. \end{aligned} \tag{9.74b}$$

Among other things, Kibel was able (also in 1955) to settle the main issue of the unsteady adjustment (towards geostrophic balance (9.63b)): $\mathbf{v}^{\tau}{}_{0}$ and $H^{\tau}{}_{1}$ – related with (9.72b) – evolves towards the geostrophic balance (9.63b), when the short adjustment time τ tends to infinity.

As a matter of fact one has a matching relation:

$$\lim_{\tau\uparrow\infty}(\mathbf{v}^{\tau}{}_{0}, H^{\tau}{}_{1}) = [\mathbf{v}_{\mathrm{QG}}(\mathrm{t} = 0,\ \mathrm{x},\ \mathrm{y}\ ,p), H_{\mathrm{QG}}(\mathrm{t}\ = 0,\ \mathrm{x},\ \mathrm{y}\ ,p)]. \tag{9.74c}$$

Finally, we obtain from (9.74b) the initial condition that must be applied to the main QG equation (9.65) with (9.66), in the following form:

$$\text{at } t = 0 : LH_{QG} = \mathbf{k}.\{\mathbf{D} \wedge \mathbf{v}^0{}_0\} + \partial/\partial p\{[p^2/K_0(p)]\partial H^0{}_1/\partial p\}, \tag{9.75}$$

where LH_{QG}, in (9.75), is the same operator which appears in the QG main outer model equation (9.65), and is given by (9.66).

Here we can observe that if it is true that to predict the field H_{QG}, in the QG approximation, it is sufficient to know only the initial value of H_{QG}! But unfortunately, this single initial value is related, according to (9.75), to the initial data in (9.73a, 9.73b), which are obtained from the unknown(!) data (9.59d) prescribed for the DH two model equations (9.54a, 9.54b), under the constraint previously mentioned in (9.73b).

As a consequence, the main unsteady adjustment problem, concerning the derivation of consistent initial conditions for the quasi-geostrophic, primitive Kibel inviscid (non-viscous and adiabatic) or dissipative hydrostatic (viscous and non-adiabatic) model equations, is the unsteady adjustment to hydrostatic balance.

In particular, for the Kibel primitive inviscid equations (DH equations (9.54a–9.54d), when $1/\mathrm{Re}_\perp \equiv 0$), the corresponding unsteady acoustic adjustment problem has been considered by Guiraud and Zeytounian [106], in which it is observed that:

> The initial conditions concerning horizontal velocity and entropy for the unsteady primitive Kibel non-viscous and adiabatic equations are simply the ones pertinent to the full Euler, non-hydrostatic equations, but shifted vertically by an amount equal to Δ. (Concerning the above-mentioned "vertical shift Δ", see our [19], Chap. V, Sect. 19.)

We also have the possibility of deriving a new initial condition (at $t = 0$) relative to $p = 1$:

$$[(H_{QG} + p[T_s(p)/K_S(p)]\partial H_{QG}/\partial p)_{p=1}]_{t=0} = (H^0{}_1 + p[T_s(p)/K_S(p)]\partial H^0{}_1/\partial p)_{p=1}. \tag{9.76}$$

The role of this "curious" condition (9.76) is to serve as an initial condition for the boundary condition on $p = 1$ (see the derived ground condition (9.87) below) that must be applied to the main QG, outer, equation (9.65) written for H_{QG}.

This condition (9.87) is derived via the formulation of Ackerblom's problem in the steady Ekman layer (region III) near the ground, and contains a time derivative. As a consequence, (9.87) may be considered as a boundary condition for our main QG, outer, equation (9.65), only if it is complemented with an initial condition – (9.76).

9.2.4.2 Ackerblom's Problem in the Steady Ekman BL Layer

Indeed, for the above derived main outer QG single equation (9.65) for H_{QG}, with the initial condition (9.75), at $t = 0$, and behaviour (decay sufficiently rapidly) condition (9.67) at $p = 0$ and at infinity in the horizontal plane, it is necessary also to write a boundary condition at the ground simulated (for low Kibel number) by $p = 1$.

For this within the region III (the steady Ekman layer) we introduce the new vertical coordinate:

$$p^* = (p - 1)/\mathrm{Ki}, \tag{9.77a}$$

and in region III we consider the following inner (BL) limiting process:

$$\mathrm{Ki} \to 0 \text{ with t, x, y and } p^* \text{ fixed.} \tag{9.77b}$$

With (9.77a, 9.77b) we consider the following inner (BL) expansion in region III:

$$(\mathbf{v},\ \omega, H,\ \mathrm{T},\ \rho) = (\mathbf{v}_0{}^*,\ \omega_0{}^*, H_0{}^*, \mathrm{T}_0{}^*, \rho_0{}^*) + \\ + \ \mathrm{Ki}\ (\mathbf{v}_1{}^*,\ \omega_1{}^*, H_1{}^*, \mathrm{T}_1{}^*, \rho_1{}^*), \tag{9.77c}$$

We then obtain from the starting Kibel equations (9.58a–9.58d) the following set of equations, for $\mathbf{v}_0{}^*,\ \omega_0{}^*, H_0{}^*, \mathrm{T}_0{}^*, \rho_0{}^*,\ \omega_1{}^*, H_1{}^*, \mathrm{T}_1{}^*$:

$$\mathbf{D}H_0^* = 0; \partial\ \omega_0{}^*/\partial p^* = 0; \partial H_0{}^*/\partial p^* = 0, \tag{9.78a}$$

$$\rho_0{}^*\mathrm{T}_0{}^* = 1; (1/\mathrm{Pr})\kappa_o\partial/\partial p^*[\rho_0{}^*(\partial \mathrm{T}_0{}^*/\partial p^*)] = 0, \tag{9.78b}$$

with

$$\text{at } H_0{}^* = 0 : \partial \mathrm{T}_0{}^*/\partial p^* = 0. \tag{9.78c}$$

Then, from matching with main QG region, when $p^* \to \propto$, we obtain

$$\omega_0{}^* = 0;\ \lim_{p^*\to\propto}\mathrm{T}_0{}^* = \mathrm{T}_0(1);\ \lim_{p^*\to\propto}H_0{}^* = H_0(1), \tag{9.79a}$$

and we also have

$$(\mathbf{k} \wedge \mathbf{v}_0{}^*) + \lambda^o\mathbf{D}H_1{}^* - \kappa_o\partial/\partial p^*[\rho_0(1)(\partial\mathbf{v}_0{}^*/\partial p^*)] = 0; \tag{9.79b}$$

$$\partial\omega_1{}^*/\partial p^* = \mathbf{D}.\mathbf{v}_0{}^*; \tag{9.79c}$$

$$(1/\mathrm{Pr})\kappa_o\partial/\partial p^*[\rho_0(1)(\partial \mathrm{T}_1{}^*/\partial p^*)] = 0, \tag{9.79d}$$

$$\mathrm{T}_0(1) = -\partial H_1{}^*/\partial p, \tag{9.79e}$$

with

$$\text{on } H_1^* = 0 : \begin{cases} \mathbf{v}_0^* = 0; \\ \omega_1^* = [1/\mathrm{T}_0(1)](\partial H_1^*/\partial \mathrm{t}), \\ [1/\mathrm{T}_0(1)][\partial \mathrm{T}_1^*/\partial p^*] = \sigma \mathrm{R}^*(1), \end{cases} \tag{9.80}$$

when we assume that the main radiative transfer does not have an Ekman BL structure.

In fact, we can assume that the flat ground in the Ekman BL layer is characterized by:

$$p^* = \mathrm{P}_{\mathrm{g}0}^* + \mathrm{Ki}\ \mathrm{P}_{\mathrm{g}1}^* + \ldots . \tag{9.81}$$

From the above relation (9.79d) and (9.80), after a matching with the main QG region I, we first obtain

$$\mathrm{T}_1{}^* = \mathrm{T}_{\mathrm{QG},1} + (\mathrm{dT}_0/\mathrm{d}p)_{p=1} p^*, \tag{9.82a}$$

where

$$\mathrm{T}_{\mathrm{QG},1} \equiv \mathrm{T}_{\mathrm{QG}}(\mathrm{t},\ \mathrm{x},\ \mathrm{y}, p = 1) = -[\partial H_{\mathrm{QG}}/\partial p]_{p=1}.$$

Then, from equation (9.79d) we obtain

$$H_1{}^* \equiv H_{\mathrm{QG},1} + \mathrm{T}_0(1) p^*, \tag{9.82b}$$

and

$$H_1{}^* = 0 : \text{ imply : } \mathrm{P}_{\mathrm{g}0}^* = [1/\mathrm{T}_0(1)] H_{\mathrm{QG},1}, \tag{9.82c}$$

where $H_{\mathrm{QG},1} \equiv H_{\mathrm{QG}}(\mathrm{t},\ \mathrm{x},\ \mathrm{y}, p = 1)$.

Now, with above results, we can formulate a consistent “Ackerblom’s problem”. First we consider equation (9.79b), for $\mathbf{v}_0{}^*$, and we set:

$$\mathbf{v}_0{}^* = \mathbf{v}_{\mathrm{QG},1} + \mathbf{V}_0{}^*, \text{with } \mathbf{v}_{\mathrm{QG},1} \equiv \mathbf{v}_{\mathrm{QG}}(\mathrm{t},\ \mathrm{x},\ \mathrm{y}, p = 1).$$

From the matching with the main QG region I, when $p^* \to \infty$, we have:

$$\lim_{p^* \to \infty} \mathbf{V}_0{}^* = 0 \text{ and } \quad (\mathbf{k} \wedge \mathbf{v}_{\mathrm{QG},1}) + \lambda^{\mathrm{o}}\ \mathbf{D} H_{\mathrm{QG},1} = 0,$$

and we obtain, from (9.82b) for $H_1{}^*$, that in (9.79b):

$$(\mathbf{k} \wedge \mathbf{v}_0{}^*) + \lambda^{\mathrm{o}} \mathbf{D} H_1{}^* \equiv \mathbf{k} \wedge \mathbf{V}_0{}^*.$$

Therefore, with the above results, we can formulate the following problem for the determination of the horizontal, perturbation, velocity vector $\mathbf{V}_0{}^*$:

$$\left.\begin{aligned} \kappa_o \partial^2 \mathbf{V}_0{}^*/\partial p^{*2} - \mathbf{k} \wedge \mathbf{V}_0{}^* = 0, \\ \mathbf{V}_0{}^* = -\mathbf{v}_{QG,1}, \text{on } p^* = -[1/T_0(1)]H_{QG,1}, \\ \mathbf{V}_0{}^* \to 0, \text{ when } p^* \to \propto . \end{aligned}\right] \tag{9.83}$$

where we have assumed $\rho_0(1) = 1$, because we work with dimensionless quantities.

The solution of the above Ackerblom's problem (9.83) is obtained in a standard way:

$$\mathbf{V}_0{}^* - \mathrm{i}\mathbf{k} \wedge \mathbf{V}_0{}^* = -[\mathbf{v}_{QG,1} - \mathrm{i}\mathbf{k} \wedge \mathbf{v}_{QG,1}]\mathrm{E}^*, \tag{9.84a}$$

where $\mathrm{i} \equiv (-1)^{1/2}$ and

$$\mathrm{E}^* = \exp\{-[(1+\mathrm{i})/(2\kappa_o)^{1/2}][p^* + [1/T_0(1)]H_{QG,1}]\}. \tag{9.84b}$$

Now it is necessary to consider the continuity equation (9.79c), and we find for $\omega_1{}^*$ the following relation:

$$\omega_1{}^* = \int_{P^*_{g0}}^{p*} (\mathbf{D} \cdot \mathbf{v}_0{}^*)\mathrm{d}p^* + [1/T_0(1)](\partial H_1{}^*/\partial \mathrm{t})_{p^*=P^*_{g0}} \tag{9.85a}$$

where

$$\mathbf{v}_0{}^* = Real\{[\mathbf{v}_{QG,1} - \mathrm{i}\mathbf{k} \wedge \mathbf{v}_{QG,1}](1 - \mathrm{E}^*)\}. \tag{9.85b}$$

A tedious computation of integral in (9.85a) with (9.85b) finally produces the value of $\omega_1{}^*$ when $p^* \to \propto$:

$$\omega_1{}^{*\propto} = \lim_{p* \to \propto} \omega_1{}^* = [1/T_0(1)](\partial H_{QG,1}/\partial \mathrm{t}) - \lambda^o \sqrt{\kappa_o} \mathbf{D}^2 H_{QG,1}. \tag{9.86}$$

Now, matching

$$\lim_{p* \to \propto} \omega_1{}^* \equiv \omega_1{}^{*\propto} = \omega_{QG,1},$$

with the main QG region I produces the following boundary condition at the flat ground for the QG model equation (9.65) with (9.66):

$$\begin{aligned} \{[1/T_0(1)]\partial/\partial \mathrm{t} + [1/K_S(1)][\partial/\partial \mathrm{t} + \mathbf{v}_{QG} \cdot \mathbf{D}]\partial/\partial p^* \\ - \lambda^o \sqrt{\kappa_o} \mathbf{D}^2\} H_{QG} = 0, \text{on } p = 1. \end{aligned} \tag{9.87}$$

The above QG model problem ((9.65) with (9.66), and (9.75) and also (9.87) with (9.76)) derived from full Kibel equations with initial and boundary conditions, for low Kibel number, is a very appropriate application of our RAM Approach, and provides a good example of the "deconstruction" of a starting problem – and its "reconstruction" – via a main, QG, model and two local, Adj and Ek, models, by matching for a Ki small parameter!

9.2.5 The Second-Order Ageostrophic G–Z Model

For the derivation of the so-called "AG" model, we need to consider three local (inner) expansions in addition to the main one. Two of them are higher approximations of those considered previously in the framework of the QG model derived above in section 9.2.4. From the outer approximation (9.61a–b) we first derive the main AG, second-order model equation, for H_{AG}.

Then, from the first local (close to initial time) approximation, via a second-order unsteady adjustment problem to the AG model equation, we obtain by matching (between regions II and I) the initial condition, at $t = 0$, for the AG main equation.

Afterwards, from the second local (near-ground) approximation, via a second-order Ackerblom's problem in steady Ekman layer, we have the possibility of matching (between regions III and I) to derive a boundary condition, on the ground p, for this AG main equation. This "ageostrophic" model is relative to the H_{AG} component in the main asymptotic expansion (9.61b).

In order to obtain the initial condition at $t = 0$, for the AG main model second-order model, the problem of adjustment to ageostrophy, in region II, must be considered. However, so as to be able to correctly formulate this problem of adjustment to ageostrophy, it is necessary to analyze the problem related to the unsteady Ekman boundary layer which develops in corner region IV! In addition, we must also elucidate the compatibility of the models between regions III and IV by analyzing the behaviour of the unsteady Ekman boundary layer when $\tau \to \infty$.

It is also necessary to consider the problem of the second approximation of the steady Ekman boundary layer, which is a necessary step for the derivation of the boundary condition in $p = 1$ associated with the AG main model equation (Detailed results for this AG second-order model are included in our monograph [12], pp. 236–262).

It seems appropriate to add a final comment to this section, because it very well shows the importance of our Approach, via the RAM, for the derivation of accurate, reliable and consistent models. This comment concerns the so-called "balance equation" which is well known in meteorological literature and often analyzed in various ad hoc theories. This balance equation was derived by Monin in 1958 (see Monin [94], pp. 42–44 for various comments), and also by Charney in 1962 [174]. In Monin this balance equation was derived as a second-order correction with respect to the Kibel number, Ki, within the framework of the hydrostatic approximation. Therefore, within the framework of the present (above) low Kibel number asymptotics analysis, we easily derive, for the second-order AG model, the following equation:

$$\begin{aligned}\mathbf{k}\cdot(\nabla\wedge\mathbf{v}_{\mathrm{AG}}) = \lambda^{0}\mathbf{D}^{2}H_{\mathrm{AG}} + 2\lambda^{02}\{(\partial^{2}H_{\mathrm{QG}}/\partial \mathrm{x}\partial \mathrm{y})^{2} \\ - (\partial^{2}H_{\mathrm{QG}}/\partial \mathrm{x}^{2})(\partial^{2}H_{\mathrm{QG}}/\partial \mathrm{y}^{2})\},\end{aligned} \tag{9.88}$$

which may be interpreted in different ways.

Here, in our RAM Approach, equation (9.88) is obviously an explicit relation for the vertical component of vorticity once H_{QG} and H_{AG} have been computed! But in Monin and Charney, confusingly, H_{QG} and H_{AG} with an mysterious(!) unknown H, (9.88) is viewed as a Monge–Ampere equation for computing this unknown H, when the horizontal velocity field ($\mathbf{v}_{\mathrm{AG}}$) is assumed known! This curious (and erroneous) interpretation of (9.88) – producing, in reality, the possibility for computing $\mathbf{v}_{\mathrm{AG}}$, when H_{QG} and H_{AG} have been computed – is obviously a direct consequence of the absence of a rigorous logical, step-by-step, RAM Approach for the derivation of sequential approximations. These erroneous results are present systematically when an ad hoc approach is used for the derivation of the second-order model, because (see the Epilogue below) it is not possible to predict the many well balanced terms in these second-order models except via a reliable method – the RAM Approach! (The Epilogue includes, among many examples, a typical example when the lack of such a RAM Approach often leads to very ambiguous results. Concerning the "Monge–Ampere equation", see Kibel [171], in which this equation is considered and analyzed).

9.2.6 Kibel Primitive Equations and Lee-Waves Problems as Inner and Outer Asymptotic Models

We mentioned in Sect. 9.2.2 that in the framework of the non-hydrostatic dominant equations (9.39a–9.39d) we have the possibility of considering two limiting processes: (9.44a) and (9.44b). In reality, when we assume that a local relief is situated around the point $\boldsymbol{P}^{\circ}(\mathrm{x}^{\circ}, \mathrm{y}^{\circ})$ in the prediction domain D with a diameter L°, a third local limiting process provides the possibility of considering the classical steady lee-waves model problem as an inner (local) model, relative to Kibel primitive (hydrostatic and non-dissipative) equations considered as an outer model problem.

First, we consider, in place of equations (9.39a–9.39d), a simplified non-dissipative ($\mathrm{Re}\rightarrow\infty$) without Coriolis force ($\mathrm{Ro}\ =\infty$) system of equations, which leads to the (classical, $\delta\rightarrow 0$) dimensionless Euler atmospheric equations

$$\rho_{\mathrm{E}}\mathrm{d}\mathbf{v}_{\mathrm{E}}/\mathrm{dt}\ + (1/\gamma\mathrm{M}^{2})\mathbf{D}\mathrm{p}_{\mathrm{E}} = 0; \tag{9.89a}$$

$$d\rho_E/dt + \rho_E\{\partial w_E/\partial z + \mathbf{D}.\mathbf{v}_E\} = 0; \tag{9.89b}$$

$$\rho_E dT_E/dt - [(\gamma-1)/\gamma]dp_E/dt = 0; \tag{9.89c}$$

$$\rho_E\{\varepsilon^2 dw_E/dt + (1/\gamma M^2)\partial p_E/\partial z + (Bo/\gamma M^2)\rho_E = 0. \tag{9.89d}$$

for $\mathbf{v}_E$, w_E, $p_E(=\rho_E T_E)$, T_E and ρ_E.

From these equations, when we write the slip condition

$$\text{on } z = 0: w_E = 0, \tag{9.90a}$$

in the case of a flat ground, we derive, in the hydrostatic limiting process

$$\lim{}^P{}_{\varepsilon\to 0} = \{\varepsilon \to 0 \text{ with t, x, y z fixed and } M = O(1)\}, \tag{9.90b}$$

the well-known Kibel primitive equations,

$$\rho_P d\mathbf{v}_P/dt + (1/\gamma M^2)\mathbf{D}p_P = 0; \tag{9.91a}$$

$$d\rho_P/dt + \rho_P\{\partial w_P/\partial z + \mathbf{D}.\mathbf{v}_P\} = 0; \tag{9.91b}$$

$$\rho_P dT_P/dt - [(\gamma-1)/\gamma]dp_P/dt = 0; \tag{9.91c}$$

$$\partial p_P/\partial z + Bo\rho_P = 0; \tag{9.91d}$$

$$p_P = T_P\rho_P, \tag{9.91e}$$

for the limit functions,

$$[\mathbf{v}_P, w_P, p_{Pr}, T_P\rho_P] = \lim{}^P{}_{\varepsilon\to 0}[\mathbf{v}_E, w_E, p_E, T_E \text{ and } \rho_E], \tag{9.92}$$

when (9.90b) is realized.

However, for these same Euler equations (9.89a–9.89d) we can also consider the presence of a local relief situated around the point $\boldsymbol{P}_O(x'_O, y'_O)$ on the ground in a local region $D_0 \in (l_0, m_0)$, inside the prediction domain D (having L° as diameter). If the local vertical length scale is h_0, then the relief can be simulated by the local dimensionless equation,

$$z' = h_0 h((x' - x'_O)/l_0, (y' - y'_O)/m_0), \text{ with } h = O(1), \tag{9.93}$$

where z' and x', y' are physical local coordinates (with dimensions) linked with the relief.

The above dimensionless Euler equations (9.89a–9.89d), however, are written with dimensionless coordinates: $z = z'/H°$, $x = x'/L°$ and $y = y'/L°$. Rewriting

(9.93), with the dimensionless coordinates, x, y and z, we obtain, in its place, the dimensionless equation of the relief in the following form (in place of: z = 0):

$$z = a\mathrm{h}(b\,\xi, c\,\eta), \tag{9.94a}$$

with

$$\xi = (\mathrm{x} - \mathrm{x_O})/\varepsilon \text{ and } \eta = (\mathrm{y} - \mathrm{y_O})/\varepsilon. \tag{9.94b}$$

In (9.94a)

$$\mathrm{a} = h_0/\mathrm{H}^\circ,\ \mathrm{b} = \mathrm{L}^\circ/l_0,\ \mathrm{c} = \mathrm{L}^\circ/m_0. \tag{9.94c}$$

Corresponding the slip condition on the surface of the relief (9.94a) is:

$$\mathrm{w_{loc}} = \mathbf{a}\mathbf{v}_{\mathrm{loc}}.\mathbf{D}_{\mathrm{loc}}\mathrm{h}(b\xi, c\,\eta) \text{ on } \mathrm{z} = a\mathrm{h}(b\,\xi, c\,\eta), \tag{9.95a}$$

where $\mathbf{D}_{\mathrm{loc}} = (\partial/\partial\xi, \partial/\partial\eta)$.

Obviously, from the limiting hydrostatic process (9.90b), where x and y are fixed, which leads to the Kibel primitive equations (9.91a–9.91e), we cannot take into account, in these primitive equations, the influence of the relief linked with the slip condition (9.95a). The reason is that

$$\mathrm{h}(\propto, \propto) \equiv 0, \tag{9.95b}$$

as a consequence of a local character of the considered relief, which is limited in D_0, inside D. As a consequence, in place of (9.90b) it is necessary to consider a new "local hydrostatic limiting process": namely,

$$\lim{}^{loc}{}_{\varepsilon\to 0} = \{\varepsilon \to 0 \text{ with t}, \xi, \eta \text{ and z fixed}\}, \tag{9.96a}$$

with, a, b, c, and M = O(1). In such a case, for the below local limit functions, when (9.96a, 9.96b) is realized,

$$\begin{aligned}&\lim{}^{loc}{}_{\varepsilon\to 0}[\mathbf{v}_{\mathrm{E}}, \varepsilon\mathrm{w_E}, \mathrm{p_E}, \mathrm{T_E}\text{and } \rho_{\mathrm{E}}],\\ &\quad = [\mathbf{v}_{\mathrm{loc}}, \mathrm{w_{loc}}, \mathrm{p_{loc}}, \mathrm{T_{loc}}, \rho_{\mathrm{loc}}],\end{aligned} \tag{9.96b}$$

we derive the following local steady non-hydrostatic Euler equations:

$$\begin{aligned}&\rho_{\mathrm{loc}}[(\mathbf{v}_{\mathrm{loc}}.\mathbf{D}_{\mathrm{loc}})\mathbf{v}_{\mathrm{loc}} + \mathrm{w_{loc}}\partial\mathbf{v}_{\mathrm{loc}}/\partial\mathrm{z}\,]\\ &\qquad + (1/\gamma\mathrm{M}^2)\mathbf{D}\mathrm{p_{loc}} = 0;\end{aligned} \tag{9.97a}$$

$$\mathbf{D}_{\mathrm{loc}}.(\rho_{\mathrm{loc}}\mathbf{v}_{\mathrm{loc}}) + \partial((\rho_{\mathrm{loc}}\mathrm{w_{loc}}/\partial\mathrm{z} = 0; \tag{9.97b}$$

$$\rho_{loc}[(\mathbf{v}_{loc}.\mathbf{D}_{loc})T_{loc} + w_{loc}\partial T_{loc}/\partial z - [\gamma - 1/\gamma][(\mathbf{v}_{loc}.\mathbf{D}_{loc})p_{loc} + w_{loc}\partial p_{loc}/\partial z\ ] = 0; \qquad (9.97c)$$

$$\rho_{loc}[(\mathbf{v}_{loc}.\mathbf{D}_{loc})w_{loc} + w_{loc}\partial w_{loc}/\partial z\ ] + (1/\gamma M^2)[\partial p_{loc}/\partial z\ +\ Bo\rho_{loc}] = 0, \qquad (9.97d)$$

with

$$p_{loc} = \rho_{loc} T_{loc}. \qquad (9.97e)$$

These local equations (9.97a–9.97e) govern the steady lee-waves problem in a non-viscous baroclinic adiabatic atmosphere over and downstream of a relief simulated by equation (9.94a) in a local domain D_0, when we use the slip condition (9.95a) for w_{loc}. Formal matching of

$$\lim_{\xi,\,\eta\uparrow\propto}[\mathbf{v}_{loc}, w_{loc}, \rho_{loc}, p_{loc}, T_{loc}] = [\mathbf{v}_P, 0, \rho_P, p_P, T_P]_{\boldsymbol{P_0}}, \qquad (9.98a)$$

$$[\partial p_P/\partial z\ +\ Bo\rho_P]_{\boldsymbol{P_0}} = 0, \qquad (9.98b)$$

between the local lee-waves (9.97a–9.97e) and Kibel primitive (9.91a–9.91e) model equations may be interpreted as providing lateral (in the horizontal plane) boundary conditions at infinity for the local (inner) steady, non-hydrostatic, dynamic lee-waves model equations (9.97a–9.97e) with (9.95a), once the prediction by primitive (hydrostatic) Kibel model equations (9.91a–9.91e) is known at the position $\boldsymbol{P_0}$ in D.

The set of local steady lee-waves equations (9.97a–9.97e) are intensively used in mesometeorology (as is the case in our *Études Hydrodynamique des Phénomènes Mésométéorologiques* [1]).

Epilogue

I hope that the reader who has become acquainted with the various parts of this monograph now has not only *une certaine idée* of my main purpose, but is also *deeply convinced* that there is really no better way for the derivation of significant consistent model problems than the RAM Approach!

It might be true that the physical approach can produce valuable qualitative analysis and a better understanding of the laws of nature related to various significant and practical fluid-flow phenomena. Unfortunately, a purely *ad hoc* physical approach would not be able to point the right way to a consistent rational obtention of approximate simplified leading-order model problems which could be used successfully for numerical simulation with a high-speed computer.

In particular, this is especially true because such an *ad hoc* approach would be unable to provide a rational, logical method (as is the case with our RAM Approach) for the derivation of a well-balanced associated second-order model problem with various complementary physical effects! A well-known example is the derivation of a second-order rational system of boundary layer compressible and heat-conducting equations (which, it seems to me, are primarily due to asymptotics) mentioned in the Prologue above. This extension of classical Prandtl boundary-layer equations, by Van Dyke (1962), from the full NS–F problem, has been a crucial practical scientific contribution to the NASA program for the atmospheric re-entry of the Space Shuttle!

Below, a typical case is considered, which is related to the modelling of the Bénard thermal convection problem (discussed in Chap. 8), up to the derivation of a second-order model associated with the classical RB shallow thermal convection model.

In general, the derivation of a second-order model problem, from the full unsteady NS–F, well-formulated problem is not an easy task! Usually, erroneous results are present, systematically, when an *ad hoc* approach is used for the derivation of such second-order models, because it is not possible to predict the many, well-balanced terms in these second-order models except via a reliable method – the RAM Approach!

R. Kh. Zeytounian, *Navier–Stokes–Fourier Equations*,
DOI 10.1007/978-3-642-20746-4,

Below, with the derivation of the second-order equation model for the Bénard convection problem, associated with the well-known RB leading-order shallow thermal convection model, we give a typical example when the absence of such a logical, step-by-step RAM Approach leads to very ambiguous results!

As starting, exact, Bénard convection equations, we choose those derived by Hill and Roberts in 1991. These dimensional equations, with physical quantities, are written in the following form (see Chap. 8 for the notations):

$$\partial u_i/\partial x_i = \alpha \, dT/dt;$$

$$\rho du_i/dt = \rho f_i - \partial p/\partial x_i + \partial/\partial x_j[\lambda d_{mm}\delta_{ij} + 2\,\mu d_{ij}];$$

$$-\alpha dp/dt + C_p[\rho/\alpha\,](\partial u_i/\partial x_i) = (\lambda d_{ii})^2$$

$$+ 2\mu d_{ij}d_{ij} + \partial/\partial x_i[k(\partial T/\partial x_i)],$$

and we assume that $f_1 = 0, f_2 = 0, f_3 = -g$, and the coefficients, λ, μ, and k are assumed constant (respectively, λ_d, μ_d, k_d, as functions of constant temperature T_d), and we write in place of $\partial/\partial x_j[\lambda d_{mm}\,\delta_{ij} + 2\mu d_{ij}]$ the term $\mu_d\{\Delta u_i + [1 + (\lambda_d/\mu_d)\nabla(\partial u_i/\partial x_i)]$; $C_p = T(\partial S/\partial T)_p$, which is related to entropy S, the specific heat at constant pressure ($C_p = C_p(T)$). For the derivation of model reduced equations, from the above (non-dimensionless) convection equations, first, Hill and Roberts consider as a limiting process (the rather exotic):

$$g \to \infty \text{ and } \alpha_d \to 0 \text{ such that } g\alpha_d \text{ fixed and } O(1)$$

which is, in fact, a "bastardized", non-formalized version of our *à la* Boussinesq limiting process (8.16). It is interesting to remark that Straughan (1992), p.279, writes: " The key philosophy of the Hills and Roberts paper (1991) is that typical acceleration promoted in the fluid by variations in the density are always much less than the acceleration of gravity!"

Then, according to the above limiting process, Hills and Roberts expand the pressure, velocity, and temperature fields, in their above dimensional equations, in $(1/g) \to 0$, such that:

$$p = p^0 g + p^1 + (1/g)p^2 + \ldots,$$

$$u_i = u_i^{\,1} + (1/g)u_i^{\,2} + \ldots,$$

$$T - T_d = T^1 - T_d + (1/g)[(T^2 - T_d)] + \ldots$$

$$\rho(T) = \rho_d[1 - \alpha_d[(T - T_d)] + \ldots].$$

– which is a rather bizarre expansion!

Now, with the above expansion, the O–B equations (for a weakly expansible liquid) are derived, via the limit:

$$\varepsilon_{H/R} = [d/C_{pd}](g\alpha_d) \to 0$$

Thus, Hill and Roberts obtain the following reduced equations (with dimensions!):

$$\partial u_i/\partial x_i = 0;$$

$$du_i{}^1/dt = \text{Ra}\ T^1\delta_{i3} - \partial p^1/\partial x_i + \Delta u_i{}^1;$$

$$dT^1/dt \ - \varepsilon_{H/R}[T_d + T^1]u_3{}^1 = (1/\text{Pr})\Delta T^1 + (2/\text{Ra})\varepsilon_{H/R} d_{ij}{}^1 d_{ij}{}^1.$$

These equations are, in fact, rather similar to our deep thermal convection equations (8.43b–d) with viscous dissipation – the parameter $\varepsilon_{H/R}$ being an analogue to our dissipation parameter Di, defined by (8.44a). When

$$\varepsilon_{H/R} \to 0,$$

we recover the RB model, leading-order, equations.

In our "deep thermal convection" equations (8.43b–d), in place of pressure and temperature, we have, respectively, the perturbations π (8.17) and θ (8.13) and $[T_d + T^1]$ in the term proportional to $\varepsilon_{H/R}u_3{}^1$does not have an explicit form! Straughan (1992), p. 274, presents the following form of our deep convection equations (a system appropriate to thermal convection in a deep layer):

$$(1/\text{Pr})[\partial u_i/\partial t + u_j\partial u_i/\partial x_j] = \ -\partial\pi/\partial x_i + \Delta u_i + R\theta\delta_{i3};$$

$$\partial u_k/\partial x_k = 0;$$

$$\partial\theta/\partial t \ + u_j\partial q/\partial x_j - Ru_3\theta = \mu(x_3)\Delta\theta + 2[\delta/R]\mu(x_3)d_{ij}d_{ij},$$

where

$$\mu(x_3) = 1/[1 + \delta(1 - x_3)]$$

and δ is a constant which represents a depth parameter, and $R \equiv \sqrt{\text{Ra}}$.

If we now consider for three functions $\mathbf{u} = (u_i), \theta$ and π, the following three asymptotic expansions, relative to expansibility parameter ε:

$$U \equiv (\mathbf{u},\ \theta, \pi) = \begin{cases} \mathbf{u} = \mathbf{u}_{RB} + \varepsilon\mathbf{u}_S + \ldots, \\ \theta = \theta_{RB} + \varepsilon\theta_S + \ldots, \\ \pi = \pi_{RB} + \varepsilon\pi_S + \ldots, \end{cases}$$

then, first derived, from the full unsteady NS–F convection equations, the RB, shallow thermal, leading-order, convection equations are:

$$\nabla.\mathbf{u}_{RB} = 0,$$

$$\partial\mathbf{u}_{RB}/\partial t + (\mathbf{u}_{RB}.\nabla)\mathbf{u}_{RB} + \nabla\pi_{RB} - Gr\theta_{RB}\mathbf{k} = \nabla^2\mathbf{u}_{RB},$$

$$\partial\ \theta_{RB}/\partial t\ + (\mathbf{u}_{RB}.\nabla)\theta_{RB} = (1/Pr)\nabla^2\theta_{RB},$$

In a second step, again from the full unsteady NS–F convection equations, we derive as a second-order, linear, well-balanced system of non-homogeneous equations for $\mathbf{u}_S$, θ_S, π_S:

$$\nabla.\mathbf{u}_S = \ d\theta_{RB}/dt;$$

$$\partial\mathbf{u}_S/\partial t + (\mathbf{u}_{RB}.\nabla)\mathbf{u}_S + (\mathbf{u}_S.\nabla)\mathbf{u}_{RB} + \nabla\pi_S -\ Gr\theta_S\mathbf{k}$$

$$-\nabla^2\mathbf{u}_S = \theta_{RB}d\mathbf{u}_{RB}/dt\ + [1 + (\lambda_d/\mu_d)]\nabla(d\theta_{RB}/dt);$$

$$\partial\theta_S/\partial t\ + \mathbf{u}_{RB}.\nabla\theta_S + \mathbf{u}_S.\nabla\theta_{RB} - (1/Pr)\nabla^2\theta_{RB}$$

$$= [(1 + \Gamma_{pd})\theta_{RB}]\ d\theta_{RB}/dt\ +\ Bo[(T_d/\Delta T) + \theta_{RB}](\mathbf{u}_{RB}.\mathbf{k})$$

$$+ (1/2)(Bo/Gr)[\partial(u_{RB})_i/\partial x_j + \partial(u_{RB})_j/\partial x_i]^2,$$

where $\Gamma_{pd} = [(1/C_p)(dC_p/dT)]_d/\alpha_d$, when we assume:

$$C_p(T) = C_{pd}[1 - \varepsilon\Gamma_{pd}\theta] \text{ and } \alpha(T) = \alpha_d[1 - \varepsilon A_d\theta]_d,$$

with

$$A_d = [(d\log C_p(T)/dT)/(d\log r(T)/dT)]_d.$$

Obviously, from the above (but not obvious) second-order equations it is possible to undertake an analysis which might produce interesting results complementary to the usual (classical, *à la* Chandresakhar, 1961) known quantitative results for the RB problem. Such analysis can, in particular, can produce specific complimentary results obtained earlier from these RB model equations.

We see that even if an ad hoc derivation is sometimes able to produce a valuable result at the leading-order, and even if a deficient ambiguous approach is chosen (as in the case of Hill and Roberts), such an approach will in no way be able to consistently derive a rational second-order approximation with well-balanced second-order terms. This precise observation is one of the main reasons for our present RAM Approach and for the publication of this book!

Actually, we have a certain idea of the intrinsic structure of NS–F equations and, via the presence of various non-dimensional reduced parameters in these equations and associated initial and boundary conditions, we begin to understand – thanks to the RAM Approach – the profound unity in the puzzle of partial fluid-flows with diverse configurations. Here in Chaps. 7–9 we have considered only some problems of aereodynamics, convection, and atmospheric motions; but our RAM Approach have a large spectrum of applications in various physical, technological, and geophysical processes. It can operate everywhere where it is possible, in analyzed problems, to detect several dimensionless significant parameters – but this is not always an easy matter. When this is the case, then the corresponding mathematical consistent asymptotic model appears under definite scaling relations (similarity rules) between these parameters, which produce diverse limitations in the use of this model! In Chap. 8 such a case was analyzed, and for the thickness, d, of a liquid layer a strong double limitation was derived for the case of RB shallow thermal convection (see Sect. 8.4).

This idea of obtaining mathematical models by proper scaling parameters in multiparametric non-linear problems is presented and very well illustrated in detail by Cercignani (kinetic theory) and Sattinger (weakly non-linear dispersive phenomena) (1998). A more recent, very interesting, but rather unusual book by de Gennes *et al.* (2004) will enable the reader to understand, in simple terms, some mundane questions affecting our daily lives – questions that have often come to the fore during our many interactions with industry (capillarity and wetting phenomena, drops, bubbles, pearls, waves, and so on). The strategy in this book is to " sacrifice scientific rigour" by an "impressionistic" approach (*à la* de Gennes) based primarily on qualitative arguments, which makes it possible to grasp things more clearly and to envisage novel situations. For me, this book is a collection of problems well suited to a RAM Approach – if, under the physics, we are able to discover the main significant small (or large) parameters.

I hope that the points of view, linked with the RAM Approach emphasized throughout this book devoted to the modelling of NS–F equations, have been well presented, and that some readers (at least) will be convinced that this technique provides very powerful tools for the derivation of consistent rational not-stiff models as an aid to numerical simulation. It is clear that at the present time a gap still exists between the Rational Asymptotic Modelling (RAM) Approach and High-Speed-Computer Numerical (H-S-CN) Simulation. In many cases the results of numerical computations (which are often fascinating) do not correspond satisfactorily with experimental/laboratory visualizations – often because of the absence of criteria by which the limit of validity of used models is known. Rarely in scientific publications do we encounter the theoretical treatment of a "full" unsteady fluid-flow problem directly inspired by the technology, and the work of producing a simple description of this fluid flow, via a model problem that can be used to explain it by using numerical simulation, is lacking (see, for instance, Sect. 6.1, footnote 2, page 119). The crucial problem of initial conditions, which strongly influence the subsequent formation and evolution of the fluid flow, is usually overlooked!

We are convinced that: "*The more computing and numerical algorithmic processes, via high-speed computers, becomes efficient, the more will be the need for conceptually consistent techniques capable of unravelling stiff fluid-flow problems. The Rational Asymptotic Modelling (RAM) Approach, among others(?), proves to be a (more!) efficient tool.*" This (rather optimistic) statement was, during 1970–80, a guideline for me when I was working on the asymptotic modelling of atmospheric flows!

Modelling of atmospheric flows – especially via approximate and consistent well-balanced models for global, short-range, and local weather-forecasting – involves vast and complex applied mathematics – as is seen clearly in Sect. 9.2 above, in the framework of the very particular QG asymptotic model. During the twenty-year period of my scientific activity at the University of Lille-I which was devoted mainly to the modelling of fluid flow phenomena (see my monograph (2002)), I published two books (1990 and 1991) and several survey and review papers (1976, 1982, 1983, 1985, 1991) concerning asymptotic models of atmospheric motions. In this scientific endeavour I was strongly influenced by my own conception of meteorology (doubtless an "unconscious inheritance" from my muscovite scientific period from 1957 to 1966, with I. A. Kibel) as a fluid dynamics discipline which lies in a privileged area for the application of the RAM Approach! My purpose (which was certainly "rather naive") was, during the 1990s, to initiate a process which does not seem to have sufficiently attracted the attention of scientists. Namely, the use of methods of formal asymptotic analysis for carrying out asymptotic modelling; that is, for building approximate consistent theoretical models based on various meteo situations and atmospheric motions for the use in weather forecasting! Conceptually, thanks to such an asymptotic (outer–inner) approach, where the matching plays a central role: "*the problem of taking into account the atmospheric initial data (using weather observations) for the used, reduced (filtered!) model is then well-posed!*"[1]

A recent operational approach at Meteo-France, working with three coupled models – Arpège-Aladin, for a numerical weather prediction model for Central Europe, and also Arome as a regional prediction model – is conceptually very

[1] In the case of a hydrostatic large horizontal-scale numerical weather prediction model, this "initialization" is consistently solved thanks to the derivation, near the (weather prediction) initial time, of a system of local unsteady equations (with a "short time" $\tau = t/\varepsilon$, where ε is the small hydrostatic parameter (Sect. 9.2) in the framework of a well-posed initial-value (*à la* Cauchy) problem. It might be true that adjustment to the state of static equilibrium is brought about (by the generation and scattering of internal acoustic waves) during only a few minutes in all (see Monin (1972), Section 6, and also our *Meteorological Fluid Dynamics* (1991), Chapter 5). Unfortunately, the theoretical analysis (for an adiabatic atmospheric motion) by Guiraud and Zeytounian (1982), *and also the numerical simulation by Outrebon (1981), show that: "The two sets of initial conditions – for full adiabatic, Euler, atmospheric equations, and the so-called Kibel primitive equations, subject to hydrostatic approximation – are merely shifted vertically by the amount of vertical displacement during the whole process (during* $\tau \uparrow \propto$ *and matching) of the vertical, one-dimensional, unsteady adjustment, atmospheric adiabatic, motion; this vertical shift being a quite significant phenomenon (Outrebon (1981))*".

different from our, "*à la* Zeytounian RAM Approach", based on asymptotics with matching. This is unfortunate, because the Arpège model, which is a "pilot model" for the Aladin model is, in fact, as large synoptic hydrostatic model, derived from the hydrostatic limiting process (Sect. 9.2.2): $\varepsilon \rightarrow 0$. In such a case we see that the problem concerning the consistent determination of the associated initial conditions for this Arpège model, in the framework of the RAM Approach, is linked with an unsteady adjustment problem where, in place of the time t, it is necessary to introduce a short time $\tau = t/\varepsilon$. Relative to τ, the acoustic-type model equations, for this unsteady adjustment problem, are derived from the full starting, non-hydrostatic, dimensionless equations for atmospheric motions.

In place of this RAM Approach – via the unsteady adjustment problem and matching, $\tau \rightarrow \infty \Leftrightarrow t = 0$ – Meteo-France uses a "statistical approach" inspired mainly by Kalnay's (2002) book, which covers methods for numerical modelling, data assimilation (for the determination of initial conditions using weather observations), and predictability, and includes a discussion of equations of motion and their approximations.

Obviously, the actual meteo philosophy is completely different from our RAM Approach, which is specifically a fluid dynamics approach using asymptotics. But perhaps this meteo philosophy is well-adapted to awkward practical weather-forecasting meteo machinery, which every day, amidst more serious business, shows, as entertainment, "what the weather will be like tomorrow or for the next few days!" Of course, weather forecasting is also an important business... but I think that atmospheric motions, as a fluid dynamics problem, poses, for our environment, numerous very interesting and specific challenges. And obviously, the applications of the RAM Approach for these specific "terrestrial" problems undoubtedly deserves particular attention!

Cited references

Hills, R., Roberts, P.: Stab. Appl. Anal. Continuous Media. **1**, 205–212 (1991)

Straughan, B.: The Energy Method, Stability, and Nonlinear Convection. Appl. Math. Sci, vol. 91, 2nd edn. Springer, New York (2004)

Chandrasekhar, S.: Hydrodynamic and Hydromagnetic Stability. Clarendon, Oxford (1961)

Cercignani, C., Sattinger, D.H.: Scaling Limits and Models in Physical Processes. DMV Seminar, Band 28, Birkhäuser (1998)

de Gennes, P.-G., Brochard-Wyart, F., Quéré, D.: Capillarity and Wetting phenomena, Drops, Bubbles, Pearls, Waves. Springer, New York (2004)

Zeytounian, R.Kh.: Asymptotic Modelling of Fluid Flow Phenomena. FMIA, vol. 64. Kluwer, Dordrecht (2002)

Zeytounian, R.Kh.: Asymptotic Modeling of Atmospheric Flow. Springer, Berlin/Heidelberg (1990)

RKh, Zeytounian: Meteorological Fluid Dynamics. Lecture Notes in Physics, m5. Springer, Heidelberg (1991)

RKh, Zeytounian: La Météorologie du point de vue du Mécanicien des Fluides. Fluid Dynamics Transactions, vol. 8, pp. 289–352. Polish Academy of Sciences, Warszawa (1976)

RKh, Zeytounian: Asymptotic phenomena in meteorology. In: Miller, J.J.H. (ed.) Proceedings of the BAIL II Conference, pp. 124–138. Boole, Dublin (1982)

Zeytounian, R.Kh.: Modèles asymptotiques pour les écoulements de fluides pesants en rotation – Application à la Météorologie. Conférence générale invitée au 6e Congrès de Mécanique, pp. 3.1–3.42. Lyon, 5–9 Sep 1983

Zeytounian, R. Kh.: Asymptotically Consistent Models for Atmospheric Flows. CISM, Von Karman session, pp. 121. Udine (Italy) Oct 1983

RKh, Zeytounian: Recent advances in asymptotic modelling of tangent atmospheric motions. Int. J. Eng. Sci. 23, 1239–1288 (1985)

Zeytounian, R.Kh.: Fluid dynamics limits of meteorological equations. Adv. Mech. **14**(3), 3–86 (1991)

Kalnay, E.: Atmospheric Modeling, Data Assimilation and Predictability. Cambridge University Press, Cambridge (2002)

References

1. Zeytounian, R.Kh.: Études Hydrodynamique des Phénomènes Mésométéorologiques. Cours à l'École de la Météorologie, pp. 163. Direction de la Météorologie Nationale, Paris (1968)
2. Zeytounian, R.Kh.: Étude des Phénomènes d'ondes dans les Écoulements Stationnaires d'un Fluide Stratifié non Visqueux. Publication ONERA N°126, 94 pages, Chatillon, février 1969. English version: Study of Wave Phenomena in the Steady Flow of an Inviscid Stratified Fluid. Royal Aircraft Establishment, Library Translation N°1404 (1969)
3. Landau, L.D., Lifschitz, E.M.: Fluid Mechanics. Pergamon, London (1959)
4. Yih, C.-S.: Stratified Flows. Academic, London (1980)
5. Batchelor, G.K.: An Introduction to Fluid Dynamics. Cambridge University Press, Cambridge (1988)
6. Zeytounian, R.Kh.: Three-dimensional rotational motion in a compressible fluid. Atmospheric and Oceanic Physics. vol. 2, p. 61. Feb 1966. English version of a Russian edition
7. Long, R.R.: Some Aspects of the Flow of Stratified Fluids: III. Continuous Density Gradients. Tellus, vol. 7, N°3, (1955)
8. Boussinesq, J.: Théorie Analytique de la Chaleur, vol. 2. Gauthier- Villars, Paris (1903)
9. Zeytounian, R.Kh.: A rigorous derivation of the equations of the compressible viscous fluid motion with gravity at low mach number. Archives of Mechanics/Archiwum Mechaniki Stosowanej **26**(3), 499–509 (1974)
10. Zeytounian, R.Kh.: On the foundations of the Boussinesq approximation applicable to atmospheric motions. Izv Atmos Ocean Phys **39**(Suppl. 1), S1–S14 (2003). English version of a Russian edition
11. Guiraud, J.-P., Zeytounian, R. Kh.: Asymptotic modelling of fluid flows. General features. J. Theor. Appl. Mech. Special Issue 1986, 1–24 (1986)
12. Zeytounian, R.Kh.: Asymptotic Modelling of Atmospheric Flows. Springer, Berlin/Heidelberg (1990)
13. Zeytounian, R.Kh.: Topics in Hyposonic Flow Theory. LNP, vol. 672. Springer, Berlin/Heidelberg (2006)
14. Van Dyke, M.: Perturbation Methods in Fluid Mechanics (Annotated edition of Perturbation Methods in Fluid Mechanics. Academic, New York/London, 1964). The Parabolic, Stanford (1975)
15. Zeytounian, R.Kh.: Les Modèles Asymptotiques de la Mécanique des Fluides I. LNP, vol. 245. Springer, Berlin/Heidelberg (1986)
16. Zeytounian, R.Kh.: Les Modèles Asymptotiques de la Mécanique des Fluides II. LNP, vol. 276. Springer, Berlin/Heidelberg (1987)
17. Zeytounian, R.Kh.: Modélisation asymptotique en mécanique des fluides newtoniens. Mathématiques et Applications, SMAI, vol. 15. Springer, Berlin/Heidelberg (1994)

R. Kh. Zeytounian, *Navier–Stokes–Fourier Equations*,
DOI 10.1007/978-3-642-20746-4, © Springer-Verlag Berlin Heidelberg 2012

18. Zeytounian, R.Kh.: La Météorologie du point de vue du Mécanicien des Fluides. Fluid Dynamics Transactions, vol. 8, pp. 289–352. Polish Academy of Sciences, Warszawa (1976)
19. Zeytounian, R.Kh.: Meteorological Fluid Dynamics. Lecture Notes in Physics, vol. m5. Springer, Heidelberg (1991)
20. Holton, J.R.: An Introduction to Dynamic Meteorology. Academic, New York (1979)
21. Gill, A.E.: Atmosphere–Ocean Dynamics. Academic, New York (1982)
22. Haltiner, G.J., Williams, R.T.: Numerical Prediction and Dynamic Meteorology. John Wiley, New York (1980)
23. Pedlosky, J.: Geophysical Fluid Dynamics. Springer, New York (1987)
24. Yang, H.: Wave Packets and Their Bifurcations in Geophysical Fluid Dynamics. Appl. Math. Sci., vol. 85. Springer, New York (1990)
25. Zeytounian, R.Kh.: Recent advances in asymptotic modelling of tangent atmospheric motions. Int. J. Eng. Sci. **23**, 1239–1288 (1985)
26. Zeytounian, R.Kh.: Asymptotic Modelling of Fluid Flow Phenomena. FMIA, vol. 64, pp. xii–545. Kluwer Academic, Dordrecht (2002)
27. Zeytounian, R.Kh.: Convection in Fluids: A Rational Analysis and Asymptotic Modelling. FMIA, vol. 90, pp. xv–396. Springer, Dordrecht (2009)
28. Birkhoff, G.: Numerical fluid dynamics. SIAM Rev. **25**(1), 1–36 (1983)
29. Zeytounian, R.Kh.: A historical survey of some mathematical aspects of Newtonian fluid flows. Appl. Mech. Rev. **54**(6), 525–562 (2001)
30. Cercignani, C., Illner, R., Pulvirenti, M.: The Mathematical Theory of Dilute Gas. AMS, vol. 106. Springer, New York (1994)
31. Serrin, J.: Mathematical Principles of Classical Fluid Mechanics. Handbuch der Physik, vol. VIII/1, pp. 125–263. Springer, Berlin (1959)
32. Newton, I.: Philosophiae Naturalis Principia Mathematica. Joseph Streater, London (1687)
33. Euler, L.: Principes généraux du mouvement des fluides. Hist. Acad, pp. 274–315. Berlin (1755)
34. Bernoulli, J.: Hydraulics (1737)
35. Bernoulli, D.: Hydrodynamica. Argentorati (1738)
36. Le Rond, J., D'Alembert.: Traité de l'équilibre et du Mouvement des Fuides (1744)
37. Zeytounian, R.Kh.: Theory and Applications of Non-viscous Fluid Flows. Springer, Berlin/ Heidelberg (2002)
38. Birch, T.: The Works of the Honourable Robert Boyle, vol. 5. A. Miller, London (1744)
39. Kirchhoff, G.: Ann. der. Phys. **134**, 177–193 (1868)
40. Navier, C.-L.: Sur les lois des mouvements des fluides, en ayant égard à l'adhésion des molécules. Ann. de Chimie **19**, 244–260 (1821)
41. Poisson, S. D.: Mémoire sur les équations générales de l'équilibre et du mouvement des corps solides élastiques et des fluides. J. Ecole Polytechnique. **13** cahier 20, 1–174 (1831)
42. Truesdell, C.: The Mechanical Foundations of Elasticity and Fluid Dynamics. Gordon and Breach, New York (1966)
43. Cauchy, A.: Ex. de math. 3. Oeuvres (2) **8**, 195–226. 227–52, 253–77 (1828)
44. Truesdell, C.: J. Rat. Mech. Anal. **1**, 125 (1952)
45. Truesdell, C.: Amer. Math. Monthly. **60**, 445 (1953)
46. Hamel, G.: Math. Ann. **66**, 350 (1908)
47. Zeytounian, R.Kh.: Theory and Applications of Viscous Fluid Flows. Springer, Berlin/ Heidelberg (2004)
48. Stokes, G.G.: On the theories of the internal friction of fluids in motion, and of the equilibrium and motion of elastic solids. Trans. Camb. Philos. Soc **8**, 287–319 (1845)
49. Saint-Venant, A.J.B. (de): C. R. Acad. Sci., Paris **17**, 1240–1243 (1843)
50. Kazhikov, A.: Some new statements for initial boundary value problems for Navier–Stokes equations of viscous gas. In: Galdi, G.P., Màlek, J., Necàs, J. (eds.) Progress in Theoretical and Computational Fluid Mechanics, pp. 33–72. Longman Scientific and Technical, New York (1994)

51. Lions, P.L.: Mathematical Topics in Fluid Mechanics, Volume 2: Compressible Models. Oxford Lecture Series in Mathematics and its Applications. Clarendon, Oxford (1998)
52. Leray, J.: La vie des sciences. Comptes Rendus Acad. Sci. Série Générale **11**, 287–290 (1994)
53. Gibbs, J.W.: On the equilibrium of heterogeneous substances. Trans. Conn. Acad. **3**, 108–248 and 343–524 (1875)
54. Fourier, J.: Mémoire d'analyse sur le mouvement de la chaleur dans les fluides. Mémoires de l'Académie Royale des Sciences de l'Institut de France (2), vol. 12, (1833), 507–530 (with editorial changes), pp. 595–616 of Oeuvres de Fourier, Tome 2 (1890)
55. Poincaré, H.: Thermodynamique. Gauthier-Villars, Paris (1892)
56. Truesdell, C., Muncaster, R.G.: Fundamental of Maxwell's Kinetic Theory of a Simple Monoatomic Gas. Academic, New York (1980)
57. Kreiss, H.O.: Initial boundary value problems for hyperbolic equations. Comm. Pure Appl. Math. **23**, 277–298 (1970)
58. Kreiss, H.O.: Initial Boundary Value Problems for Hyperbolic Equations. Lecture Notes in Mathematics, vol. 363, pp. 64–74. Springer, Berlin (1974)
59. Belov, Yu.Ya., Yanenko, N.N.: Influence of viscosity on the smoothness of solutions of incompletely parabolic system. Math. Notes **10**, 480–483 (1971)
60. Oliger, J., Sundström, A.: Theoretical and practical aspects of some initial boundary value problems in fluid dynamics. SIAM J. Appl. Math. **35**, 419–446 (1978)
61. Majda, A.: Compressible Fluid Flow and Systems of Conservation Laws in Several Space Variables. AMS, vol. 53. Springer, New York (1984)
62. Kreiss, H.O., Lorenz, J.: Initial-Boundary Value Problems and the Navier–Stokes Equations. Academic press, Boston (1989)
63. Greenspan, H.P.: The Theory of Rotating Fluids. Cambridge University Press, Cambridge (1968)
64. Nayfeh, A.H.: Perturbation Methods. Wiley, New York (1973)
65. Beirào da Veiga, H.: Singular limits in compressible fluid dynamics. Arch. Ration. Mech. Anal. **128**, 313–327 (1994)
66. Gresho, P.M.: Advances in Applied Mechanics. vol. 28, pp. 45–140 (1992)
67. Gustafsson, B., Sundström, A.: Incompletely parabolic problem in fluid dynamics. SIAM J. Appl. Math. **35**, 343–357 (1978)
68. Viviand, H., Veuillot, J.P.: Publication ONERA, N° 1978–4 (1978)
69. Viviand, H.: Publication ONERA, N° 1983–4 (1983)
70. Velarde, M.G., Zeytounian, R.Kh.: Interfacial Phenomena and the Marangoni Effect. CISM Courses and Lectures, vol. 428. Springer, Vienna/New York (2002)
71. Whitham, G.B.: Linear and Nonlinear waves. Wiley-InterScience, New York (1974)
72. Zeytounian, R. Kh.: Nonlinear long waves on water and solutions. Physics-Uspekhi, English edition of the Russian Uspekhi Fizicheskikh Nauk, **38**(12), 1333–1381 (1995)
73. Dobrokhotov, S.Yu., Safarevich, A.I.: On the behavior of an incompressible fluid velocity field at infinity. Fluid Dynamics, **31**(4), 511–514 (1996)
74. Serrin, J.: On the uniqueness of compressible fluid motion. Arch. Ration Mech. Anal. **3**, 271–288 (1959)
75. Batchelor, G. K.: J. Fluid. Mech. **1**, 177–190 (1956)
76. Wood, W.W.: Boundary layers whose streamline are closed. J. Fluid Mech. **2**, 77–87 (1957)
77. Guiraud, J.-P., Zeytounian, R.Kh. (eds.): Asymptotic modelling of fluid flows. J. de Mécanique Th. et Appl. Special Issue 1986, 313 pages (1986)
78. Doering, C.R., Gibbon, J.D.: Applied Analysis of the Navier–Stokes Equations. Cambridge University Press, Cambridge (1995)
79. Zeytounian, R.Kh.: Well-posedness of problems in fluid dynamics (a fluid-dynamical point of view). Russian Math. Surv **54**(3), 479–564 (1999)
80. Riley, N.: J. Inst. Math. Appl. **3**, 419–454 (1967)
81. Strikwerda, J. C.: Comm. Pure Appl. Math. **30**, 797–822 (1977)

82. Valli, A.: Mathematical results for compressible flows. In: Rodrigues, J.F., Sequeira, A. (eds.) Mathematical Topics in Fluid Mechanics. Longman Scientific, New York (1992)
83. Temam, R.: Some development on Navier–Stokes equations in the second half of the twentieth century. In: Pier, J.P. (ed.) Development of Mathematics 1950–2000, pp. 1049–1106. Birkäuser, Basel (2000)
84. Leray, J.: Etude de diverses équations intégrales non linéaires et de quelques problèmes que pose l'hydrodynamique. J. Math. Pures Appl. **12**, 1–82 (1933), and see also: Acta Math., vol. 63, 193–248, 1934
85. Zeytounian, R.Kh., Guiraud, J.-P.: Evolution d'Ondes Acoustiques dans une Enceinte et Concept d'Incompressibilité. Publications IRMA, vol. 2, fasc. 2, n°4, IV–1 à IV–35. Lille (1980)
86. Guiraud, J.-P., Zeytounian, R. Kh.: Proceedings of the BAIL I Conference, pp. 297–300. Boole Press, Dublin, June 1980. See also: Zeytounian, R.Kh., Guiraud, J.-P.: Comptes rendus Acad. Sci., Paris, 290, B, 75–78 (1980)
87. Wilcox, C.: Scattering Theory for the d'Alembert Equation in Exterior Domain. Lecture Notes in Mathematics, vol. **442**. Springer, Berlin (1975)
88. Zeytounian, R. Kh.: Int. J. Eng. Sci. **38** (17), 1983–1992 (2000)
89. Stokes, G. G.: Trans. Cambridge Phil. Soc., vol. ix (8), Papers iii, 1, 1851
90. Oseen, C.W.: Arkiv für Mat., Astr. och Fysik, vol. **6**, N° 29, 1910; 7, N° 1 (1911)
91. Lagerstrom, P. A.: Laminar Flow Theory, In: Moore, F.K. (ed.) Theory of Laminar Flows, pp. 20–285. Princeton University Press, Princeton, Section B (1964)
92. Coulouvrat, F.: J. Acoustique. **5**, 321–359 (1992)
93. Kevorkian, J., Cole, J.D.: Multi Scale and Singular Perturbation Methods. AMS, vol. **114**. Springer, New York (1996)
94. Monin, A.S.: Weather Forecasting as a Problem in Physics. MIT, Cambridge (1972). English translation
95. Miles, J. W.: J. Fluid Mech. **33**(4), 803–814 (1968)
96. Kozhevnikov, V.N.: Isv. Akad. Nauk SSSR, Atmosph. Ocean Phys. **4**(1), 16–27 (1968)
97. Guiraud, J.-P., Zeytounian, R. Kh.: Geophys. Astrophys. Fluid Dyn. **12**(1/2), 61–72 (1979)
98. Zeytounian, R.Kh.: Comptes rendus Acad. Sci., Paris, **290**(Série A):567–569 (1980)
99. Howarth, L. N.: Q. J. Mech. Appl. Math. **4** (2), 157–165 (1951)
100. Rozhdestvenskii, B.L., Yanenko, N.N.: Quasi-linear Systems of Equations and their Applications to Gas Dynamics. Nauka, Moscow (1968) (Russian edition)
101. Stewartson, K.: The Theory of Laminar Boundary-Layers in Compressible Fluids. Clarendon, Oxford (1964)
102. Oleinik, O.A., Samokhin, V.N.: Mathematical Models in Boundary Layer Theory. Chapman & Hall/CRC, Boca Raton (1999)
103. Van Dyke, M.: Nineteenth-century roots of the boundary-layer idea. SIAM Rev. **36**, 415–424 (1994)
104. Antonsev, S.N., Kazhikhov, A.V., Monakhov, V.N.: Boundary Value Problems in Mechanics of Nonhomogeneous Fluids. Nauka, Novosibirsk (1983)
105. Solonnikov, V.A., Kazhikhov, A.V.: Existence theorems for the equations of motion a compressible viscous fluid. Ann Rev. Fluid Mech. **13**, 79–95 (1981)
106. Guiraud, J.-P., Zeytounian, R.Kh.: A note on the adjustment to hydrostatic balance. Tellus **34**(1), 50–54 (1982)
107. Germain, P.: The "New" Mechanics of Fluids of Ludwig Prandtl. In: Ludwig Prandtl ein Führer in der Stromungslehre. Meier, G. (ed) pp. 31–40. Vieweg, Braunschweig (2000)
108. Germain, P.: My discovery of mechanics. In: Maugin, G.A., Drouot, R., Sidoroff, F. (eds.) Continuum Thermomechanics. Paul Germain's Anniversary Volume, pp. 1–27. Kluwer Academic Publishers, Dordrecht (2000)
109. Eckhaus, W.: Asymptotic Analysis of Singular Perturbations. North-Holland, Amsterdam (1979)

110. Temam, R.: Navier–Stoles Equations, a talk at the Special Session. On the Diverse Mathematical Legacy of Jean Leray. American Math. Soc., Austin, Texas, 8 Oct 1999
111. Chemin, J.-Y. Jean Leray et Navier–Stokes. In Jean Leray 1906–1998, pp. 71–82. No. spécial de la Gazette des Mathématiciens, SMF, Paris (2000)
112. Ladyzhenskaya, O.A.: The Mathematical Theory of Viscous Incompressible Flow. Gos Izdat Fiz-Mat Lit., Moskva, 1961 (in Russian). English translation by Gordon and Breach, New York (1963)
113. Quartapelle, L.: Numerical Solution of the Incompressible Navier– Stokes Equations. Birkhäuser, Basel (1993)
114. Feireisl, E.: Dynamic of Viscous Compressible Fluids. Oxford Science, Oxford (2004)
115. Novotny, A., Straskraba, I.: Introduction to the Mathematical Theory of Compressible Flow. Oxford Lecture Series in Mathematics and its Applications, Oxford (2004)
116. Guiraud, J.-P., Zeytounian, Kh: A double-scale investigation of the asymptotic structure of rolled-up vortex sheets. J. Fluid Mech. **79**(part 1), 93–112 (1977)
117. Guiraud, J.P., RKh, Zeytounian: Vortex sheets and concentrated vorticity: a variation on the theme of asymptotic modelling in fluid mechanics. In: Hormung, H.G., Müller, E.A. (eds.) Proceeding of the colloquium "Vortex Motion". Vieweg and Son, Braunschweig (1982)
118. Guiraud, J.-P.: Ann. de Physique France. **5**, 33–59 (1980)
119. Stewartson, K., Stuart, J. T.: J. Fluid Mech. **48** (3), 529–545 (1971)
120. Guiraud, J.-P., Zeytounian, R. Kh.: J. de Mécanique. **17** (3), 337–402 (1978)
121. De Coninck, F., Guiraud, J.-P., Zeytounian, R. Kh.: Q. J. Mech. Appl. Math. **36** (1), 1–18 (1983)
122. Feuillebois, F., Lasek, A.: Une méthode de calcul des dégénérescences significatives et quelques applications en mécanique des fluides. J. Theor. Appl. Mech. N° spécial 1986 (Guiraud, J.-P., Zeytounian, R. Kh., eds) 295–313 (1986)
123. Godts, S., Zeytounian, R. Kh.: ZAAM. **70** (1), 67–69 (1990)
124. Eckhaus, W.: Fundamental concepts of matching. SIAM Rev. **36**(3), 431–439 (1994)
125. Lagerstrom, P.A.: Matched Asymptotic Expansions – Ideas and Techniques. AMS, vol. **76**. Springer, New York (1988)
126. Germain, P.: Méthodes asymptotiques en mécanique des fluides. In: Balian, R., Peube, J.L. (eds.) Fluid Dynamics, les Houches Lectures, pp. 3–147. Gordon and Breach, New York/ Paris (1977)
127. Guiraud, J.-P.: In: Asymptotic modelling in fluid mechanics. Bois, P.A., Dériat, E., Gatignol, R., Rigolot, A. (eds.). Lecture Notes in Physics, vol. **442**, pp. 257–307. Springer, Berlin/ Heidelberg (1995)
128. Robinson, A.: Nonstandard Analysis. North-Holland, Amsterdam (1966)
129. Lightstone, A.H., Robinson, A.: Non-Archimedean Fields and Asymptotic Expansions. North-Holland, Amsterdam (1975)
130. Lutz, R., Goze, M.: Nonstandard analysis. A practical guide with applications. Lecture Notes in Mathematics, vol. **881**. Springer, Berlin/Heidelberg/New York (1981)
131. Eckhaus, W.: In: Lecture Notes in Mathematics. **985**, Verhulst, F. (ed.), Springer, Berlin (1983)
132. Braaksma, B.: Phantom ducks and models of excitability. J. Dynam Differ Equat **4**, 485–513 (1992)
133. Germain, P.: Progressive Waves. 14th Ludwig Prandtl Memorial Lecture (Mannheim, 16 April 1971). Sonderdruck aus dem Jahrbuch 1971 der DGLR
134. Guiraud, J.-P.: Focalisation dans les ondes courtes non linéaires. Application au bruit balistique de focalisation. Colloque AGARD. ONERA, TP **686** (1969)
135. Sanchez-Palencia, E.: Comportement local et macroscopic d'un type de milieux physiques hétérogènes. Int. Eng. Sci. **12**, 331–351 (1974)
136. Mei, C.C., Auriault, J.-L., Nig, C.O.: Some application of the homogenization theory. Adv. Appl. Mech **32**, 277–348 (1996)

137. Stewartson, K.: Multistructural boundary layers on flat plates and related bodies. Adv. Appl. Mech **14**, 145–239 (1974)
138. Meyer, R.E.: A view of the triple deck. SIAM J. Appl. Math. **43**(3), 639–663 (1983)
139. Neiland, V.Ya.: Towards a theory of separation of the laminar boundary layer in a supersonic stream. Isv. Akad. Nauk. SSSR. Mekh. Zhidkosti i Gaza **4**, 33–35 (1969)
140. Veuillot, J.-P. La Recherche Aérospatiale. vol. **6**, p. 313, ONERA (1971)
141. Guiraud, J.-P., Zeytounian, R.Kh.: La Recherche Aérospatiale. vol. **2**, pp. 65–87, ONERA (1971)
142. Veuillot, J.-P.: Calculation of the Quasi Three-dimensional Flow in a Turbomachine Blade Row. Trans. ASME, J. Eng. Power, **76-GT-56**, 1–10 (1976)
143. Wu, Ch-Hua.: Trans. ASME, **74**(8), 1363 (1952)
144. Guiraud, J.-P., Zeytounian, R.Kh.: La Recherche Aérospatiale. vol. **5**, pp. 237–256, ONERA (1971)
145. Guiraud, J.-P., Zeytounian, R.Kh.: Application du concept d'échelles multiples à l'écoulement dans une turbomachine axiale. Int. J. Eng. Sci. **12**, 311–330 (1974)
146. Guiraud, J.-P., Zeytounian, R.Kh.: Cascade and through-flow theories as inner and outer expansions. Int. J. Eng. Sci. **16**, 515–526 (1978)
147. Sirotkin, Ya.A.: Fluid Dynamics (Translation of Izv. A. N. SSSR, Mechanika Zhidkosti i Gaza), vol. **2**, pp. 64 (1967)
148. Chester, W.: J. Fluid Mech. **18**, 44 (1964)
149. Rott, N.: Rimrot and Tabarrok (ICTAM, 163), North Holland (1980)
150. Müller, B.: Low Mach number asymptotics of the Navier–Stokes equations. J. Eng. Math. **34**, 97–109 (1998)
151. Meister, A.: Asymptotic single and multiple scale expansions in the low mach number limit. SIAM J. Appl. Math. **60**, 256–271 (1999)
152. Ali, G.: Low mach number flow in time-dependent domain. SIAM J. Appl. Math. **63**(6), 2020–2041 (2003)
153. Zeytounian, R.Kh.: C. R. Acad. Sci., Paris, série I, **297**, 271–274 (1983)
154. Zeytounian, R. Kh.: Int. J. Eng. Sci. **27**(11), 1361–1366 (1989)
155. Zeytounian, R. Kh.: Int. J. Engng. Sci. **35**(5), 455–466 (1997)
156. Zeytounian, R. Kh.: The Bénard–Marangoni thermocapillary instability problem. Uspekhi Fizicheskikh Nauk, **41**(3), 241–267 (1998) (English edition)
157. Davis, S. H.: Ann. Rev. Fluid Mech. **19**, 403–435 (1975)
158. Parmentier, P.M., Regnier, V.C., Leblond, G.: Nonlinear analysis of coupled gravitational and capillary thermoconvection in thin fluid layers. Phys. Rev. E **54**(1), 411–423 (1996)
159. Joseph, D.S.: Stability of Fluid Motions, vol. II. Springer, Heidelberg (1976)
160. Takashima, M.: J. Phys. Soc. Japan, **50**(8), 2745–2750 and 2751–2756 (1981)
161. Turcotte et al., J. Fluid Mech.**64**, 369 (1974)
162. Crowley, D.G., Lawrence, C.J., Wilson, S.K.(eds.): The Dynamics of Thin Liquid Film. J. Eng. Maths. Special Issue, **50**(2–3) (2004)
163. Rayleigh, L: On convection currents in horizontal layer of fluid when the higher temperature is on the underside. Philos. Mag., Ser. **32**(6), 529–546 (1916)
164. Pearson, J.R.A.: On convection cells induced by surface tension. J. Fluid Mech. **4**, 489 (1958)
165. Block, M.J.: Surface tension as the cause of Bénard cells and surface deformation in a liquid film. Nature **178**, 650–651 (1956)
166. Guyon, E., Hulin, J. P., Petit, L.: Hydrodynamique Physique. Inter Edition/Editions du CNRS, Paris 1991. English edition (with Mitescu, C.D.): Physical Hydrodynamics, Oxford University Press (2001)
167. VanHook, S. J., Swift J. B.: Phys. Rev. E. **56**(4), 4897–4898 (1997)
168. Oron, A., Davis, S.H., Bankoff, S.G.: Long-scale evolution of thin liquid films. Rev. Mod. Phys. **69**(3), 931–980 (1997)
169. Dorodnitsyn, A. A.: Trudy Tsentral Inst. Prognoz. **21**(48), 3–35 (1950)

170. Zeytounian, R.Kh.: Criterion for filtering parasite solutions in numerical computations. Phys. Fluids. **12**(Suppl. II), II46–II50 (1969)
171. Kibel, I.A.: An Introduction to the Hydrodynamical Methods of Short Period Weather Forecasting. Pergamon, New York, 1963. Translated from the Russian edition, Moscow (1957)
172. Norbury, J., Roulstone, I. (eds.): Large-Scale Atmospheric–Ocean Dynamics. I. Analytical Methods and Numerical Models. II. Geometric Methods and Models. Cambridge University Press, Cambridge (2002)
173. Guiraud, J.-P., Zeytounian, R.Kh.: Asymptotic derivation of the ageostrophic model for atmospheric hydrostatic flows. Geophys. Astrophys. Fluid Dyn. **15**((3 + 4)), 283–295 (1980)
174. Charney, J.G.: Proceedings of International Symposium Numerical Weather Prediction, p. 131. Tokyo, Meteorological Society of Japan (1962)

Index

DOI 10.1007/978-3-642-20746-4,

MIX
Papier aus verantwortungsvollen Quellen
Paper from responsible sources
FSC® C105338

If you have any concerns about our products,
you can contact us on
ProductSafety@springernature.com

In case Publisher is established outside the EU,
the EU authorized representative is:
Springer Nature Customer Service Center GmbH
Europaplatz 3, 69115 Heidelberg, Germany

Printed by Libri Plureos GmbH
in Hamburg, Germany